Frederick Billings

Frederick BILLINGS

A Life

ROBIN W. WINKS

New York   Oxford
OXFORD UNIVERSITY PRESS
1991

Oxford University Press

Oxford   New York   Toronto
Delhi   Bombay   Calcutta   Madras   Karachi
Petaling Jaya   Singapore   Hong Kong   Tokyo
Nairobi   Dar es Salaam   Cape Town
Melbourne   Auckland

and associated companies in
Berlin   Ibadan

Copyright © 1991 by Oxford University Press, Inc.

Published by Oxford University Press, Inc.
200 Madison Avenue, New York, New York 10016

Oxford is a registered trademark of Oxford University Press

The monogram FB, used here to adorn the main title and each chapter title, was selected by Frederick Billings to be used on his carriage harnesses in Woodstock.

Dust jacket: The medallion of Frederick Billings is from an oil portrait attributed to Robert M. Platt, apparently painted at the time of Billings's departure from Vermont for California in 1849. (By permission of Mrs. John H. McDill, Woodstock, Vermont.) The background is an adaptation of William Bradford's painting *Sunset in the Yosemite Valley* (1881), the original of which hangs in the Billings Mansion in Woodstock. (By permission of the Billings Mansion Archives, Woodstock, Vermont.)

Frontispiece: Charcoal and chalk portrait of Frederick Billings, derived in 1873 from a photograph by William Kurtz, fashionable New York photographer whose studio was located in Madison Square. The portrait hangs in the Billings Mansion. (By permission of the Billings Mansion Archives.)

Library of Congress Cataloging-in-Publication Data
Winks, Robin W.
Frederick Billings : a life / Robin W. Winks.
p. cm.   Includes bibliographical references and index.
ISBN 0-19-506814-9
1. Billings, Frederick, 1823–1890.   2. Businessmen—United States—
Biography.   3. Northern Pacific Railway Company—History.
I. Title.
HE2754.B5W56   1991
385'.092—dc20   90-22367

9 8 7 6 5 4 3 2 1

Printed in the United States of America
on acid-free paper

To Vermont and California,
States of Mind

ACKNOWLEDGMENTS

THIS BOOK COULD NOT HAVE BEEN WRITTEN without the assistance of many people. Archivists in over one hundred libraries and archives I visited helped to ferret out relevant documents, while librarians and archivists in fifty more depositories responded to my cries for aid with a flood of photocopied materials. In particular, the staffs of the Bancroft Library at the University of California, Berkeley, and of The Huntington Library in San Marino, California provided a great quantity of original resources and answered many questions. The staff of the Minnesota Historical Society, which holds the papers of the Northern Pacific Railroad, guided me through that vast treasure trove, while the assistants at the library of the University of Vermont and at the Vermont Historical Society in Montpelier combed their collections repeatedly. The Yale University Library purchased, or obtained on interlibrary loan, hundreds of titles across a range of subjects.

Janet Houghton, curator of decorative arts at the Billings Museum and archivist of the Billings Family Papers, kept in the Billings Mansion Archives in Woodstock, Vermont, established the most ideal of working arrangements for a writer. Without her meticulous organization of that extensive range of family correspondence, this study would not have been possible. She also researched and prepared the Billings genealogy that appears in this book. Peter Jennison of Taftsville, Vermont, together with Jane Curtis of Hartford and Frank Leiberman of Woodstock, authors of an exemplary short popular biography of Frederick Billings, time and again demonstrated that all scholarship is a cooperative endeavor. Thomas Debevoise of the Woodstock Foundation and the staff of the Woodstock Historical Society were also consistently supportive.

No biography of Frederick Billings could be written without the encouragement and support of Mr. and Mrs. Laurance Rockefeller. Mary Rockefeller is Frederick Billings's granddaughter, and the Billings Archive is the possession of the Rockefeller family, without whose permission one may not consult the records. Far more than mere permission was given, however, for they showed a supportive and steady interest in the project from its inception, meeting travel and research costs either personally or through the Woodstock Foundation, though at no time did they attempt to shape research priorities or change my interpretations. An historian can ask for no greater assistance than the expressions of interest, curiosity, and warm neutrality shown by Mr. and Mrs. Rockefeller throughout research and writing.

Finally, I wish to thank Yale University and my department for granting me leave so that this book might be written, and those scholars who reacted to chapters in draft form so that errors might be rooted out. Gunther Barth of the University of California, Berkeley, read all the material on California and helped crystallize my conclusions about Billings's reactions to his experiences there. Gerald Stanley, of California State College, Bakersfield, helped me to understand the complexities of California politics, while Peter E. Palmquist of Humboldt State College in Arcata saved me from potential errors concerning Carleton E. Watkins. Kevin Graffagnino, curator of the Wilbur Collection at the University of Vermont, provided a lively critique of my chapters on Vermont, as did Samuel Hand, also of the University of Vermont, on Chapter 24. Albro Martin of Bradley University, author of a superb biography of one of Frederick Billings's chief rivals, James J. Hill, clarified many points on railroad history with good humor and pungent phrase. Janet Houghton helped to verify and correct citations to the Billings Papers in Woodstock, in the process of reorganization during my research, and Julia McDill, also of Woodstock, reacted with a critical eye to numerous excesses of prose. Above all, Fred Smith, a former fellow member of the National Park System Advisory Board, read every line of the manuscript, offered cheerful and amusing support at crucial moments, and gently prodded the project to completion. To all I am most grateful.

New Haven, Connecticut   R. W. W.
June 1991

CONTENTS

Frederick Billings

THE BILLINGS FAMILY

Oel Billings   m.   1817   Sophia Farwell Wetherbee
(1788–1871)   (1796–1870)

Edward Horatio
(1818–1844)

Charles Jason   m.   1849   Sarah Towne
(1822–1896)   (b. 1827)

Laura   m.   1845   Bezer Simmons
(1820–1849)   (1810–1850)

Sophia Farwell   m.   (1) 1853   Goldsmith Fox Bailey
(1826–1895)   (1823–1862)
                      (2) 1876   Rodney Wallace

Richard Oel
(1831–1854)

Oliver Phelps Chandler   m.   1868   Charlotte Lane
(1836–1894)   (b. 1842)

[Another child
who died in
infancy]

Elizabeth Sprague   m.   1853   George Washington Allen
(1833–1905)   (1827 or 1828–1864)

Franklin Noble   m.   1859   Nancy Swift Hatch
(1829–1894)   (1822–1904)

Frederick   m.   1862   Julia Parmly
(1823–1890)   (1835–1914)

Parmly
(1863–1888)

Frederick Jr.   m.   1912   Jessie Starr Nichols
(1866–1913)

Laura   m.   1901   Frederic Schiller Lee
(1864–1938)   (1859–1939)

Elizabeth
(1871–1944)

Mary Montagu   m.   1907   John French
(1869–1951)   (1863–1935)

Ehrick
(1872–1889)

Richard   m.   1898   May Merrill
(1875–1931)   (1874–1965)

By permission of the Billings Mansion Archives, Woodstock, Vermont

Introduction

THE LAND RISES ALMOST IMPERCEPTIBLY from the south toward the Vermont border. Today, great patches of green blanket the area in the summer, for heavy forestation has returned to the Massachusetts Berkshires and the ascending slopes of the Green Mountains and, across the Connecticut River, to the White Mountains as well. Much of this land was bare in the middle of the nineteenth century, denuded of its timber cover to feed the industrial revolution on whose fringe Vermont lay. The traveler journeying from Long Island Sound, paralleling the Connecticut River, or traversing Massachusetts from the Atlantic Coast, notes a clear steepening of the gradient a few miles south of that border and thereafter a steady ascent into the rugged hills of central Vermont.

There lie the Green Mountains, a plain-spoken name from a plain-speaking people. The mountains are really not green at all, but grey, or a misty blue, and a dozen varieties of subtle shadings in spring and fall. Snow comes early and rests heavily upon these hills, lingering well into May in the clefts and rills and deep in the woods, under the fallen oaks where the morels grow. Suddenly, for two spring months, a house—almost any house—is likely to be embattled by mud. Winters are, it would appear, less severe today than in the 1840s, for Vermonters know March as the Mud Season, while a century and more ago the mud oozed up in late April and continued through May; correspondence and the press of the time are filled with routine references to snows in mid-April or late August. During Mud Season even the grand homes (of which there are few) are entered through the kitchen door, so that one may deposit great cakes of mud on the bare floorboards.

Even now Vermonters live in a kind of mountain time, and except on the modern thruways, distance is measured by answering "how long" rather than "how far." Today a snug town such as Woodstock, nestled a few fortunate miles off those thruways, is three hours from Boston; in the 1850s, with the best available transportation, that distance took eight hours. In the worst of times one did not travel. Mountain time and lives are different from sea coast time and lives and different again from plains and prairies, islands and marshes— not better or worse, just different.

Vermont's people—city or country—are independent, not so much of each other, for family ties are close, but independent certainly from the rest of the country. Vermonters did not get around to joining the new United States until 1791, eight years after the first thirteen states came together, and fully two centuries later that sense of independence, though sometimes exaggerated or romanticized, is still very real. The land is huge in stature and feeling, though the state is one of the nation's smallest, and is as different politically and economically from its neighbor, New Hampshire, as New Mexico is from Arizona. Vermont always has been the most purely rural of all the older states: it is a land where the wet winds and high mists, once lifted, reveal the tops of trees and an occasional steeple and often not much else. The people boast of little, and have small cause to blush, though as Thoreau remarked, in autumn the hills will blush for them. Some think it a land of ostentatious understatement.

Here on a late September Saturday in 1823 Sophia Wetherbee Billings bore her fourth child, to be named Frederick. On the Vermont piedmont, in Royalton, the weather was hot, the hills enveloped in a shimmering haze. Here on the last day of September 1890, some fifteen miles over the hill in Woodstock, Frederick Billings died. During this interval, significant history occurred to create the character of the United States and of Vermont and Frederick Billings had an influential hand in much of it.

1

Woodstock Is Where
the Sheriff Lives

FREDERICK BILLINGS WAS BORN TO Sophia Wetherbee Billings and Oel Billings on September 27, 1823, in Royalton, Windsor County, a small town on the banks of the White River, some nineteen miles above its confluence with the Connecticut. Though family tradition holds that Royalton was an isolated Vermont village, it was in fact on the main stage route between the New Hampshire border, Montpelier, and Burlington. In the census of 1820 Royalton ranked 27th in population of Vermont's 230 towns. There were thirteen schools, all quite tiny, and lively signs of the woollen boom, then at its height. Cattle, sheep, and horses played an omnipresent role. Four saw mills and three grist mills also helped make Royalton a community of significance in 1823.[1]

The sheep era was still young: William Jarvis of Weathersfield brought the first Merinos to Vermont in 1811. Windsor County ranked fourth in Vermont in sheep per square mile, and until the price of wool spiraled downward in the 1840s as protective tariffs were lowered and grazing lands further West were brought into production, land was laid out and managed, to the extent that it was managed at all, for growing wool. As sheep spread, so did support industries: carpenters to build barns for the sheep, loggers to cut trees for the barns, mills to saw raw timber, tanneries to care for the skins, fulling (or cleansing) mills to wash the cloth brought in from handlooms worked in dozens of private homes.[2] Anyone growing up in such a community knew that the thigh bone was connected to the hip bone and understood almost intuitively the interrelatedness of human activities.

Dominated by an imposing Congregational church, the village had two attorneys, a physician, four stores, two taverns, and a flourishing academy. Royalton was also famous: settled in 1771, it had been burned in October 1780 by a band of three hundred Indians under British direction. The Indians had killed four people and taken twenty-six prisoners. This was the most serious of Vermont's Indian raids and the focus of a substantial memoir litera-

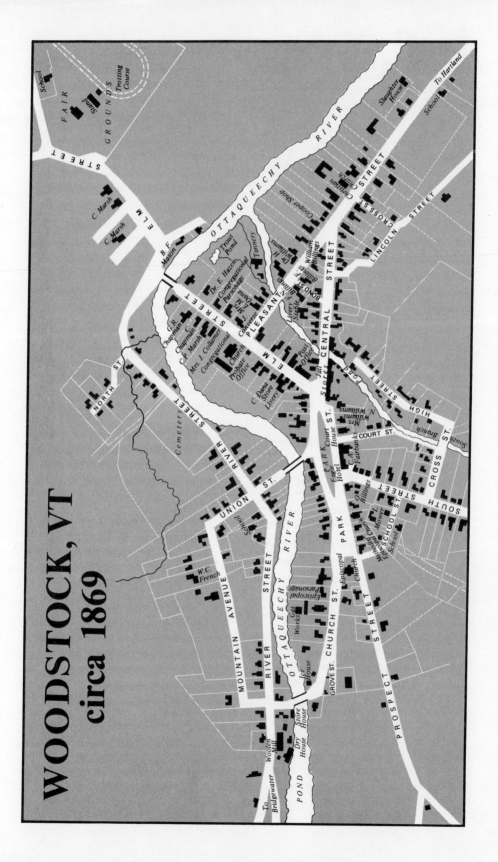

WOODSTOCK, VT
circa 1869

ture, a sub-genre of the late eighteenth and early nineteenth century, "captivity narratives," which Frederick Billings would collect in his Woodstock retirement during the 1880s.

Family legend holds that Oel Billings was improverished and driven to leave Royalton for Woodstock because of debt.[3] This is only partially true. Billings was of solid yeomenry background and well regarded in Royalton. Oel and Sophia were able, through close saving, to send their first child Edward (born in 1818) to the University of Vermont, where he read law, and their second son Charles (born in 1822) to the Royalton Academy. When the Marquis de Lafayette made his triumphal tour of America in 1824–1825 and progressed from Windsor to Burlington, Oel Billings was marshal for the turning-out at Royalton, an honor that hardly would have been bestowed upon the nondescript. As Lafayette journeyed to Royalton from Woodstock on the afternoon of June 28, 1825, a concourse of citizens formed in procession under Billings's direction, and he and several assistant marshals escorted Lafayette to the community's leading home, where one of the two attorneys read a welcoming address. Twenty Revolutionary War soldiers then shook Lafayette's hand, stepped back a few paces, and fired their muskets in salute. All proceedings were supervised by Oel, who during the less formal part carried a young son (Charles, then three?—more likely Frederick, not quite two) on his shoulder.[4] This was not the duty of a poor man. And in 1832 Oel apparently was considered for the superintendency of the state prison and was for a time a justice of the peace.[5]

Oel Billings had made a respectable living from the good soil along the White River, growing potatoes and hay, and from selling a bit of maple sugar. He had married a bit later than usual for the time—at twenty-nine—and by 1835, when the move to Woodstock occurred, he and Sophia had had nine children, with eight surviving (a tenth, Oliver, would be born the next year in Woodstock). Four were girls; none of the boys was yet of an age to supplement the family income, and Oel was paying tuition for the two in school. Money was running out. In May 1833, Oel sued his older brother Asa to recover a loan made in better times. He apparently won $75, payable over three years, in settlement, but he also acquired some of Asa's indebtedness to others. To pay, Oel borrowed from his father-in-law, Jason Wetherbee, who lived in Charlestown, New Hampshire. Two years later the litigious Wetherbee went to court to get his money back. Oel, who was trying his hand at running a store, had no ready cash, and when the court found against him, he was told that the law required him to live within a mile of a jail, to whose sheriff he must report regularly.*

---

*The previous year, after four years of public agitation against the state's debtors' act, which required imprisonment for debt, the legislature opened a loophole in an otherwise barbarous law. The reformers argued that to imprison for debt was to "confound innocence with guilt, by inflict-

Though the circumstance was embarrassing for Oel Billings, one may assume that his neighbors understood and sympathized. He was a respected member of the community, a trustee at the Royalton Academy, a representative to the General Assembly, the son of a deacon of the Baptist Church, and through his wife and brothers was linked to distinguished families. He, like three-quarters of Royalton residents, owned a story-and-a-half frame house and a hay meadow, and this Oel now had to sell.

Family tradition, perhaps further enhancing the rags-to-riches notion, holds that the Billings family went to Woodstock with their tails between their legs, but this seems unlikely. America was, after all, the land of beginning again, of the second chance; Woodstock was a growing, sophisticated town that would afford more opportunities than Royalton; and Oel had acquired his debt honorably and attempted to deal with it equally so. Woodstock was, simply, where the sheriff lived.

One of the most enduring scenes of family legend is, however, almost certainly true: that of the trek from Royalton to Woodstock. With his sister Sarah and brother-in-law Calvin Skinner helping him, Oel piled the family possessions into his farm wagon and set out early one hot September morning in 1835 to walk the Woodstock turnpike. Charles was to shepherd the cow, while Frederick, not yet quite twelve, made certain that the family pig reached its destination. The road rose steeply out of Royalton and by the time Woodstock was reached, fifteen miles later, everyone was dirty and exhausted. The toll, thirty cents for the wagon and one cent for the cow, proved another embarrassment. Frederick later confessed that he had never forgotten that day: "as we climbed yonder hill, the pig gave out and would go no further and, to my unsophisticated heart, when that pig stopped the whole world stopped."[7] Family legend also has it that as Frederick trudged past the Charles Marsh mansion on the edge of Woodstock, not then the present grand estate but nonetheless spaciously comfortable, he vowed he would never be poor again. After all, Marsh had founded the turnpike company by which Frederick, with pig trotting errantly ahead, was ushered into his new home.[8]

"Two movings are equal to one fire." Frederick may have felt this Vermont adage too true, for though the family had changed houses once in Royalton, nothing could have so wiped out the sense of continuity as the enforced move to Woodstock. Still, the move was eased by friends and acquaintances, in both Royalton and Woodstock, who helped settle the Billings family in their new home.

Woodstock was the seat of Windsor County, by now a bustling, growing

---

ing the same penalities upon the unfortunate as upon the fraudulent," while depriving the former of any means of raising money to pay the debt. While imprisonment for debt was not abolished until 1838, Oel thus benefited from the more lenient interpretation of the law.[6]

town of about 3,250 people, in transition from being the fifth to the third most populous city in the state. By 1840 it was exceeded only by Burlington and Montpelier—indeed, Woodstock was larger in 1830 than in 1980. Though larger than Royalton, it was not yet greatly so: it had seven saw mills rather than Royalton's four, six grist mills, three carding machines, and five fulling mills. It had far fewer sheep than Royalton: that is, Woodstock already was in transition between rural and urban frontiers. There were eight attorneys, attracted by the County Court, five physicians, two taverns, two printing offices, a weekly paper, and a stone jail that Zadock Thompson, indefatigable chronicler of nineteenth-century Vermont, called "the best in the state."[9]

Two early sources of moral support were Norman Williams, who with his wife Mary called on the Billingses soon after arrival, and Oliver Phelps Chandler. Williams was a happy man with a dour face, and he enjoyed himself and those around him in equal measure. Born in Woodstock in 1791, he attended the University of Vermont and had been practicing law in his hometown since 1814. There was no one he and his wife did not know, both in Woodstock and in the state at large, since he had served for eight years as secretary of state.

O. P. Chandler was a twenty-eight-year-old lawyer who already had made a success of himself. A Dartmouth graduate, he had opened his law office in Woodstock only three years before, and was doing well. He liked the Billingses at once, and shortly after Edward Billings graduated from the University of Vermont, Chandler took him into his office to read law and, when Edward was only twenty-two, made him a partner.

Frederick Billings would remember both Williams and Chandler with affection. He was invited into the Williams home, where he especially enjoyed examining some scenic wallpaper, then fashionable, that showed episodes from history, of places he would one day visit. Mrs. Williams in particular was welcoming and, as Frederick declared in 1884, when he commemorated that friendship at a new public library built on the site of the Williams's house, there was hardly a day he did not go to visit her in what became "a kind of second home."[10]

These two friendships never wavered; others underwent moments of strain, especially when Frederick's mother, who, though not formally educated was both intelligent and well-read, felt she was being condescended to. People were attracted to warm-hearted energetic Sophia, who had seen a bit of the world outside Vermont. Born in Charlestown, New Hampshire in 1796, she had lived in Montreal at the home of her uncle, Levi Mower, learned some French, as well as how to dress as well as she could afford, and honed the lively sense of humor that lay behind her often intense gaze. She was determined that her children would do well—college education for her sons, good marriages for her daughters—and she saw Woodstock for all its opportunities, joining its social organizations and, as family finances recovered, doing her share of entertaining. Oel, orderly and dour, was respected for his undoubted hard work and obvious desire for self-improvement—he founded the town's literary society—but it was Sophia's strength and humor that attracted people to help the family. Both saw the future through their children: as Oel wrote, "all my hopes are bent upon my children, that they may have an education and make Vermont proud of her Billings."[11]

Education was no simple matter, however, for it was not to be had without cost. There is no record of Frederick's schooling in his first year in Woodstock. In the second his parents placed him, with his brother Charles, in a new school opened by William Bates, a son of the president of Middlebury College. The class met in a room over the Bank of Woodstock in the fall, and in the winter term in a smoke-filled and unpleasant cellar of the local Episcopal church, "better fitted for the habitation of bats and owls than for a school-room."[12] The wandering school then moved into a room on the second floor of a nearby hall, and at the end of the second term Bates abandoned it.

The school served to give Frederick a circle of friends, including young Charles Marsh (from whom Billings would, in 1869, realize his ambition to purchase the estate he had trudged past in September 1835), William Collamer, son of Jacob Collamer, a future postmaster general of the United States,[13] and Benjamin S. Dana, subsequently a successful farmer in the area. These and others were joined in a larger school the next year taught by a student on vacation from college in Burlington. Classes met in a small office, attached to a woodshed; when the tiny room became too warm in summer, the class sat out under the maple trees, and it is likely that it was during one of these sessions, rather than a hot and bedraggled moment while tramping past on the long march from Royalton, that Frederick decided he must one day own Marsh's hill and house.

The third year saw the school moved once again, this time to a barn, reached through a cow-yard. After one term the school died, and Frederick's parents had to cast about for another tutor. There were other schools, but they were not better. Teachers were young boys often not five years older than their pupils, and by no means always educated. Frederick had discovered that he liked to study, unlike many of the boys, who were disruptive and rude.

Frederick wanted more than this, however—a good deal more—and he asked his parents whether he could attend Kimball Union Academy. The academy, which took boarders, was twenty-six miles away, across the Connecticut River in Meriden, New Hampshire. There he could expect to learn enough Greek and Latin to win entrance into a university, or so the principal of the academy, Cyrus L. Richards, assured him when Frederick wrote to inquire. Frederick, even as a youth, knew how to ask the right questions—What do you teach that I need and cannot get elsewhere?—and back came the answer: Xenophon, Virgil, Sallust.

Chartered in 1813 "to train young men for leadership in the ministry," Union Academy, as it was first known, was in good hands. Richards was near the beginning of a long and productive reign, serving as principal and later as headmaster from 1835 until 1871. There was a staff of five, and in Frederick's class twenty-four scholars pursued the classical course. Years afterwards he still corresponded with six of these boys; one, Henry S. Dana, was also from Woodstock; another, a friend from Royalton, E. W. Hutchins, offered to share a room with Frederick. Tests were rigorous and for graduation covered twenty-seven books in addition to separate examinations in English grammar, arithmetic, geography, and American history. In 1838 tuition was $4.00 a term— there were four terms each year—and room and board was an additional $1.75 per week.[14]

Frederick found the course demanding and the costs worrisome. The academy made it clear to prospective students that three years was "quite short enough to meet the present academical requirements," yet he proposed to complete the work in two. He had not been required to apply himself to his studies in the motley schools he had previously attended, but now he found that he had to awake at four o'clock each morning and stay up until ten o'clock at night to make up for lost time. "I have never worked so hard," he reported, almost joyously. At Kimball Union he learned to believe in hard work for its own sake, a virtue much associated with the Victorians, English or American, but in fact far less practiced then myth would suggest. Further, his budget often was only fifty cents a week, which provided a filling, if stodgy, diet of brown bread, molasses, codfish, and potatoes. That throughout his adult life he would be known as a gourmet, loving good food and fine wine, was surely in part a reaction to these two spartan years.

They were also productive, exhausting, very exciting years. A fire had destroyed the old academy building the previous spring, and Frederick participated in building a new hall. A female department was introduced in his second year, and young ladies might be seen in the library and at lectures. A new minister took over the village church: Amos Blanchard, an outspoken abolitionist, who brought with him something of a reputation, as it was said that in 1835 no fewer than 135 of his meetings had been broken up by mobs. (When the Civil War erupted, nearly two hundred graduates of Kimball Union served in the northern armies, and a quarter of these died.) The wave of abolitionism was followed by an intense religious revival in which many of the students participated. Work, poor food, anxiety over money, and the enclosed and overheated atmosphere of the school often had Frederick in bed suffering from or fighting off a cold. Since Oel Billings's response, whenever he learned that Frederick was sick, was to scold him for allowing them to hear of it through others and tell him to come home, Frederick took to referring to all illnesses as a ''a bit of cold,'' a practice he would continue in letters to his parents throughout his life, even when seriously ill.

Frederick was learning more than Greek and Latin. He was discovering that he did not want to make his life in Woodstock, at least not immediately upon graduation from the university he confidently expected to attend, and he was discovering, as most people do when away from home for the first time, how to dissemble with his parents. They were anxious for him, and he seemed unable to find quite the right tone with his father, who was always worried about money.

Sophia wrote frequently, long newsy letters filled with Woodstock gossip, while Oel added perfunctory, cranky short messages in whatever space was left, invariably remarking that nothing important had happened. Frederick was three weeks into his first term when Oel scolded him for not writing often enough (and for not dating his letters, so that his parents might know whether they had missed one): ''You seem to think that you have much more to do in order to keep up with your class than you should have. . . . I have indulged you, and curtailed my own ease and convenience to send you to school as much as I have, and yet I get no thanks for it.''[15] Oel instructed Frederick to come home at the end of the term; as his anger at his own misfortunes grew—Sophia was bedridden, Oel was spending $1 a week for an Irish girl to help in the house, Oliver had whooping cough—Oel concluded that he might send for Frederick in twelve days; if so he was to come without fail.

To be sure, to his stern father young Frederick did seem a bit pampered: he was at study, which Oel seemed never to believe was arduous work, and he gave too much thought to his appearance (throughout his life Billings would take great care with how he dressed, frequenting the best tailors, and always noted—especially by the ladies—for his elegance). Things were very tight, for Charles was in his last year at Middlebury College, and final-year expenses always exceed earlier ones. Little wonder, in the face of stern reproof, that Frederick took solace in letters from his sister Laura, which he read over and over, and from his younger brother Oliver, whose lionization of Frederick provided a needed boost to his ego.

After the warning that he might be summoned in twelve days, Frederick was told that his mother had prevailed upon Oel so that he might stay to the end of the term, though he would have to come home then for good. By the end of October Oel had relented, however, and told Frederick he could stay for a second term. Then in the spring Frederick received a maddening letter, abruptly telling him that his father would come in five days to return him to Woodstock, as both older brothers had to leave for Boston, and casually remarking that he regretted Frederick would not be able to stay for his examinations, scheduled for three days later. Frederick was having none of this, for he had studied too hard. He must have spoken to the principal, for Richards wrote Oel immediately, stating flatly that Frederick must remain with them, whatever Oel's wishes might be, as there was only one time when exams were given with the board of trustees present. Richards concluded, either as justification or in mollification, that Frederick was one of the best students in the school. [16]

By the time Frederick completed his two years at Kimball Union he knew Greek and Latin, qualified for entry to university, developed a lifelong tendency to come down with colds, and had established some independence from his parents. At the end of the first year his mother began calling him Fred instead of Freddy (except when he was ill) and his father no longer threatened him with sudden withdrawals. Frederick had also developed his habit, from which he never deviated, of keeping all his correspondence, and of using letters to both maintain contact and put distance between himself and others.

The voluminous Billings family correspondence, which would swell over the years to fill many boxes, is deeply informative of nineteenth-century values. Sophia Billings writes Frederick a distressed letter, not having heard from him in three weeks, and reminds him that it is his duty to write since theirs is a family of only nine children, requiring a greater sense of closeness than a large family might. He is, she tells Frederick repeatedly, studying too hard; if he cannot handle it without endangering his health, he is to come home. There is news of a steady flow of hired girls: one proves unable to milk or draw water and is dirty besides; another is colored, ignorant, but hard-working; a third proves excellent and sadly does not stay. Sophia is frequently confined to her bed. There is much worry about liquor and who is consuming it. From such correspondence one can reconstruct the social fabric of a representative Vermont community and, perhaps without realizing it at the time, Frederick was learning from such letters much that Virgil could not teach him. [17]

Frederick had gone to Kimball Union to qualify for the University of Vermont and, though most of his classmates chose Dartmouth College instead, Frederick held firm. He did write to a Dartmouth professor, posing a somewhat pretentious question concerning meteors, no doubt because he genuinely wanted an answer but also from a youthful desire to impress, and perhaps to test. If so, Dartmouth failed its examination, for though the professor replied, the answer was slow in coming. Frederick felt he would receive more attention at UVM, as the university was affectionately known, and in 1840 he proudly and promptly accepted the place offered him. [18]

2

The University Makes
All the Difference

THE UNIVERSITY OF VERMONT of the 1840s was the intellectual construction of a quite remarkable group of men. It was regarded then and has been judged since by historians as an institution of exceptional ferment and vigor. Perhaps more social éclat was attached to attending Harvard or Yale—Frederick would worry a little over this later—but neither gave their graduates a better education. Frederick literally loved his time at "good old UVM": he preferred Burlington to Woodstock, enjoying the company of the "many real, good hearted fellows" he found there, studying and talking with his professors, working with the debating team, arguing politics, religion, and philosophy with his classmates. He found a reason each year but one not to go home for Thanksgiving, and financial need made it necessary for him to teach during the eight-week winter break, so he was not much in Woodstock then either. The intellectual and social pleasures of a thriving university town so captured him that he thought of Woodstock as home in a nostalgic rather than an actual sense.[1]

James Marsh was the university's overarching intellectual force. He and his cousin, George Perkins Marsh, who would also deeply influence Frederick, were Dartmouth graduates; James had studied further at the Andover Theological Seminary, where he had been steeped in orthodox Calvinism that was later diluted by a heavy dose of Samuel Taylor Coleridge and a subsequent diet of German literature. In 1826 Marsh had become president of the University of Vermont, then passing through a rough patch intellectually and financially, and had turned it into a front-rank exponent of the kinds of arguments that would come to be associated with American transcendentalism, with Coleridge, Kant, and Emerson. Under his guidance professors gained permanent positions, students were encouraged to mix with their teachers socially, and great emphasis was placed on writing and speaking well so that the university's graduates might play future public roles. When Billings arrived, Marsh was no longer president, but until his death in 1842 he was a force on the faculty, making certain that students received a good grounding in contemporary poetry, hard reasoning, and in a Christianity that was both muscular and argumentative.

The Reverend John Wheeler, the new president, was a foil to Marsh, a man who kept the university solvent and left the more exciting teaching to others, delivering rather long sermons in which he mixed Christianity and political economy, and presiding over morning and evening prayers. His home was always open to students with the courage to visit it, and Frederick was there often, both because he genuinely liked to argue issues of revealed religion with Wheeler and because he quite enjoyed a cheerful banter with the president's attractive daughter.

Frederick did his best work for the teacher who was the most feared: Ferrand N. Benedict, professor of mathematics, called behind his hunched back Little Ben. By the end of his first year Frederick was regularly attaining 20s (the highest grade) in algebra and geometry, and for a moment entertained the thought of being an engineer before his pleasure in arguing politics and metaphysics turned him toward the law.

Frederick admired these men, and like many students then and since, consciously set out to emulate them. Wheeler was an eloquent speaker, and Frederick did not miss a chance to attend his lectures; he set a tone of courtesy and slightly distant graciousness evident in much of Billings's later social behavior. Marsh told the students to reach for the stars: hard work, courtesy to others, and faithfulness to oneself would lead to success that was spiritual as well as material. Courtesy, Frederick thought, was the highest form of efficiency; yet to be efficient for the benefit of others was also a form of courtesy. From Marsh he acquired a taste for conundrum and riddle, and he read steadily—how deeply one cannot tell—Plato and Aristotle, Kant and Coleridge, Lamb, Locke, Bentham, Hobbes, Hume, and John Stuart Mill, and across a strong if narrow range of natural science, beginning with von Humboldt.

There was time for friends, of course, for the university was, in 1840, quite small: 109 students in all (Harvard had 409). During his four years the usual undergraduate escapades occurred—students stole wood, broke windows, were disorderly in chapel, violated the honor code, cheated at examinations, got drunk, and slipped girls into their rooms—and nine members of Frederick's graduating class were failed and three others expelled. Nonetheless, these were the golden years at the university, filled with both good deeds—students worked to improve the public common, built fences, cleaned the chapel, and kept four literary or speaking clubs vigorously alive through their own efforts—and intellectual growth. The year Frederick graduated, the university applied for its own chapter of Phi Beta Kappa (even then a sign of intellectual distinction), to which he was later elected.

Frederick was not fond of any of the four societies, thinking them frivolous, and with seven other students founded his own in March 1841. The societies claimed to support debate and literary effort; still, one could find other means of sharpening one's rhetorical skills and an abundance of opportunities existed to practice writing. From the record the societies do appear somewhat undisciplined, and attendance at their meetings was not invariably elevating: frequent complaints against public spitting, fining members for swearing, eating apples, whittling, smoking, or making "subterranean noises" (about which one may only guess) were common. As for improving his writing skills, Frederick found the best means possible: in January 1842, Professor Joseph Torrey, his classics teacher, became quite ill, and Frederick sat up with him through the night. This led to an invitation to live in the household and act as the professor's assistant, and in exchange for his board and room, Frederick devoted four hours a day to recording Torrey's translation of J. C. Neander's *History of the Christian Church*. Frederick's prose style assumed an added elegance and, as a bonus, he had (as he confessed to a friend) a very good excuse not to go home for the holidays. Frederick's complaint was common to the scholarly: he loved to study but his parents would not let him "do anything that [looked] like studying" at home.[2]

Frederick had entered the university with a strong recommendation from Cyrus Richards and a warm letter from Joseph Marsh, and he fully lived up to their expectations. His marks were consistently high, he never received a reprimand, he was awarded a certificate for perfect attendance, and he was chosen to deliver an oration—an honor reserved for top students—at his commencement ceremony. He was later awarded a Master of Arts degree on the basis of the excellence of his record and his reading of law in offices in Boston, Woodstock, and Montpelier.

Billings's reputation for speaking his mind was formed while at the university. He did not hesitate to pronounce one student "a little empty fop" or a noted visiting lecturer "a humbug." He remarked of a family friend whose business had failed that he had "always looked upon him as spreading rather too many sails for his boat," and of another failure he noted that he had "supposed that the money he had cheated out of every person he met, had made him perfectly independent." When his mother, who for years would worry that Frederick appeared disinclined to marriage, thrust a young woman at him, he responded that she was "a very clever girl, but her upper story has not been taken care of enough to suit me." Of another, he firmly told his mother that he would consent to nothing more than being friends; his time was, after all, taken up with writing his senior essay.

The rich, gossipy correspondence of the Kimball Union years grew, spreading its leaves of wicked speculation, witty observation, and often unconscious revelation, especially between mother and son.[3] Oel Billings continued to write brief but happier notes, for the scarcity of money, though still present, was no longer so acute, Charles having graduated in 1839. Still, Oel could never bring himself to sign his letters more warmly than "with affection, your father." He sends Frederick a life of Henry Clay to read; Frederick replies that he already has one. The parents ship a garment they have had made; it does not fit, and they are distressed, while Frederick simply goes to a tailor and for fifty cents has it made right, instructing his father to get his money back from the Woodstock maker. The parents tell him he studies too hard, and Frederick replies that he must have a winter break filled with study. His mother advises him not to teach, for it is better to have less learning than poor health; Frederick teaches. She becomes ever more interested in finding a proper wife for him; he tells her not to press the matter. Oel tells him that he must board at a cheaper house; Frederick coolly replies that he cannot, as he has made a commitment to his present one, adding that he knows he must ease the family's financial burden and that he will move at the end of term. But he will not, he says, "make mincement of a business."

Frederick thought quite a bit about money, and intended not to be a burden on his parents in the future. He took odd jobs, conducted sabbath schools, and, drawing on his Kimball Union education, preached in small town churches in the absence of the pastor. By his senior year he had put away enough cash to be able to lend a little to those in need. "I would rather have others be in debt to me, than be in debt to them," he wrote his mother when she expressed mild surprise that her son was already becoming an entrepreneur. (He was, in fact, becoming a philanthropist as well, for he made his first gift—$5—to a charitable cause in the spring of his final year at college.) The last time he asked for a large sum of money—$25, mid-way through the fall term of his second year—he remarked that he was sorry to have to ask; still, he promised he was almost "at the end of this hill, at which I have been tripping, & when I am at the top then the stone will roll t'other way." He kept his promise, for by the second year after graduation, he was sending small sums home to help his parents. He would send them money all his life, in increasing amounts as he prospered, usually with warmly firm instructions that some of it be spent to bring his mother pleasure: help in the house, a new piece of kitchen equipment, a trip to a nearby hot springs.

He also made clear his incipient independence: he would come home for Thanksgiving, he wrote, having missed the last five, to "discuss the turkey." He allowed that this was his last winter at home, so, he gently urged his parents, they should all work to make it an agreeable one.

The correspondence also reveals Sophia's sense of hurt about life in Woodstock and her confidence that Frederick would ease it. Putting little Oliver, the youngest child, to bed at night, she heard him say that Freddie would one day become a great man and take the entire family to Burlington; she liked the thought. The wedding of John Dunbar in Benjamin Swan's front parlor still rankled, for no Billings had been invited, not even Charles, who had been very close to Dunbar. After the ceremony Sophia received part of the wedding cake and sent it back, causing "some sensation." Sophia was then told that she had not been invited because of her speech, and she wrote Frederick that it was best she have nothing more to do with "Elm Street society." (One wonders about her speech: her letters are quite literate, though Oel's were not always so, and they contain frequent flashes of unsuppressed wit.) Sophia clearly understood the need to be friendly with the Marshs, and Jacob Collamer, and the Dana family, but she nonetheless was clear-eyed about them: Collamer, she feared, drank too much, and "Lady Dana" overestimated her literary ability. Oel tended to dismiss them as the "White House gang," a reference to the old Congregational White Church they attended.

These strikingly frank, gossipy letters about who was in whose "set" trained Frederick Billings well in understanding the importance of human frailty. He kept the letters faithfully, and they read today with a candid, straightforward grace characteristic of both mother and son. From them we learn that by 1844 Oel, who was appointed town clerk in October 1840, was doing quite well, off to Albany on business, improving the family home, buying a piano, while Sophia was attending all (well, nearly all) the parties, "large & brilliant" affairs, and having people in to tea in steady streams. Her letters, virtually without blot—a suggestion that she cared enough to compose them first—flow on with news of the children, of sister Laura teaching in faraway South Carolina, of Charles helping his father prepare the Grand List and then, in 1842, leaving home to work in Lowell as a clerk, or of at last finding an Irish servant willing to shine Oel's boots every night without a word. Oel reports on town politics—there are persistent squabbles over who shall have the post office, and Oel is convinced that son Edward has been done out of a job as secretary for civil and military affairs to the governor in Montpelier—and hews to the Whig line. Oel gives the bad news as well: divorces, blindness from alcohol, the burning of his barn.

If Oel favored Edward, Sophia favored Frederick, and usually answered his missives on the same day, or at the latest on the next (leading Frederick to twit her on occasion about breaking the sabbath by writing to him). Frederick, in turn, was clearly the letter writer in the family for all of the sons, as Laura was for the daughters, and both parents often wrote to their children with the expectation that it would be Frederick who would reply if they were together. At times all this letter writing appeared a substitute for affection rather than an expression of it.

Letter writers have a different world view than non-correspondents, often using missives to put space between themselves and others, or to create an orderly sense of their world that may not fully reflect reality. They have, or come to have, rather more sequential minds than most, and—as with any act of writing—they need to select language enabling clarity of expression. Those who may take refuge behind the merely spoken word, however, need not be as meticulous in their use of language, for their expressions can always be disavowed or claimed to have been misunderstood. To reply promptly to letters, and especially to inconsequential ones, requires and reinforces discipline. It may also reveal ego and certainly a sense of responsibility. And the orderliness with which mother and son preserved their correspondence over the years suggests a sense of attachment, to each other and also to the past, to the need for continuity and accountability, that went beyond the norm.

In politics, a subject usually introduced by Frederick, he and his father at last found common ground. Woodstock is the most Whiggish of all Vermont towns and both are pleased. Frederick thinks politics "rather interesting," and, in midst of writing his senior piece, is invited to attend the Whig State Mass Convention (not a major distraction since it will be held in Burlington), finding himself in charge of the students in the procession. He helps to build an arch over the entrance to the university—it is, he reports, equal to anything in Baltimore, a reference to the fact that brother Edward is away in 1844 at the National Whig Convention in Baltimore to nominate Henry Clay—and Frederick seizes on the occasion to try his hand at a bit of journalism, hurrying off a short piece with the names of the Woodstock electors to his father with instructions to get it into the local paper if possible. He cannot imagine being other than a Whig, Frederick reports, and Oel is happy.

Throughout his four years at the University of Vermont, Frederick's affection for his sister Laura deepened. It was her letters from South Carolina that he most enjoyed and she whom he most missed, and they confided in each other as only brothers and sisters do. Frederick shaped these letters most carefully, for he clearly wanted Laura to think well of both them and him; much of the gaiety that his mother wished he would show to any of the several young women she suggested he lavished instead upon his sister.

This close-knit family sustained its first paralyzing loss that summer of 1844: Edward, Oel's and Sophia's first-born, the apple of Oel's eye, died in Baltimore while at the Whig Convention. His death was little spoken of afterwards, though to Frederick it was clearly instructive. For a year or two earlier the correspondence hints of a growing worry on Sophia's part that Edward was drinking too much. Much of his energy had gone into Whig politics and into the militia, in which he was a colonel. He persisted in hoping for a minor appointment in government but failed to attain it. Handsome, gay, in demand for singing groups, Edward had not married and his mother thought he was adrift. In Baltimore, after an evening of strong drink, he and a group of young men set out into the harbor in a rowboat, apparently to look back upon the city's lights from the water, and Edward fell overboard and drowned. He might have lived had any of his companions been sober enough to apply artificial respiration, for apparently his body was recovered quickly; none did. Edward's funeral in Woodstock was a major event: a large procession, with fifty friends mounted on horseback, followed by members of the bar, medical students, and the citizenry generally, made its way through heavy rain to the cemetery. The cause of death was never mentioned. Frederick did not attend the funeral; it was examination week at the university.[4]

The university had clearly become the center of Frederick Billings's life. He had worked to persuade the state assembly to provide direct assistance to it—for it was not yet state supported and in Frederick's freshman year the faculty had gone without salary—and he had labored to clean its rooms, master its ways, devour its books, and debate its teachers (as when, in 1844, he took on the historian and naturalist Zadock Thompson, who was in charge of astronomy at the university, on an astronomical matter).[5] His mother wrote that she hoped he might go on to study theology, his father mentioned the law, he hinted at a life of scholarship. Though he did not remain at the university, he never forgot it: a quarter century later, in 1869, he would accept appointment by Governor Julius Converse to the board of trustees, in 1871 he would take an interest in the possible purchase of a collection of specimens for the university, in 1885 he would dedicate a new library, at the time the finest in the East, that he had provided, and he would testify then of how dear to him old UVM had been. In 1890, very ill and unable to attend the ceremony, he would receive an honorary LL.D. from the university. Clearly the university was more than the seat of his education and the promise of his independence: at least in 1844, at the time of his graduation, it was his spiritual home.

Frederick was casting about for a job well before commencement, which found him marginally in debt and anxious for his parents' welfare (travelling to Baltimore to claim Edward's body and paying for appropriate mourning clothes had used up their small savings). Sophia was urging Frederick to be godly and to show it by turning to the ministry, while his father began a quite understandable crusade against alcohol by refusing to license any rum sellers in Woodstock.

Accordingly, various family accounts state that Frederick's "one ardent desire" was to go into the ministry. There really is little evidence for this beyond one letter to his mother, which clearly was intended to placate her, and a statement many years later to his wife Julia. While Kimball Union boasted that it prepared young men for the ministry, this was not the reason Frederick had chosen it, and at the university he did not pursue subjects especially intended for a ministerial career. His faithful attendance at chapel is not evidence, for attendance was required. He did enjoy discussing religion with his mother, and he kept an open mind, attending churches of various denominations in Burlington. But his behavior was not that of a person bent on a life in religion. The law, and perhaps a career in politics, clearly appealed to him most, and O. P. Chandler was prepared to take Frederick on as an apprentice.

3

First Steps into the World

FREDERICK DID NOT WANT TO FOLLOW in his brother Edward's footsteps. He was, perhaps, of two minds, for he strongly desired to be his own person and, reasonably enough, had not been pleased when his father had tended to favor his eldest son. Yet there is more than a hint in Frederick's actions that he wanted to prove that he was at least as worthy of his parents' regard as Edward had been, and this might best be served by taking on something Edward had done, or had hoped to do, and excelling in it.

The chance came in the summer of 1846. During the previous autumn Frederick had spent some time in Montpelier, the state capital, reading a bit of law, writing an essay or two for the university, and closely watching the political scene. He was working for Chandler, who had wanted a man in the capital and paid enough to meet Frederick's needs while he considered what direction his life should take. Oel's hold on his own position, first as register of probate and later as a pension agent in Woodstock, was precarious, for though the Whig party was in utter dominance in Windsor County, it was beginning to fragment into various groups, breaking on the rock of abolitionism. Edward's death had removed a prime source of funds, for he had lived at home and turned over most of his earnings to his parents. Oel feared, one suspects, both loss of income and the attendant loss of the prestige that had come to him through Sophia's social success.

To Oel, as register, it mattered above all who the judge of probate would be, and in October 1845, he learned that a strong campaign had been mounted on behalf of a lawyer, Walter Palmer, who would, Oel knew, remove him from his position. Oel asked Frederick to go at once to the town representatives from Sharon and Hartford, good friends through temperance connections, to tell them of the risk a Palmer appointment posed and ask them to press for the appointment of George E. Wales, the Hartford Town Clerk. Failing Wales, Oel hoped for John Porter, whom he knew through banking connections. Because the register had to live in Woodstock, it made sense, Oel concluded, that the judge should be taken from somewhere else. Oel requested a report by the night stage and promised that if necessary he would start for Montpelier the next day.[1]

Oel Billings was not displaced and his star would continue to rise in the Woodstock community, culminating in his election in 1847 as a trustee, and then secretary, of the Ottauquechee Savings Bank. It appears, however, that during the course of making inquiries on behalf of his father, Frederick met Horace Eaton, who shared the temperance persuasion, and learned enough about the realities of patronage to play the game for himself.

Eaton was a doctor with long experience in state government—sixteen years as a state representative and senator. He was a member of the 1843 constitutional convention and was now lieutenant governor. Frederick discussed education with Eaton—perhaps the problems of the university and almost certainly, from the evidence of Eaton's later correspondence, the need for reform at the secondary level—and at some point Frederick clearly suggested that he would like to be Eaton's secretary if Eaton were to become governor, and that he wished in particular to work with him on maintaining and strengthening popular education. Thus in September 1846, Eaton, now the governor, informed Frederick that he intended to appoint his secretary from Windsor County and urged him to get all his supporters to send letters on his behalf. He did not need a great number of names, Eaton cautioned Frederick—to solicit too much support would alert opponents unnecessarily—though he required enough to "show that I was acting upon public principles." The letters followed, and Eaton appointed Frederick secretary of civil and military affairs, with an annual salary of $100.

The historian must be careful of indulging in cheap psychology. Still, it cannot be without some significance that Frederick had set his cap for and won the very position his brother Edward had wanted. Frederick certainly was more qualified for the position than Edward, but qualifications often played a small role in lesser government offices in the nineteenth century. He confessed to Laura that he would never have gone into the law had Edward lived, for there would have been no place at Chandler's firm, and he believed that a good bit of life was governed by luck. Perhaps he wished to demonstrate that he knew his own mind and could serve his own interests without conspiratorial pressures from his father—and ineffective ones at that. If a sense of competition with Edward for his father's regard did play a part here, Edward's ghost must now have been laid to rest.

Sophia was delighted, though Edward's ghost was not yet exorcised for her. She put Frederick under heavy pressure, telling him he was her "precious jewel," her only son who professed godliness, the member of the family of whom "the most was expected" and who would, she knew, meet all those expectations. But she had heard that, in celebration of his selection, Frederick had been seen playing cards and drinking. Nothing was worth such an evil, and she sounded the alarm: Frederick above all the children must stand for God. He was not again to touch "the accursed thing that made shipwreck of beloved Edward," and that was, she hinted, now destroying some of the most honored figures of Woodstock society.

Frederick replied by discussing education. Public support for the common schools was still new and uncertain, and Eaton sensed that the alliance of prison reformers, abolitionists, anti-Masons, and others who had formed behind the common school movement was likely to be broken soon. The revival of the schools owed much to an educational innovator from Massachusetts, Horace Mann, whose chief disciple in Vermont was Thomas H. Palmer, author of an influential essay on education that had won first prize in a contest run by Mann's *Boston Common School Journal.* In 1841 Governor Charles Paine had appointed a committee to investigate the state of education in Vermont, and his successors, John Mattocks and William Slade, also favored Palmer's ideas.[2]

As lieutenant governor, Horace Eaton worked to help persuade the General Assembly to pass a comprehensive education act in 1845 that created three new offices to begin the process of bringing order to Vermont's 2,750 school districts. There would be town superintendents, county superintendents, and a state superintendent of education. Eaton became the first state superintendent and continued in this role while governor. Increasingly preoccupied with rifts in the Whig party, the governor came to rely heavily on his young secretary to read and respond to the county superintendents' reports and draft gubernatorial statements on education. Convinced that there should be far fewer school districts—on average each had only thirty-seven pupils—Eaton and Billings campaigned for consolidation, new schoolhouses, and better equipment. They wanted free, tax-supported common schools throughout the state (a goal not achieved until 1864), and gathered supporting statistics that such a system was entirely feasible. Much of Eaton's path-breaking first annual report drew upon materials assembled by Billings, who proved an alert researcher. Eaton happily reappointed him, without needing letters of recommendation, to a second term.[3]

By late spring 1848, however, Eaton's faction within the Whig party was falling out of public favor as the slavery and alcohol issues grew, and Frederick realized his patron would soon be looking for another job. Eaton made a peregrination through Vermont in June, getting in touch socially with "the upper ten thousand," as he wrote satirically to Billings, and reported that those who were spreading "systematic meanness" were in the ascendency. Eaton was suffering from depressing headaches, Montpelier had accentuated his asthma, and he turned to his young assistant for good cheer. But as he watched Eaton decline, Frederick concluded that the anxieties of public office were not for him, or at least not yet.[4]

In 1848 the Whig candidate, Carlos Coolidge, won the statehouse as a minority governor only by moving well out in front of the national party to advocate total abolition. George Perkins Marsh, a five-year member of the national House of Representatives on whom Billings looked with enormous respect, won reelection, and Woodstock remained firmly in Whig hands. Statewide, however, disaster was narrowly averted, with the Whigs claiming victory only because the Democratic and Free Soil parties, which took 57 percent of the vote, split the opposition. Abolitionists were now in the majority and the resulting fragmented and shifting factionalism would dampen any temporary enthusiasm Billings might raise for pursuing elective office.[5]

By now Frederick knew that he wanted to move on. He had read enough law under O. P. Chandler to qualify for a shingle at a time when qualifications were not yet precise. A friend working on Wall Street had written in May 1848 to encourage Billings to come to New York City (since Frederick had confessed a growing boredom with Montpelier) to work for the law firm of Horace L. Clark. Frederick responded that while he was interested, he would not accept a clerkship. His friend replied that he would be a clerk in name only, that his destiny would be in his own hands, and that he would be paid $300 for the first year, a fine sum even given higher living costs.[6] But Billings's mind was turning elsewhere. In May, apparently for the first time, he had spoken to a friend about the prospect of going to California.

4

Laura

CONVENTIONAL ACCOUNTS—histories of the Northern Pacific Railroad, brief sketches of Frederick Billings as businessman—suggest that he went to California because of the gold rush. However, family tradition holds that he journeyed West primarily to accompany his sister Laura and incidentally to try his hand at the law. Family tradition is nearer the mark.

Frederick and Laura were good friends: chums enjoying a loving, open relationship. Three years Frederick's senior, Laura was his only older sister. They exchanged confidences of the "don't tell mama" kind, and at least once Laura wrote so frankly she asked Frederick to burn her letter (he did not). Attractive, high-spirited, and altruistic, Laura possessed a good intellect. She also liked to do the right thing, to be well organized and disciplined, and she hoped for a little glamour and excitement in her life, though she was uncertain where to find it. Frederick was enthusiastically ready for every possible adventure, full of zest, and so too, within the limits permitted by contemporary strictures, was Laura. They were kindred spirits.

At some point in 1838 or 1839, while Frederick was away at Kimball Union, Laura went to Pendleton, South Carolina, to work as a teaching assistant to Mary Bates, a close friend who had taught in Woodstock the previous year and had been appointed head of the Pendleton Female Academy.[1] The academy had opened in 1827–1828, and in the somewhat condescending words of Pendleton's nineteenth-century (male) historian, R. W. Simpson, it taught "the young ladies etiquette and French, graceful attitudes and high-falutin notions, modern manners, to walk daintily and to scream fashionably at a bug or mouse."[2] (In fact, Laura taught music and French.)

The academy, under the patronage of South Carolina's noted Senator John C. Calhoun, was very much an institution of its time. Pendleton was in the full tide of prosperity, a town astride fertile fields serviced by the old slave regime, its wealth in agriculture and land and not yet in trade. Simpson wrote (sounding a bit like an early-day Garrison Keillor) that the community "was famed for the beauty and gentleness of her women as well as for the high-tone and pluck of her men'' (and, he added, for the happiness of its slaves). Laura reported otherwise and had mixed feelings about her work; she taught in Pendleton until 1844, each year agonizing over whether she would return, and then loyalty to Mary and Anna Bates would bring her back again. She confided to Frederick of the scandalous elopement of a student, admitted to several gentlemanly suitors from the nearby military academy, and referred gently to the growing agitation over abolitionism, which she and her family would discuss when she was at home in the summer. She enjoyed teaching—and advised Frederick that whatever he ultimately chose to do, he should teach for a time—and remained steadfast to the Bates sisters until, at last, a successful suitor kept her in New England.[3]

The suitor was Bezer Simmons, New Hampshire-born and Woodstock-bred, now a ship's captain out of New Bedford. Laura and Bezer met in Woodstock, when he was visiting a relative, and he took to calling on her whenever he was in town. Good looking, apparently determined, ten years older than Laura, he and his prospects seemed promising. Still, his long absences at sea on his whaling vessel, *Magnolia*, did not endear him to Sophia Billings. And gossipy hints of possible minor financial improprieties further worried her. There was a stiffness when he was in the house that kept any warmth from developing between Simmons and the rest of the family, and Laura confessed to Frederick that Edward, who as the eldest was entitled to a formal opinion in the matter, was very opposed to her marrying Bezer. Charles "raves at the bare mention of a Simmon's [sic] name." She admitted to being very fond of someone else, of whom, however, Oel was suspicious, and it seemed clear her parents would disapprove of any marriage contracted in the South.

Then, in January 1844, Bezer presented his ultimatum by mail, and Laura surrendered, agreeing to the marriage. There were repeated delays, however, for Sophia wanted the captain to remain at home and renounce his seafaring ways, and he seemed in no hurry to return from the Pacific. When Simmons announced that in June he would be off on another long voyage, perhaps for two years, Sophia declared the arrangement "disappointing." At first Bezer talked of taking Laura away to live for five years in Brazil, but when he failed to sell the *Magnolia* on a trip to Pernambuco, this idea had to be given up, to the joy of the Billings family. Simmons called on Frederick in Boston, mindful of the desirability of his approval, and apparently it was given, for Frederick wanted Laura to have whatever she had decided upon. By slow fits and starts, with Oel approving and Sophia now silent, Laura and Bezer moved toward marriage, which took place on March 26, 1845. Bezer left again three months later.[4]

The marriage was not a particularly happy one and there were no children. Bezer was often away in pursuit of whales. In 1846 Laura returned to South

Carolina. She wrote to Frederick that the captain was away and had disappointed her by not returning; because he had not "accounted for his four hundred sperm," he must go on to the Sandwich Islands, from which he returned in 1847, only to set out again a few weeks later. No one was happy with the situation, and when Simmons returned in October 1848, having purchased property in California, the family made it clear they found him more devoted to his ship than to his wife. He was about to be put into drydock.

But Bezer Simmons had come back full of the news of an exciting discovery in California. In January 1848, James Wilson Marshall had found gold on the land of Johann Augustus Sutter, and by May news spread that a major strike had occurred. At the end of the year the gold yield exceeded $10,000,000 and the gold-hungry were pouring in from Oregon, Mexico, Peru, and Chile. Ambiguities arose concerning land ownership in California, and lawyers as well as miners would be needed. By the early summer of 1848 the full force of the news hit the East Coast, and Frederick then confided to his college friend that he was thinking of going to California.[5]

This became a firm decision when Bezer and Laura announced that they were leaving for the West Coast, according to Bezer, "for life." A ship captain could expect to do extremely well in California, even if he did no prospecting himself. A fortune was to be made in carrying passengers to the isthmus of Panama or up the Pacific coast to California, another in running gold seekers from the coast to the Sacramento River. Simmons might well find gold too. In any event, unlike so many hurrying to the West, he already held property in Yerba Buena,* so he would have a base from which to start. Laura was unhappy with Bezer's decision—to her journal she confessed her heart was heavy "at the prospect of going so far away & not a word of sympathy from my husband"—but determined to go through with it; seeing this, Frederick decided that he must join them. If he did not have a try at the California madness, he might well regret it, and with Laura to talk to, he could enthusiastically look forward to a year or two of new experiences in one of those lands that, if not on Mrs. Williams's wallpaper, would soon deserve to be.

Frederick researched the trip, and he and Bezer agreed they would go by the isthmian route. Leaving Woodstock the day after Christmas, 1848, they journeyed to New York to make final arrangements. Frederick's spirits were high: "O this California," he wrote to his parents, "what a madness there is about it [;] everywhere it is California, in Congress & on the streets, in the pulpit, in the press, out doors & in doors."[6] His next younger brother, Franklin Noble, six years his junior, also wanted to try his hand in California.

* So read Simmons's indenture, since he purchased the property in 1846. In January 1847, Yerba Buena was rechristened San Francisco. After the change became official in March, the former name remained attached to an island in the bay where the "good herbs" of the original Spanish name did grow.

Sophia and Oel were frightened at the thought of losing three of their children so quickly to the madness, but Frederick urged them to let Frank go, for there would be thousands of young men there like him.

Frederick actively solicited advice from friends and considered his course of action. He determined that he would not dig gold; rather, he would practice law and buy real estate, for in so prospectively litigious a society, the law would be "better than gold." He would take only one or two partners and would avoid joint stock ventures, so that he might stand essentially on his own. Legal fees would provide leverage—he did not use the word, of course—for the real estate. And he would take an immediate interest in politics so that the New England Whig stamp might be impressed upon California. Horace Eaton wrote cautioning Frederick not to believe that gold would add to his wealth: it had little intrinsic value. Rather, he must be a force for good in the boisterous, booming land, so that people would look up to him as a model; he must remember that he would, one day, return to Vermont and must be worthy of his state's respect.[7]

The Simmons-Billings party engaged second-class passage on the Atlantic Steamship Company's *Falcon*, scheduled for departure from New York on February 1, 1849, for Chagres, on Panama's Atlantic coast. (The party did not include Frank, who followed on his own eight days later.) Frederick obtained letters of introduction from Governor Coolidge and from Eaton, from members of the bar, and other well-placed friends. (Bezer, in turn, obtained a letter from Senator Stephen A. Douglas of Illinois.) Billings arranged that he should be a commissioner (that is, a notary public able to execute documents) in California for legal matters relating to Vermont, Connecticut, Massachusetts, and later for New York and Maryland. He borrowed $1,000 from Albert Catlin, a family friend in Burlington, and paid off Bezer's debts. He had a lawyer's shingle made that read FREDERICK BILLINGS, ATTORNEY AT LAW, and packed it away carefully with a large box of texts on contracts. When Laura came to New York to join the men, Frederick thoughtfully had portraits of the three of them painted as farewell presents for Oel and Sophia. The elder Billingses regarded the three as "the young people," breaking from home for the first time, with every chance that they might not return. They were not, by the standards of the time, quite so young, and though Laura grew more depressed as the sailing date approached, Bezer—at thirty-nine embarking on perhaps his last chance to achieve success—and Frederick, twenty-five, grew more and more enthusiastic.[8]

The party, pressed and shoved by the crowd, boarded *Falcon* at noon in the rain on February 1, exhausted from staying up very late the night before packing trunks. A group of friends saw them off, a small knot of people obscured by a great throng. As the vessel weighed anchor, deafening cheers went up from the wharf and were answered from the ship—"death cheers to me,"

Laura wrote in her diary—and at one o'clock *Falcon* slipped out of port. Frederick rushed to Laura's stateroom, a tear in his eye, and caught her hand: "This is the last of home," he said, and they sobbed together. Within half an hour the ship was plunging and Laura, seasick, took to her bed.[9]

The journey to Chagres was typical of the era. Drizzling rain followed the pilgrims past Cape Hatteras. Each day passed into another, the monotony broken by counting ships. Laura grew more ill and nearly went into convulsions, so a fellow passenger who was a doctor gave her doses of ether. She lay on deck during the day, wrapped in dressing gown and cloak; there was little space, for *Falcon* was grotesquely overcrowded, with 314 passengers in space designed for 100 and bunks rigged in the dining saloon and hold. A substantial crew, cooks, and an infant added to the crowd—Frederick estimated there were "upwards of 400 mouths to feed"—and many passengers never dined at table during the entire voyage. Only one other woman was on board. Frederick hovered near Laura throughout the journey, his earnest brown eyes ever alert to her needs.

For Frederick the voyage was exhilarating. He found the passengers agreeable, his stateroom companions pleasant, a school of porpoises exciting. On the first Sunday out, at last under a scorching sun, the passengers sang hymns together and an Episcopal service was read. A German declaimed from Isaiah, "Remember ye not the former things, neither consider the things of old." Off Charleston, South Carolina, a mail packet came alongside and Laura received six letters and three gift books from the Misses Bates.

On the sixth the passengers knew they were entering a different world, and for the first time, the sun bright, all came on deck to observe the Florida Everglades, so vast that *Falcon* was all day in passing them. Laura, having recovered, dutifully began to study Spanish as the ship was nearing Havana, where the passengers spent a "joyously happy day" on shore, dining on fresh fruit and attending the opera in the evening. "Everything," Laura wrote, now in good spirits, "was new & strangely beautiful."

The crossing of the Gulf of Mexico was extremely hot to the Vermonters, the sea like glass, the food disagreeable. A second sabbath passed, with all quietly in prayer and no sermon or singing, and at two o'clock in the morning of February 14, with the smell of land permeating the coastal fog, *Falcon* arrived off Chagres. The passengers chose a committee of six, two representing each class, to make arrangements for the isthmian crossing. Frederick was elected secretary.

At dawn the captain went on land with the committee to learn that *Falcon* could not get in to shore and no lighter was waiting. He would have to move the ship eighteen miles along the coast to Limón Bay to discharge his passengers. In the meantime *Crescent City*, which had set out from New York five days behind *Falcon*, passed at a distance, and by the time *Falcon* dropped anchor in Limón Bay many of the passengers off *Crescent City* had hired the larger canoes for passage up the Chagres. The *Falcon* committee thereupon engaged the *Orus,* a 250-ton side-wheeler out of New Orleans and the first river steamer to begin plying the Chagres. Unhappily it could accommodate only half the clamoring passengers. All made the decision that the two ladies should be in the first group to lessen the time they would be exposed to Panama fever, and thus Frederick and Bezer also joined it.

The captain of the *Orus* engaged fresh canoes, called bungoes, rowed by totally naked boatmen. The previous December, when the first Americans had put ashore on the way to California, the native villagers had no notion of why they were in such a frantic hurry and had charged only a dollar for the journey. Now, seven weeks later, they had learned enough about supply and demand to charge $15 or more. While the committee negotiated, the passengers walked about on shore, visiting with a white woman who had been there for two months, swatting mosquitoes as they talked, unmindful of the causes of malaria. In the early afternoon of February 15 *Orus* chuffed up-river towing fifteen small bungoes which constantly became entangled with lush vegetation along the river banks and periodically grounded on gravel bars. Still, the prospect was beautiful. The river, Laura noted, was so winding and so colorful it was "like walking through a green house with labyrinthine walls attached."

After a stop at Gatún, a town of mud huts, *Orus* drew up at sunset at Los Hermanos, twelve miles from Chagres. The town boasted two huts and no accommodations. The vessel could go no further, and that night the two women rested on board on two narrow settees while the men lay down on a cabin floor slippery with tobacco washings. There was no water and most passengers drank from the river, despite the firm advice of a minority that they ought not. No one slept. The *Orus* moved on at daylight since the river level changed, but after three miles could go no further, and the two women and their husbands scrabbled into a bungo. Five natives could pole and paddle one of these crafts—a hollowed-out tree trunk that held four to six passengers with their baggage—ten to twelve miles a day in the rainy season even against the famed currents of the Chagres, which were known to shift and rise and fall almost without warning, and if offered double pay some would pole in shifts through the night. The polesmen had created small roofs of bamboo to cover one end of each bungo, under which two passengers could take shelter. Again there were not enough bungoes, so Frederick remained behind to contract for one of those that would be coming back the next day.

The slow journey up river was exhausting. The distance from Chagres to the town of Panama, in a straight line, was not quite thirty-eight miles, yet the twisting, urgent river, the hard, muddy crossing over the heighth of land, and the rough, jouncing descent by mule down the Pacific slope might well take seven days. It was unsafe to sleep on the shore—Laura observed two six-foot alligators well up among the trees—and so the women rested in the bungoes while the men, guarded by a servant, slept in a cleared space high on the river bank, since the river might well rise eight feet in the night. Two canoes upset, while the Simmons-Billings bungo was turned twice around by the force of the current, though it did not capsize. The passengers were always wet and cold, apprehensive of striking one of the large timbers that drifted down the river toward them, or of being driven against one of the trees that seemed to rise suddenly out of the water. A little cold meat and some biscuits, nothing more, provided nourishment, for February was at the edge of the rainy season and the coffee sold on shore stank and the rice was maggoty. Darkness came suddenly, and at twilight the Indians refused to go further, drawing the bungoes in toward the bank.

The party was warned that smallpox was rampant in the provision huts above their first night's camp, and Bezer went alone to verify this. Finding everyone with fever, he returned to say they must remain on the river. By 1851 Chagres was well known as an unhealthy place, the very soil supposedly poisoned, and the terms "Chagres fever" and "Panama fever" were household phrases for pernicious malaria and yellow fever. By 1850 some insurance companies would void a life insurance policy on any person who remained overnight in Chagres. But in February 1849 there was as yet no established record for the pestilential passage. The crucial role of the mosquito was unknown, and passengers, little perceiving that there were hidden dangers, feared only those dangers they could see, such as alligators. A passenger died on *Orus* the day before the *Falcon* group boarded it, and another had died in one of the tents on shore at Limón Bay; no one commented on possible causes.

Refreshed by coffee supplied by a Virginian who had the foresight to bring his own, the enlarged group—passengers from *Crescent City* having camped the night at the same spot—set out with the American men at the poles, convinced that they could do better than the "lazy oarsmen." Though he lost his hat, Bezer stayed with the poling until noon, when the convoy—for such it now was, with several canoes strung out in a line to the extent that the vagaries of the current would permit—stopped at an island on whose high bank sat a small ranch. The advance bungoes bought some milk and a chicken, and the party had hot broth. That night, exhausted yet filled with wonder and admiration for the incandescent vegetation surrounding them, the tunnel of river parting it like a moving sword, the group reached Gorgona, at last on relatively high ground.

At Gorgona mud huts, without floors or windows, had signs proclaiming them hotels. Laura opted for the Hotel Française, where she and Bezer obtained the only room. It was hot and insect-ridden, however, and Laura quickly accepted an invitation from the officers of the Isthmus Survey Company to come to their encampment and pass the next day in a tent. There, in a dirty and torn calico wrapper and broad-brimmed straw native hat, she received callers. After the long days at sea, the smell of vomit, the wash of tobacco juice, the mud and heat and hovering mosquitoes, the tent was "so nice & the green grass so clean."

The next day Laura visited a man who had been shot while coming upriver, since he had sent word that the sight of an American woman would do him much good, and she returned to find that Frederick had arrived. Happy that she was safe, he sank into an exhausted sleep on the ground by her cot, not having undressed for three days. This was the first sabbath without even a religious prayer service in a very long while, but neither worried.

On Monday morning the group began the mountain portion of the isthmian crossing. The Survey Company had constructed a palanquin for Laura. It was to be carried by four men who between them would be paid $28, a good price since a riding animal cost $10, and double that if the traveler wanted someone to lead it along the rocky path. Bezer had a horse and Frederick, having arrived later, a rather bone-weary mule. Two locals were hired to carry the trunks of money and provision, with the rest of the baggage sent along in a few days. With three hearty cheers from the survey crew, the explorers were off, almost certainly not prepared to agree with Robert Louis Stevenson's notion that to travel was better than to arrive.[10]

The mountain crossing was completed in two days, though with difficulty. The path was so steep that the makeshift palanquin had to be abandoned after a mile lest Laura be tossed out, and she rode on Frederick's mule as he walked beside her. The trail often was so narrow that no one could pass, and if anyone were encountered in the opposite direction it was necessary to back up, mule and horse clumsily skittish in the drying mud. There was neither steady ascent nor steady descent; the terrain consisted of one seemingly interminable ridge after another. The night was spent just beyond the summit, in the tents of a group of surveyors, and the following day the band, grown to twenty as others pitched their tents near the survey camp, descended steadily to Panama, there to await their baggage and the arrival of their steamer, *Oregon*, on which they hoped to depart quickly for California.[11]

Panama City was crowded and in panicky turmoil, with hundreds of passengers demanding space on the next vessel. In January, when *Oregon*'s sister ship *California* had anchored at Panama ready to take on 150 passengers (who would be the first "Forty-Niners"), five to six times that number had fought and bribed their way aboard. Everyone feared a similar scene, for the town bulged with "bungo artists." *Oregon* arrived on the night of February 23, and a member of the *Falcon* party went along to confirm their passage, only to find that there were problems. *Oregon* had been ordered to take no more than 250 passengers, though it soon became apparent that the captain, R. H. Pearson, would have to increase the limit if he were to get away at all. Eventually

Pearson agreed on three hundred, but this meant finding provisions at outrageous prices where few were to be had. Another delay occurred when Pearson reluctantly agreed to double up and accommodate four hundred passengers, and seventeen long, very hot days dragged by before *Oregon* sailed. While the Vermont party had been warned that it was unhealthy to pass many days in the village, nothing could be done about it. Bezer, Frederick, and T. L. Chapman, a doctor from Castleton, Vermont, with whom they had struck up a lasting friendship on the crossing, called round daily to the shipping office of Zachrisson & Nelson, to the vessel itself, and to the American Hotel to guard their places, for they were in no position to bribe their way aboard.

Laura studied Spanish, wrote letters, made a riding skirt, and suffered from a steady headache. One evening she braved the night air to see an old church by moonlight. In a few days the party's baggage descended the mountain to them—by the next year travelers dared not leave their belongings behind, for even in the American Hotel trunks were forced open by the natives who, in 1849, were noted for taking nothing and by 1851 were suspected of taking everything—and they enjoyed the comforting ritual of repacking. They made many friendships during this time, all useful to Frederick at his destination— John B. Geary, who would become mayor of San Francisco, and whose wife was the other woman on *Falcon*, the Reverend Albert Williams, to whose Sunday sermons at the American Hotel they thrice went, and a dashing Virginian (he of the fresh coffee?), Archibald Peachy. Late on March 12 they were taken in bungoes to sleep on *Oregon*, which captured the breeze that blew offshore, and the next day set out on the third leg of their arduous adventure.

By now Laura was very tired. She had abandoned her lengthy, observant journal for brief entries in a small diary. She seldom noted her husband in either, and did not mention him again until reaching California. *Oregon* made its way through terrible heat, the sea like a mill pond, the sun "almost dissolving the deck." Laura took to a settee again, diverted now and then by sighting a large whale or flying fish. The drinking water was stale, the ship stank, the food was, as another passenger remarked, "terribly bad" and unhealthy. On March 29 Laura dreamt of home for the first time. Two days later she stopped writing in her diary.

Oregon entered San Francisco Bay on April 1, but did not anchor until the next day, and not until April 3 did passengers go ashore. The captain had briefly to put down a mutiny amongst his crew, who wanted to leave ship instantly for the gold diggings, and he could get no one to unload the baggage. *California*, though it had reached harbor a month before, had not been able to sail again because of wholesale desertion, and Captain Pearson was determined to get back to Panama as quickly as possible for more passengers. Rather than anchoring, Pearson steered *Oregon* under the guns of the USS *Ohio*, whose pennant was flying off Sausalito, and marines from *Ohio* restored order. Pearson put several men in irons and tried to hold the rest of his crew by posting a new payroll: it was said he simply added a zero to each rate from the previous schedule (and then passed the increase on to the next load of Forty-Niners). The weather remained unexpectedly cold and the Vermont party was unprepared both for it and for the chaotic arrival. The passengers (and two of the crew, disguised in dress coats) were warped ashore on a floating stage.[12] The next day Laura and Mrs. Geary created a sensation walking in the streets, there being so few women in San Francisco.[13] A day later Laura fell ill, confined to bed with what the doctors diagnosed as Chagres fever.[14]

Frederick and Bezer, as well as the three doctors in attendance, did not realize Laura was seriously ill until it was too late. The once-neglected Bezer rose to the occasion, sitting with her night after night. Frederick was almost constantly present, "as kind and attentive as a woman": one of the physicians wrote that he had never known "such devotion between brother & sister[;] they seemed like lovers."[15] On April 24, holding tightly to Frederick's hand, Laura told him he had been a precious brother to her and that it seemed she had known him for forty or fifty years. She then handed him her Bible and died.[16]

5

No Small Affair

THE CIRCUMSTANCE OF LAURA'S DEATH wove for Frederick a network of
California friends that he never broke. He remained in touch with Dr.
George F. Turner, one of the attending physicians who had sailed with them,
through Turner's peripatetic career on the Texas frontier, acting as Turner's
lawyer and purchasing land with him in Benicia when they concluded it was
likely to become California's administrative center.[1] Billings corresponded with
Dr. Chapman long after he had returned to the East Coast, and in 1874 Chapman reminded Frederick that he still treasured a daguerrotype of Laura. Billings stayed in contact with Dr. Augustus J. Bowie, the third physician (and
the only one with tropical experience), for Bowie had written a full medical
note for Frederick to include in his letter to his parents. Laura's last request
had been that Frederick should write to Anna Bates Butler, and Mary replied
at length for the family: the Bates, Butler, and Billings clans were even more
closely knit by this Victorian bond of shared grief, and though the friendship
cooled as sectional tensions increased in the late 1850s, Laura remained a link.
Each of the ministers who visited Laura while she was ill would command
Frederick's faithful financial support for life, and Frederick would remain in
touch with many he had met on *Falcon* and *Oregon*: of 161 names traceable
to *Falcon*'s departure from New York harbor, Frederick was still in contact
with at least 38 when he left California for the last time, in 1863.[2]

There was the grim thought that Franklin, his twenty-year-old brother,
unaware of the disaster, would arrive soon and have to be told, deepening
Frederick's sense of grief. At the time of Laura's death Franklin was actually
somewhere in the rough Atlantic seas south of Rio de Janeiro. On February 8
he left New Bedford on Bezer's converted whaling ship *Magnolia*, with Bezer's brother Benjamin as captain, traveling before the mast on a palm-leaf
mattress, fortified by a bushel of Woodstock apples. The voyage was extremely rough; knowledgeable sea captains aboard declared they had never

experienced such gales, and the ungainly *Magnolia* performed as though it were a floating loaf of bread.

He did not reach San Francisco until August 29; frequent delays occurred along the route and the *Magnolia* had been driven by high winds far into the Atlantic. He was shattered to learn that his sister had been dead for over four months. Frederick and Franklin, sharing their sense of misery, agreed that they "only want[ed] a little gold" and would then go home.[3]

Bezer, in these four months, had expanded his business substantially. He had promised to accompany Laura's body to Vermont, and had wanted to send Oel and Sophia a variety of mementos of Laura, but had done neither as yet, for he was fully engaged in business. His land in Novato, north of San Francisco, was escalating at a dizzying rate, as were his lots in Sacramento (he eventually owned 132), while prospects for profit in using *Magnolia* to ply the route to Sacramento were too attractive to give up. His store in Sacramento—Simmons & Hutchinson, partnered with Titus Hutchinson, Jr., who had come out from Woodstock on *Falcon*—already required eight clerks. In September he joined several investors in further developing steamer traffic to Sacramento. He had become "a great man," Franklin reported, and was looked up to: indeed, he had just been elected to the *ayuntamiento*, or town council. Still, Franklin predicted grief for Bezer Simmons: he had taken in as a partner a Roman Catholic, John Francis Pope, who had come out on *Magnolia* as supercargo, and was allowing his brothers Frank and Nathan too much leeway; further, their wives, Maria and Julia, were not happy in California.

By now relations between Frederick and Bezer were strained. Frederick felt Bezer had been insufficiently attentive on the isthmian crossing, and he was inclined now to remember how his mother had been against the marriage. He opposed Bezer's decision to take in Pope as a partner (though not on religious grounds), and thought Bezer a little too inclined to resort to guile to achieve his ends. He also thought Bezer rather imprudent in his investments. Oel waspishly declared that anyone having anything to do with Bezer would lose money.[4]

Such proved to be the case. *Magnolia*, put on the Oregon rather than the Sacramento run, did not prosper. Money sunk into the Rancho Llano Seco was not giving any returns. With his partners, Simmons had invested in the town of Vernon, on the east bank of the Sacramento River opposite its junction with the Feather River. In the winter of 1849 Vernon had been at the head of navigation for vessels to unload, and warehouses and stores had popped up along the river banks. But the Feather River was seldom so dry as in 1849, and heavy rains in the winter made the rivers rise, so that ships could go on to Marysville. Vernon died, utterly. Thus the "needful" (as money often was referred to then) had to come almost entirely from the trading company of

Simmons & Hutchinson. The Sacramento branch was hard-hit by a flood early in 1850, and June 13 witnessed the destructive blow, as a great fire swept through San Francisco and reduced the warehouses of three major mercantile houses—S. H. Williams & Company, Macondray & Company, and Simmons, Hutchinson & Company—to heaps of ashes, high pyramids of lumber burning for days and emitting great crackling sounds, a funeral pyre for Bezer's dreams. With the warehouse went most of his merchandise, recently arrived and waiting for transport to Sacramento and the mines.

On September 8, 1850, Simmons, Hutchinson & Company closed. The last indignity had occurred the day before, when there was a run on its banking and commission counters. The company was forced to make an assignment of all its property to meet debts of over $230,000. Simmons, bed-ridden at the time, found he could pay only thirty cents on the dollar. Frederick called in money to help but apparently raised only $3,000, for he too was quite ill and confined to his room. Franklin, now the firm's single clerk, sold the goods and fittings, and Simmons was able to pay off $80,000 of his indebtedness. Whether from grief, for both his mother and his brother Benjamin had recently died, from loneliness after Laura's death, from rheumatism, or from an undiagnosed wasting disease—Franklin reported that Bezer had been reduced to skin and bones and the contemporary record tells us that cholera was rampant at this time—he died on the night of September 26, 1850. All shipping flew at half mast for Bezer Simmons the next day, and the procession to his grave drew over half the city's population.[5]

Despite such personal tragedy, Billings did not waver in his belief that in coming to California he had done the right thing. He was, in fact, markedly prescient about his new home, predicting the day would come when California would be more powerful and rich than France, that it would be the largest and most productive state in the Union. He said that its citizens would sway national politics, anchor the continent, might even produce a president of the United States. California's true resources were its people, its grass, its grain, fruit, timber and cattle, not its minerals, he wrote, and soon a railroad would link this vast new land with the rest of the nation of which it had so recently become a part. The day also would come, he later wrote, when he could fly to friends on the East Coast across wires, with distances brought to little account. Because he never forgot the long sea voyages and the horrid bungo and mule trips he, Laura, and Bezer endured to reach San Francisco, he was constant to his dream that the crossing could be, must be, reduced to a matter of hours. The surveying teams in whose tents Laura had slept on the crossing were harbingers of the application of steam to the isthmus, by either rail or canal.[6]

Billings was intrigued by the idea of a trans-isthmian canal in Central America. While not particularly sentimental on the subject, he always believed that his sister's death would not have occurred had there been no long wait in Panama City for transportation up the coast to California. His un-happy memories fueled his interest in an alternative route through Nicaragua, for which there were sound diplomatic and business reasons.

Frederick agreed with those who thought the Nicaragua route might be quicker (and thus healthier as well as cheaper). Three weeks after *Falcon* had departed New York, the brig *Mary* had sailed with 130 passengers for San Juan del Norte, near the mouth of the San Juan River. Though the voyage had taken twenty days—longer then the passage to Chagres—the crossing by land was faster and far more pleasant. Billings reasoned that if West Coast shipping could be made dependable, the Nicaraguan passage was a better choice. However, with the completion of a railroad across Panama in January 1855, reducing the isthmian crossing to three hours under favorable conditions, Bil-lings's interest in a canal route was put aside and not taken up again until the 1870s.

Frederick had passed his first day in San Francisco walking about, observing the frenzy. The harbor was like a forest burned over and resembled a vast graveyard, so clogged was it with abandoned ships lacking replacements for crews that had deserted for the gold fields. The vessels were jammed so close to each other, some still in the water, others drawn up on dry ground to serve as makeshift hotels, that a man in a small boat could hide from the law simply by maneuvering through the intricate channels between those ships still afloat. The tiny shanty village of the year before had become a sprawling, ramshackle city with a population that would reach twenty-five thousand by the summer of 1850.[7]

On that first day Frederick confirmed three conclusions. First, he would not go to the diggings—indeed, he had not really expected to—for while one could bring out an ounce of gold a day, at $16 an ounce, a laborer in the city could earn $15 a day without risk. The gold rush was a placer rush—that is, the flakes of gold lay exposed in river beds or dry gravel and did not require sophisticated equipment or experienced miners. Thus many gullible, inexpe-rienced, and quite greedy men rushed to the gold fields—and there would be, as a result, many contested claims. Second, much money was to be made in the inflated real estate market. Third, the city's most immediate need was for building materials, and the best at hand were the planks from the ships in the harbor, which would sell more readily for their timber content than as sailing vessels.

Frederick immediately went into business: at the waterfront he purchased an abandoned canal boat for $200, selling its timbers at a profit. With this money he made an initial payment on what he understood would be a "sand lot"—a lot above the level of high tide along the harbor—though in fact it quickly proved a "water lot," below the water line at high tide. He used the rest of his money to rent a tiny office on Brenham Place, where he slept on a cot until he collected his first fees, and then took a room above a nearby restaurant (where he discovered that breakfast for two cost $27.50). On April 14 he hung out his shingle, the first lawyer in San Francisco to do so, and for the next ten days—until Laura's death—and for the following two weeks or so thereafter, he worked frantically, buying and trading lots, handling any legal matter that walked through his door, using every spare moment to pass out his business cards.

Luck struck almost at once: through Billings's door came John Augustus Sutter, on whose land James Marshall had discovered gold the previous year. By 1852 Sutter would be a ruined man, with squatters settling on his lands, his workmen and herders deserting him, but at this moment he still felt in control of his destiny. He wanted a legal paper prepared and Billings obliged; when Sutter asked his fee, Frederick changed $10 (which was five times what the simple task would have cost in Vermont). Sutter cried, "Why, young man, you'll never get rich in this country on $10 fees," and threw down on the table fifty Mexican silver dollars, warning the young lawyer never to take less. (If this family tradition is true, it explains both why Frederick prospered so quickly and why Sutter was soon bankrupt.) By November Billings had purchased from Sutter, who was beginning to feel the pinch, twenty lots in Sacramento (to which, in the following year, he added an entire square in what would prove to be the center of the city).[8]

These first months were a blur of hard work, with Frederick taxing his bookkeeper, William H. Stowell, with dozens of transactions each day. Caspar T. Hopkins, a classmate from the University of Vermont, looked in on Frederick only a week after *Oregon* arrived, and Billings grabbed him, asking Hopkins whether he could make a town plat from the field notes of a dead surveyor and produce a dozen copies within the week so that he could sell lots. Hopkins sat to the task at once when Billings offered him $250—as much as Hopkins earned in a year in Vermont—for the job. When Hopkins discovered the prices of supplies—drawing paper at $20 a sheet, India ink at $10 a cake— he reported to Billings, who at once told Hopkins that he would meet any costs to get on with the work. The maps were for the town of Vernon, Bezer Simmons's ill-fated venture on the Feather.[9]

In these first months, Billings was willing to look at any honest means of making money: speculation in sugar, investing in a freight line between Placerville and Sacramento, building a fleet of barques for the China trade, taking his fees in gold dust. None would yield a fortune, though within three years Frederick Billings by hard work, prudence, integrity, and luck, would be a member of the most highly respected and prosperous law firm in California. Although Billings had been the first to hang out his shingle, merely being first had not led to success: after all, there were nineteen lawyers in San Francisco by the end of 1849.

Hard work, a keen mind, and luck were joined, however, by another significant factor: Billings became, in effect, the first attorney general of California, and through this office expanded the circle of those who were impressed with him. Late in April his friend Dr. Turner introduced Billings to Brigadier General Bennet Riley, who had just arrived as commander of the Pacific Division and seventh military governor of California. Riley was to assist the organization of civil government. On May 1 he appointed Billings commissioner of deeds for the district of San Francisco. A month later Riley issued a proclamation, calling on the residents of ten specified districts to select delegates to a convention in Monterey on September 1 to draft a state constitution. To certify elections, Billings was appointed chairman of the Board of Inspectors and Judges, and on August 1 informed those who were elected of that fact. At a stroke, at twenty-five, he knew the most important political figures in the proposed new state.[10]

Two other individuals accounted for Frederick's quick appointment to office: Colonel John Geary, who had ascended the Chagres with Laura and arrived with his appointment as San Francisco's first postmaster in his pocket, and Captain Henry Wager Halleck, secretary of state under military governors Richard B. Mason and Bennet Riley. Geary was elected the city's last *alcalde* (a combination of mayor and justice of the peace) on August 1, 1849, and thirteen days later Riley, at Halleck's suggestion, appointed Billings territorial legal adviser and attorney general (initially under the old *ayuntamiento*, or town council, the title was *procurador fiscal*); the appointment was retroactively effective as of the first of the month. Billings's annual salary was $4,000, which to his parents seemed a fortune.

Frederick Billings's tenure as attorney general was short but important: it was, as his mother concluded, "no small affair."[11] His most important action was to enforce the federal government's claims to land assigned to it, thus establishing the precedent that in time would protect the Presidio and, today, Golden Gate National Recreational Area, modern San Francisco's most important public land. A government reserve had been set aside by General Stephen Watts Kearny, "conqueror" of California and the first military governor. Billings read the scant documentation on the reserve and in October 1849 wrote to Riley to draw his attention to the "immense value" of any land set aside for public purposes. Already there were two encroachments on it, Billings warned: a significant part had been claimed at the end of 1848 under a purported grant of 1839, and one of Riley's predecessors, Colonel Mason, had neither inquired into the validity of the title nor intervened against the claimants. In the summer of 1849 a second encroachment had taken place, and

Riley had followed Mason's precedent, concluding that the claimants had a prima facie title and should be given the opportunity to test that title in the courts. Billings sternly told Riley that there was now a third precedent, for in September a building had been put up on Rincon Point, a detached portion of the reserve.[12]

Billings was determined that the people of San Francisco were to have their public reserve as park, parade and mustering ground, and as a hedge against the future when the federal government might need building space. He reported to the governor that he had already taken action, going with Captain Erasmus Darwin Keyes, the officer in charge of the Presidio—and yet another fellow *Falcon* passenger—to order the removal of the Rincon Point structure. If it was not taken down voluntarily, Billings informed the squatters, he would instruct Keyes to have his men remove it for them: the squatters tore it down. Now came an attempt by Wright & Co., brokers, using the name of the Miners' Bank, to place a building in the midst of the gnarled oak and brush that ran back from the point; Keyes and Billings demanded to see their title. If Wright did produce a title, Billings said, he felt bound by the Mason-Riley precedents, and he hoped that Riley would make a ruling to set those precedents aside. But in the meantime A. S. Wright decided on a bold game, apparently having no valid documentation, and began building. Billings confronted him, declaring that if Wright did not stop, he would either remove the building or take it for use by the military. A second person then began to erect a building, while Wright, in order to involve as many individuals as possible in any future litigation, began selling portions of the land that he claimed.[13]

The government, Billings concluded, had been asleep in two instances—a sharp remark to make to Riley, since it was Riley who was asleep on the second occasion—and if this new trespass were not stopped now, Billings predicted, it would grow ever bolder, to the point that there would be no reserve. This in turn would decrease the stature of the federal government in favor of purely local interests.

The solution, Billings said, was to take the bold course themselves and forcibly eject the trespassers. To do nothing and wait for clarification of the law would permit the practice of fraud against innocent third parties who were buying up lots from Wright in the belief that he had the right to sell them. To use an injunction would bring the matter to court; but this would be hazardous in the present transitional state of the courts and in the absence of detailed knowledge of the Mexican statutes or any word from Washington on the validity of Kearney's creation of the reserve. If Riley would act at once, the problem would be brought before the courts and the trespassers would have to sustain their titles. There was a further urgency: Riley's authority to act on behalf of the rights of the federal government would end soon, since the constitutional convention in Monterey on September 3 rose on October 13 with a finished document, partly intending to end military government.[14]

Keyes, with Billings accompanying him, apparently did oust Wright & Co., for at the end of the year they were operating from rented quarters on the plaza. The second person who had been building was also removed. Rincon Point would soon become more attractive, too, for it would no longer be cut off from the peninsula at high tide, when anyone wishing to reach it had to wade in water waist-deep, since contractors were at work moving the high sand hills in the area into the cove.

On November 13 Californians ratified their state constitution by a nearly unanimous vote and Peter H. Burnett, a friend of Billings, was elected governor. San Jose became the temporary capital, and on December 20 Burnett was inaugurated. General Riley resigned his powers as governor even though congressional approval had not yet been granted for California's admission to the Union. In the interval between Burnett's assumption of authority from Riley and California's admission as the thirty-first state, which came on September 9, 1850, matters remained legally ambiguous, though Burnett proceeded as if he had full and clear powers.

The matter came to a head on February 28, 1850. Rincon Point had continued to attract temporary settlers, for it was sandy and an ideal location for pitching a tent. Still, most settlers had left in short order. By February, however, it was evident that a large group intended to stay. To compound public anxiety, the squatters were largely "from Sydney"—that is, they were Australians, and "Australian" was, to many San Franciscans, synonymous with "criminal." Judging that Burnett would be on their side, Keyes and Billings decided to get the matter into the courts before statehood might further confuse the issue.

The strategy worked. Keyes marched to the Point with twenty soldiers, tore down the shanties and tents as the squatters looked on, hooting and threatening, some thirty of them surrounding Billings, and informed the crowd that the land was a public reserve and could be used only with the express consent of the federal government. To the squatters' cry that they were the public, Keyes responded that the public meant all of the people, who would determine their own interest, and not merely a portion of the public. As Billings had anticipated, one of the squatters brought civil suit for damages against Keyes, and the court dismissed the case out of hand, with the principle of public reserves deemed clear. Billings was delighted, for the Rincon reserve, and by extension the Presidio and similar future reserves, now enjoyed protection of legal precedent. Soon after, with full Americanization of municipal government, the abolition of the *ayuntamiento* carried with it the termination of Billings's office.[15]

Founding a Law Firm: HPB

W HILE ADVISING THE MILITARY GOVERNMENT of California, Frederick
Billings had made another fundamental decision: he would specialize
in contested claims, especially those arising from pre-conquest Spanish and
Mexican grants (as the creation of the Presidio had). He felt the early land
grants were generally valid and believed that if so, the United States had the
obligation to confirm them, rather than allowing squatters to occupy the lands
of the old *Californios* (as the native-born Spanish-speaking population was
called) much as bands of squatters had attempted to take over Rincon Point.
He would take as clients many *Californio* families now threatened with loss
or severe dimunition of their land. Further, many *Californios* had sold their
land—legally in Billings's opinion—to newcomers, and in defending the validity of these sales he would be at the cutting edge of the developing legal
market, able to discover and acquire properties that no one held valid titles to
while earning steady fees through litigation.

Billings quickly realized there was more work than he could possibly do
and he would need partners even sooner than expected if he were to become
a man of property. Whether from calculation or friendship, it seemed obvious
that the first partner should be the Virginian who had befriended Laura on
the isthmian crossing, Archibald Cary Peachy. Three years Billings's senior,
Peachy had defended Keyes in the suit against the Rincon Point expulsion and
Billings had been favorably impressed. Cavalier, gracious, and handsome, Peachy
bore, in speech and in his knowledge of legal literature, the signs of learning.
A Southerner, Democrat, and Episcopalian, Peachy would balance Billings, a
Yankee, Whig, and—in California—a Presbyterian. He was prepared to work
hard, for he was the chief support for his father and family in Virginia; he
could be depended upon for integrity, Billings thought, both because Frederick
trusted his own judgment in such matters and because Peachy seemed very
touchy about his personal honor, a sensitivity Billings put down as a southern
trait. (On at least three occasions Peachy challenged people who he felt had

insulted him to duels.) The men enjoyed each other's company, a cigar and brandy together, and the presence of spirited and intelligent women. Neither drank heavily or frequented San Francisco's many gambling halls (though Peachy was known to make a wager every now and then). Peachy was a bit slower than Billings to conclude that the gold fields were not for him—he went with five officers of the U.S. Navy to Sacramento, where he met with Sutter, whom for a time he represented, until the irascible Sutter discharged him[1]—but when Peachy opted for San Francisco, Billings offered him a partnership. Peachy accepted at once, and the two men opened an office in the City Hotel.

Peachy may have felt he had something to prove, for though he did not tell Billings, he had left Virginia under a cloud. In California Peachy let it be known that he had been professor of moral philosophy at Virginia's College of William and Mary for five years, though in truth he had never succeeded in filling that chair. An 1841 graduate of the institution, Peachy had been put forward for the professorship by members of the faculty who were seeking someone to support an extreme position on states' rights, and Peachy's age (he was only twenty-seven), the fact that his chief promoter was Nathaniel Beverly Tucker, professor of law and a leading secessionist, and Peachy's intemperance in challenging the college president to a duel discredited him. The appointment led to a split in the faculty, a strike by a majority of students, charges of financial improprieties, and vilification by the Richmond press, which at the time was taking a more Whiggish line. Peachy struck out for California to put all this behind him, arriving stony broke.[2]

Peachy proved to be the right partner. He was rather less careful of research than Billings but more effective in the courtroom, where his magisterial manner and professorial language were convincing in case after case. Like Billings, he sent money home regularly, helping his father to pay his debts, and the two gave moral support to each other. Peachy was often incapacitated by intermittent fever, probably malaria, and was happy to represent the firm in Los Angeles, sparing Billings the long journey, since the warm weather agreed with him. He enjoyed late breakfasts, reading novels, and a romantic sense of place—he liked Los Angeles, he wrote to Frederick, because he could stroll through a vineyard eating grapes and, in the evening, could "fall in love with beautiful white skinned blue veined girls of 18, dressed in light pink or white"—and did not grow impatient, as Billings would have done, at the fact that everything was "very slow" in the south. The two complemented each other well.[3]

The second partner, added at the end of the year, was Henry Wager Halleck, and as he was by far the more experienced and senior of the three, Billings insisted that Halleck's name go first in the expanded firm. Born in Westernville, New York, in 1815, Halleck had attended both Union College and West Point, from which he had graduated third in the class of 1839. A scholar, he had taught military history and French, and published an influential book, *Elements of Military Art and Science*. Halleck had worked on coastal defenses, and at the time of the Mexican War he was assigned to California. There he read law, learned administrative skills, and acquired much land and many mineral rights. Appointed secretary of state by Governor Riley in 1847, Halleck became intimately acquainted with the archives, and especially with the Spanish land records, first in Los Angeles and then in Monterey. (Indeed, he was the first to recognize the need for an archive.) At the constitutional convention that convened in Monterey in September 1849, he was both a delegate for Monterey and liaison officer between the convention and the military government. Widely viewed as one of the founders of the new California—when the Society of California Pioneers was established he was an irresistible choice for its board of directors—Halleck was known for his advocacy of the rights of settlers, yet was respected by the native-born Californians. Obviously he thought well of Billings, since he had helped him to become attorney general. Halleck was the ideal senior partner.[4]

The firm of Halleck, Peachy, and Billings, Attorneys and Solicitors, thus opened its doors on January 1, 1850. Halleck, with his detailed knowledge of all land grants and *expedientes* (collections of documents of title), of which he had made a list while secretary of state, and his understanding of Spanish and Mexican law, of *diseños*, surveys, and maps, was the scholar of the firm, the man who could decipher even the fine, antique Spanish script used in many records. It was Halleck who worked up the facts of a case with such care that in 1859 he published a then-definitive *Collection of Mining Laws of Spain and Mexico*. Peachy, fiery orator, active politician—he would serve two terms as a state legislator—was superb in backroom negotiations. Billings, social, quick, and open, known to everyone and knowing everyone, "brought in the business," as one commentator somewhat tartly observed.[5]

When the United States government established a Land Commission, which began to meet in San Francisco early in 1852, HPB found itself so quickly and so heavily involved that it was soon overwhelmed in its need for new clerks, researchers, and advisers. Further, cases in the Los Angeles area were taking far too long, and the judge there was proving dismayingly careless, so that the firm had always to be at hand to protect its clients' interests. HPB needed another lawyer, not just a young man read in the law—it had three of these in addition to the partners—but someone with weight. Hiland Hall, the chairman of the Land Commission, a Vermont Whig who had served in the House of Representatives in Washington, and had been appointed to the commission by President Millard Fillmore, suggested his son-in-law, Trenor Park. Billings had known Hall in Vermont, he needed his good will in California, and he was mindful of the adage that it was better to know the judge than the law.

Born in 1832 in Woodford, Vermont, Park was a driven, difficult, self-taught lawyer who had a practice in Bennington. Twenty-eight years old, with a formidable wife and an attractive daughter, he was not going anywhere. Hall had written to tell Park that he ought to bring his family to California, and on April 26, 1852, Park set out from New York harbor on the steamship *Illinois*, crossed Panama by the Chagres route, and arrived aboard *Golden Gate* on May 22.

On the Chagres Park met Fernando Wood, a New York investor and former congressman who, having made a small fortune, was on his way to buy land in San Francisco before returning to New York (where, in 1854, he would become mayor). Wood liked Park, and particularly liked his connection with the land commissioner: he asked Park to be his agent in San Francisco. Park immediately began his law practice, and two weeks after his arrival won a court victory against James McDougall, a highly regarded lawyer with considerable clout. Hall then pressed Park on Billings; Halleck, though particularly overtaxed, resisted. He worried that Park, as a Whig, making three in the firm, might lock out the Democrats as clients; that as a married man, he would be distracted and unable to put in the necessary hours; and that Judge Hall, out of a desire to appear impartial when dealing with a firm that included his son-in-law, might well bend over backwards against HPB. Billings persisted, however, and with the understanding that Park would be a full partner in all business except that dealing directly with the Land Commission (and the New Almaden Mine, which Peachy thought ought to be excepted because the bulk of the work on it was believed nearly done), Park was brought into the firm in late September 1852.[6] Halleck, Peachy, Billings and Park would continue until August 1855, when Park withdrew from the partnership.[7]

Park was ambitious, hard-working, and very hungry. He had not enjoyed his first months in San Francisco, living in a tiny rented house so small "a span of donkeys could move it anywhere in the world," with blankets hanging across the room to create private sleeping quarters. Hall had moved in with the entourage as well, adding to the discomfort. Admission to HPB meant that Park could move his family to a substantial house on Tehama Street, near, if not on, the fashionable Rincon Hill. He would soon win a reputation for sharp practices, and one historian of the period, and of mining cases in particular, has concluded that Park was "a pirate in a frock coat."[8]

At first Park proved to be a good partner. He was more eager for money and the firm needed someone who would hustle. True, HPB charged higher fees than any other law firm in the city; true, HPB represented more land claims than any other firm; true, Halleck usually charged all the traffic could bear, and in the many Napa Valley cases where acreage was small and legal work for one claim would do for a dozen, he made no concession on costs. But Halleck was given to charging no fees from ladies in distress, and Billings demanded the best fees only of those who could clearly pay them, or where the land in question was well developed (in which case he sometimes pressed for as much as 25 percent of the value of the land if the case was successful and 5 percent if not). So the firm needed someone unreservedly in search of material rewards. Park was soon buying lots, warehouses, wharves, ranches, and even a dairy farm, and he quickly became a rich man.

All three partners drew friends and acquaintances into the firm as advisers, senior and junior clerks, bookkeepers, and draftsmen. Lonely men naturally enough preferred the company of people from their own states, and Billings was happy to favor Vermonters with a temporary job while they were finding their feet or until they proved ineffective. His bookkeeper William Stowell and his first draftsman Caspar Hopkins had been hired simply because they walked through his door and he knew them. Though he did not think highly of the legal knowledge of several lawyers whose company he kept, he remained on good terms because they were from upper New England, or knew his parents, or were a friend of one of his brothers. Two men he did think highly of, Lafayette Maynard and Gregory Yale, frequently worked for the firm on commission, while William E. P. Hartnell provided invaluable research into the old Spanish records, usually for Halleck but on occasion for the other partners. Hartnell was married into the influential de la Guerra family, which held vast properties in southern California, and, until his death in 1854, was a valued adjunct.[9]

Even more important to HPB during Park's tenure with the firm was Oscar Lovell Shafter, who was appointed senior legal adviser in 1853. Shafter was a shrewd, honest lawyer of considerable skill and experience, who had more formal legal education than any of the partners—Wesleyan University and Harvard Law—and until 1855 was often in charge of key cases. Born in 1812, Shafter was the son of a long-time member of the Vermont legislature and in 1848 was a candidate for governor for the Young Liberty Party. The land of steady habits had not produced many fees for him, however, and as Park was urging him to give up his law practice of eighteen years in Wilmington, Vermont, Shafter concluded that he ought to seek out the land of profligate consumerism and rampant materialism, knowing that both brought on easy attacks of litigiousness. The offer HPB made him was for two and a half times what he was making in Vermont, and so, leaving behind his wife and children, he arrived in search of his fortune. Shafter would succeed handsomely, in time bringing his family to California and becoming one of the leading land holders on the Marin Peninsula, a candidate for the U.S. Senate, and a distinguished judge.[10]

Halleck, Peachy, and Billings was the most thriving legal house in San Francisco throughout most of its business life, from January 1850 to April 1861. It handled over half of the land claim cases on the dockets of the California courts, over eight hundred of fourteen hundred cases.[11] HPB rarely lost. Though the firm experienced some decline, as land cases were settled and as land values fell, especially after 1858, the proceeds from it, at least for Halleck and Billings, represented only a portion of their income. Both bought heavily in real estate, Halleck was deeply involved in the richest quicksilver mine in the world, and Billings was engaged in a dozen entrepreneurial activities. If legal fees fell off, real estate might hold up, and when real estate declined, there were a variety of cushions.

This is not to say that HPB did not experience, to some degree, the fluctuations all Californians faced.[12] By 1858 San Francisco had gone through four major business cycles, each quite independent of the national economy. When Billings arrived the city had been in a boom stage that continued until January 1850. The city suffered through a recession that began with a downturn in February 1850 and did not bottom out until April 1852. San Francisco experienced another boom from May 1852 until December 1853, and during this time HPB became the premier law firm, with all partners prospering. In 1853 HPB completed the construction of the Montgomery Block, the largest group of commercial offices in the city, and the project's rental income thereafter played a major role in the growth of Billings's fortune.

From January 1854 the city sank into a prolonged depression, brought on by an oversupply of goods in the San Francisco market, which did not end until January 1858. This affected HPB fees and Montgomery Block rents, and in 1856 Billings was short, but he held to most of his properties while others were selling and benefited greatly from an upturn prior to the Civil War. Throughout the decade well over half of all merchants failed; of the seventeen lawyers listed in San Francisco in the directory of 1850, most had either moved elsewhere, gone into politics, or died and HPB continued as the firm to turn to if one wanted to win.

HPB's services were expensive. The firm knew that in the end many clients would not pay, so fees were adjusted to take losses into account. The paperwork essential to a good case was enormous, and clients often were unmindful of how quickly fees could mount under the constant routine of correspondence to obtain documents, solicit witnesses, persuade land agents to a more favorable point of view, conduct title searches, draft statements, and prove documentation. In one case involving thirty-two documents, HPB also felt compelled to produce thirty-eight witnesses to testify as to the documentation's legitimacy. But HPB's business records, read consecutively across the three archives in which they now rest, do not suggest fees that were out of line.

This is also the view of the leading historian of nineteenth-century American land law, Paul W. Gates.[13] In examining the hundreds of California cases of the 1850s, and using HPB records, Gates concluded that the firm's fees were not unduly high. Taking forty-five claims, ranging from one acre to 62,000-

acre Santa Rosa Island, Gates noted fees running from nothing to $50 (for a near-hopeless claim) to $1,500 for the largest, and he calculated that the fee per acre was two-and-a-half cents. For example, Gates found that HPB confirmed 474,000 acres in the Southern District for $12,050, and described such a fee as "small." To be sure, these fees were for carrying the cases to the Land Commission only; costs on appeals were probably the same, which would double those that were appealed. Halleck and Peachy routinely submitted claims to the Land Commission to protect their share of a fee, whatever the decision, while Billings did this less frequently. Still, in one instance, the Rancho los Coches, he would own two-thirds of the property and, in a second, the Peninsula of San Diego, he would take a quarter interest, though this he acquired in good measure as a result of handling Bezer Simmons's indebtedness.

HPB had good strategists: to build up their record of winning, they brought those claims on which documentation was sound and evidence of occupation and improvement were clear to the commission as early as possible before taking on more questionable claims. They managed to delay their cases at any given session so that the judge's current reading on various matters would be a bit clearer and they would have time to study the impact of any precedent. They also saw to it that they came up late enough in a session so that the beleagured and underpaid U.S. district attorney was overloaded and unable to do his homework properly.[14] Thus HPB made the law's delays work for them, their fees, and—at least most of the time—their clients.[15]

Billings remembered his two long-term partners with affection to the end of his days. Halleck, who left the firm in April 1861, eventually to become commander-in-chief of the armed forces under President Lincoln, thus effectively dissolving HPB, died in 1872. For years Billings provided free financial advice to his widow, Elizabeth Hamilton. When heart disease brought death to Peachy in 1883, Billings reached out to contact his widow and adopted daughter and offer his help, notwithstanding the division the Civil War had temporarily brought between the New England Yankee and the Southern cavalier. Erasmus Darwin Keyes, the commander of the Presidio who with Billings had cleared Rincon Point, wrote that HPB was one of the finest law firms in the state: it worked, he said, because its members were incongruous and dissimilar, the laborious Halleck, aristocratic Peachy, and philanthropic Billings composing—one wonders whether Keyes really thought about his words—"a conjunctive disjunctive continuance."[16]

7

A Man of Property

FOR BILLINGS, THE MOST PRESSING LEGAL ISSUE in 1850 concerned water lots. From them he would make his first substantial income; they also contributed to the founding of HPB, and to some extent clouded Billings's reputation, since several observers believed many of the sales and their surveys were fraudulent.

In March 1847, General Stephen Kearney had ordered a sale of lots for commercial facilities under a survey that had originated in 1839 and been altered as the definition of the shoreline at Yerba Buena cove was refined. A second sale followed in mid-1848. These lots generally were between the high and low water mark, and most were covered with water at the time of sale. Structures built upon them required stilts, but if a program of landfill, or "made land," existed, they would command very high prices since they were directly on though not in the water. Kearney had conveyed the title to this land to San Francisco and, believing they had the authority, the Alcaldes had proceeded with their sale. In 1849 a third survey of beach and waterfront property led to a public auction in January 1850.

The 1849 survey, especially away from the water, formalized the usual lot size at fifty *varas*, a Spanish unit of measurement not given a formal definition in American law until 1852, and until 1851 San Francisco's municipal government financed its operations by the sale of these lots at auction and on petition. Many questions had arisen about the validity of titles, whether from a fundamental doubt over Kearney's authority to have permitted sales, from issues directed to the surveys, or over the basic matter of whether the municipality was entitled to four square leagues of land as an original Mexican *pueblo*. Social issues, relating to squatters particularly, were also part of the problem.

Billings was both an owner—that is, a prospective litigant—and the lawyer for many clients who had purchased water lots. Some lots were legitimate, clearly some were not, and since all were under a cloud, there was much speculation in them, with individuals buying up lots from those who feared their titles would, in time, be proven invalid. These individuals then worked behind the scenes as well as through their lawyers to have the matter either clarified, if that was in their interest, or postponed if use of the lot was, in the short run, producing substantial income. In short, the situation was entirely normal. Billings was among the speculators, though he never owned more than thirteen San Francisco lots, taking his income more from his clients who, knowing he was riding in their boat, trusted him to be as vigorous in their interests as possible.

In 1849 Billings purchased four water lots so as to be in the running for whatever might flow from these properties. At a town sale in January 1850, he acquired three more in a pattern that gave Billings and Peachy between them one side of a full city block. In 1852–53 Billings acquired four more.[1]

These lots were more than a legal problem: they made for severe unrest. Buyers wanted to fill them in; wharf owners objected. As one historian remarked, "a species of guerilla warfare ensued, featuring court injunctions, midnight and Sunday pile driving and extracting, the sinking of sand-filled hulks by sea-borne squatters, and even armed clashes between claimants to the same property."[2] Fire was an obvious hazard, as these jerry-built structures sat amongst the vessels and along the wharves that connected with planked streets. Billings was not happy with the chaotic situation, though he was part of it.

Although Billings had met Peachy on the Chagres crossing, and had achieved his first official appointment in California through Halleck, giving him good reasons for taking both as partners, he might not have considered partnership with them had it not been for the water lot cases. Not only was Halleck the acknowledged authority on Spanish land law, he had traced the *vara* measure back to its first official use in 1802, knew its Castilian origins and its Mexican variants, and knew how to find equivalents between poles, rods, feet, and links, on the one hand, and a *vara, cordel* or *pulgada* on the other. He was excellent when a dispute involved a question of adequacy of survey. He also owned twelve water lots. As the newly-appointed City Attorney, Peachy was good on local Anglo law: it would be Peachy, on behalf of the town council, who would begin legal proceedings against Gardner Q. Colton, JP, who late in 1849 had made grants of city lots to individuals contrary to the council's understanding of his powers. Many councilmen owned water lots too, so that their views would be parallel to Billings's on the larger questions.

On March 26, 1851 the state legislature granted use and occupation of the lots in question to the City of San Francisco for ninety-nine years. On June 2, Samuel D. King, the first U.S. surveyor general of California, arrived and set about establishing measurement equivalences. There would be challenges, of course, to the law and to King, though not until March 1855 would the

Supreme Court of California, in two decisions, declare that new surveys must, in those instances, be made under the authority of the United States; previous surveys were held inaccurate and thus invalid.

After the passage of the state law of March 1851, numerous individuals attempted to secure confirmation for water lot titles that were patently illegal. Purchasers of fifty-three voided water lots appealed to the common council. When speculators coerced the council into passing an ordinance that had the effect of confirming them in their property, the mayor, Stephen R. Harris, vetoed the ordinance and had Peachy further investigate the claims. The situation was complicated by the influx of squatters and by an effort on the part of the state of California to assert ownership over the lots to the detriment of the city. Firms like HPB soon had more business than they could handle, and were adding clerks and messengers as quickly as they could find them.

Whether the sale of water lots was legal or not, there was no question that the lots were of great value; lots that sold in July 1847 for $100 skyrocketed in price, and at a sale in December 1853 lots half the size of the original ones and well beyond the low water mark brought up to $16,000. The lots held by Billings and Peachy, and perhaps by Halleck, appear to have been worth at least $8,000 each by this time. Selling, which of course depended upon an uncontested title, would yield a substantial profit and the leverage the partners needed for larger purposes.

Peachy appears to have used the information he gained as city attorney to try to shape water lot litigation so as to favor Halleck, Billings, and himself as owners. He won election to the State Assembly on September 3, 1851, as a Chivalry Democrat (a follower of Senator William Gwin), and at once tried to get favorable legislation passed on the water lots that he and Billings held, causing the opposing faction to charge him with bribery and fraud, both to win his seat and to gain votes for his proposed bill. The Senate subpoenaed Peachy, who pleaded constitutional grounds in refusing to attend any hearings. His opponent also owned water lots—indeed, 362 of them—and was intent on taking advantage of the confusion to speculate his way into a substantial fortune, and he did not want Peachy's far smaller interest to get in the way.[3]

Billings felt unhappy about Peachy's insider trading, but he moved quickly to sell most of his lots, from which he realized roughly $50,000 in clear profits. He soon put his gain to work by purchasing property in the Napa, Sonoma, and Petaluma valleys. He also bought land on the Stanislaus River, in Butler City on the Sacramento River, and, in concert with Peachy or Dr. Turner, additional shorefront lots in Benicia, hoping that it might yet become the state capital.

Late in 1850 two booksellers, William B. Cooke and Josiah J. Lecount, published a booklet with the rather saucy title *A "Pile," or A Glance at the*

Wealth of the Monied Men. Billings was placed in the top 125 residents of San Francisco with $35,000. Cooke and Lecount were wrong, however: Billings was worth at least $100,000 by the end of 1850. In modern terms he had become more than a millionaire eighteen months after his arrival.[4]

Though Billings's wealth came largely from real estate and lawyer's fees, he was always willing to put a bit of money into a growth area: mining, railroads, shipping, wine, or whatever came recommended to him by a trusted friend and that appealed to his sense of adventure. He would buy fifty thousand shares in the Merced Mining Company, which did moderately well but gave him much grief. In 1854 we find him a director of the Bear River & Auburn Water & Mining Company. He invested in the Spring Mountain Coal Company, and he helped Owen Taft, the son of one of his father's old friends, who had come to Butte County and was faltering with his California Coal Company. Later in the 1850s Billings acquired slightly less than a half interest in the Hermitage Mill and Mine in El Dorado County, and was lucky to get out with a small profit by selling the machinery to a foundry at Placerville. In 1857 he wrote excitedly to his parents about a scientific discovery that might make him a fortune, dashing off to Vacaville to check on it, though it is not clear what the discovery was. He put $1,000 into the Comstock Lode as soon as silver was discovered in June 1859, and in 1860 sank a good bit more into the Ophir Silver Mining Company in Carson City, losing badly as Ophir stock fell 80 percent before he was able to get out. He also acquired substantial properties near Austin, Nevada.

One profitable venture was in borax. With Peachy as partner, Billings bought Borax Lake, near Lower Clear Lake, in 1858, getting in just ahead of a survey party on a tip from a friend. Peachy and Billings hired a well-known scientist, Benjamin Silliman, to report on this and other mineral sites for them; Billings converted his share into California Borax Company stock later and did very well, for the lake produced the finest natural crystals ever discovered, 99.94 percent pure. Protected by an import duty against foreign borax, the lake provided nearly the nation's entire supply until 1873, when the discovery of "marsh" deposits of borax in the deserts of California and Nevada ended production.[5]

Over the years he dipped into a mixed bag of other ventures as well. Billings gave serious thought to the China trade, obtaining estimates for the cost of building vessels of southern white oak and provisioning them for six months, with advances to their crews; in the end, however, he put this money into F.W. Macondray & Company, run by two of his friends. Billings also invested several thousand dollars in land in Port Townsend, Washington Territory (especially after he and Halleck were successful in getting an old friend appointed to the Court of Claims there), and bought into the Central Wharf Company in San Francisco, holding on through the severe depression of 1857–

1858. He was a stockholder in the San Francisco Gas Company, which put coal gaslights on the city's streets in February 1854; he purchased a sixteenth share in a clipper ship, *Crystal Palace*, and by acquisition from Bezer Simmons had a part interest in the Sacramento River steamer *McKim*; as he prepared to leave California for good he would invest heavily in W. C. Ralston's new and very ambitious Bank of California. He thought, Frederick told his parents, that he ought to try everything. He was, however, "a pretty safe operator," and no one in San Francisco had better credit.

Among the "everything" Billings was involved in was the fledgling wine industry. He liked wine, a good cigar, and an occasional brandy (even the raw aguardiente that often had to be substituted), and believed that the California climate ought to make quality wine production possible. With his usual sense of precision, he felt strongly about the careful labeling of wines and declared that Americans were too inclined to ignore matters of origin and vintage in favor of showing off by going for the most expensive bottles. He thought that Ageston Haraszthy, a Hungarian who had come to California in 1849, was likely to find a way to put the state on the wine map, especially after he shifted his efforts from around the Mission Dolores in San Francisco to Crystal Springs in San Mateo. Billings was right in the main: the climate in Crystal Springs did not prove hospitable, but soon after Haraszthy went to Sonoma in 1857 he was producing wines that Billings pronounced excellent. The Vermonter was less prescient about a German, Charles Krug, who began as Haraszthy's assistant and opened his own winery at St. Helena, though Billings rather liked Krug's experimental Zinfandels. And in due course two of California's best wines would bear the names of two of the old ranchos for whose claims Billings was the lawyer: Simé and Guenoc. Indeed, the latter was the second land claim case HPB took up.[6]

Though Billings put his money into a variety of enterprises, he preferred to invest through good friends who, even if investments turned sour, usually remained good friends. In addition to his law partners, he was particularly close to three businessmen, Robert Sedgwick Watson, S. Griffiths Morgan, and Carlile P. Patterson. All three returned to the East Coast before Billings, and he corresponded with them, expressed pleasure at their successes, doted on their children, and, when he returned East himself, was a frequent guest in their homes. He called them his "little band of brothers."

Watson was almost forty when he arrived in California on August 18, 1849, and already was a partner in the commission house of Macondray & Company, determined to make his fortune and return to his home in New Bedford, Massachusetts, where his family waited for him. Billings thought him balanced, wise, and attractively introspective, and later when Watson brought his family out, Frederick was often a guest in their home. Watson encouraged Billings when he was depressed and listened to him when he was most lonely for Vermont. They shared an interest in the outdoors and attended the same church, but most of all they shared a sense of home. In 1852 Watson returned to the East a rich man, drawing $150,000 out of Macondray & Company and leaving $60,000 behind for Billings to manage for him, and remained Billings's firm friend thereafter, as did Mary Watson, who joined the long line of those worried about Frederick's lack of interest in marriage. Mary's twin sister married John Murray Forbes, who was on his way to a vast fortune from railroads, real estate, and the China trade, and in the 1880s Billings would become well acquainted with the Forbes at their estate on Naushon Island.[7]

S. Griffiths Morgan ("Griff") was a second member of the little band, close to both Watson and Billings, though here the initial tie was through marriage, as his wife was Carrie Simmons, one of Bezer Simmons's sisters. He too was a member of Macondray & Company, but late in 1851 joined Halleck, Peachy, and Billings as a chief clerk and legal adviser. He and Billings also attended the same church, went to the same political meetings, and liked the same people. Morgan returned to Massachusetts in 1853 and remained a constant correspondent until his death, having put Billings in charge of his California properties. Carrie continued a series of lively, rather loopy letters which gave Frederick great pleasure long after he returned to Woodstock.[8]

Most important, there was Captain Carlile P. Patterson ("Kline") and his wife Elizabeth, in whose home Frederick passed many of his happiest hours. Patterson came to the bay area in 1851 as master of *Golden Gate*, a popular vessel built expressly for the Panama–San Francisco run. This steamer, able to accommodate 850 passengers, had been put into service by the Pacific Mail Steamship Company, and at once commanded the affections of San Franciscans, named as it was—or so they felt—for them. Patterson was as strikingly handsome as his ship, and the city doted on the captain of "its vessel," especially after he left it to pursue various dryland investments and his immediate successor as captain ran it aground. The Pattersons and Billings met through Peachy and soon found they had common acquaintances through Laura's sojourn in South Carolina. They invested in the Hermitage Mine and Mill together and, when it failed, Billings took over the mill from Patterson in payment of his debt and arranged for Patterson's creditors to operate the mill. The two men also shared an interest in railroads and often discussed the changes a Pacific coast line would bring to California.

This friendship had an added dimension, for Lizzie Patterson came to mean much to Frederick. When the captain was traveling, Billings often passed his evenings at the Patterson home in San Francisco and later in Oakland, where he cheerfully washed dishes, swept floors, and relaxed into a routine of domesticity that gave him a sense of family. He enjoyed looking after Lizzie and her little nephew, who lived with the Pattersons, and whenever he returned from their company would write introspectively to his mother about how, one day, he would like to be married. After Patterson moved to Washington, D.C., he often wrote Billings, urging him to leave California, where "every species of atrocious depravity" flourished, and return to the more benign East Coast. When that benignity was interrupted by the Civil War, they remained friends, despite the Pattersons' leanings toward the South, and whenever he was in Washington Billings was a regular visitor at their estate, Brentwood. Some years after the war, when Patterson had been rehabilitated in the eyes of northern administrations, he was appointed chief of the U.S. Coastal Survey. When in 1881 Lizzie Patterson—always, as her correspondence shows, just a little *distrait*—lost her husband and then Brentwood, and had to move in straitened circumstances to far smaller lodgings, Billings persuaded his old friend and University of Vermont classmate, Senator John Kasson of Iowa, to sponsor a private bill to relieve her of her indebtedness.[9]

These four men shared a life of laughter, and good cigars, and a sense of youthful elegance in a city often rampantly ugly. More important, Billings found in the three households the sense of closeness he so missed. The three wives twitted him about marriage but did not press candidates on him, giving him the feeling that in time he would find someone to make him as happy as they had made their husbands. The women as well as the men shared a sense that the West was a grand adventure, the countryside full of beauty, the future a land where a calm old age lay just out of reach yet was attainable. They all liked good gossip and when Frederick did marry, he was particularly delighted to find his quick-witted wife Julia admitted to this inner circle.

The four men shared a common business interest as well: railroads. Patterson was treasurer of the Atlantic and Pacific Railroad (a rather grandiose title for a line meant to run from San Francisco to San Jose), of which Halleck was president and HPB the legal representatives. Patterson hoped to draw Billings in more, so that they might make a greater fortune in order "to found colleges, asylums, and hospitals and to build churches." Watson invested in railroads shortly after his return East, beginning with the Hannibal and St. Joseph, and urging Billings along the same track so that they might have money "to live comfortably with the daily grind of business." He would, he wrote Billings, fix him up as president of a first-class railroad if he would come home.

Still, railroads remained in the future—land was the staple. In combination with Peachy, less often with Halleck, and increasingly with fellow lawyer and Vermonter Lafayette Maynard, Billings acquired more prime properties in San Francisco, Sacramento, and elsewhere. He and Maynard guessed that, once Benicia faded, Sacramento would be the permanent state capital, and the guess paid off handsomely after 1854. The 1853 Sacramento tax list shows Maynard, Peachy, and Billings jointly owning 147 lots, with Billings holding the majority interest in at least 80, spread over thirteen increasingly significant downtown blocks. Through a shrewd real estate agent and title searcher, A. K. Grim, Billings exchanged lots, trading up and down as advised, sometimes disposing of property to Bezer Simmons's brother Nathan, who was in the hardware business, and at other times selling to squatters who had proved stable and untroublesome. When he was told that he owed Sacramento some six thousand in taxes, Billings arranged to let squatters remain on several of his lots in exchange for rents that would meet the taxes, waiting for property values to rise, as they did. He might have cantered happily, with Tennyson's Northern Farmer, to "proputty, proputty, proputty."[10]

 8

The Montgomery Block

B ILLINGS'S ASCENT TO WEALTH AND SECURITY did not spring primarily
from the water lots or his diverse holdings but from his part ownership
in the Montgomery Block, which when finished in December 1853 was San
Francisco's largest, most expensive, and most desirable office property.[1]

The only "absolutely fireproof" building in the city—and so it proved
when put to the test—the Montgomery Block was the commercial and social
center of San Francisco. It brought in rentals of nearly $12,000 a month in
the mid- to late-1850s, almost never experienced a vacancy, and offered all
the services of a small city within its walls: twenty-eight ground floor shops,
including the best bar in the entire bay area, two banks, a superb tobacconist,
a bookseller and stationer, a dining and oystering saloon, an undertaker, a
barber shop, portrait studios, a shirtmaker, a tailor, a hatter, and two newspaper offices. Above were a law library, the three rooms of the Mercantile
Library Association, and a swank billiard room that ran nearly the entire length
of the second floor. A variety of real estate and other commercial offices, and
fully half of San Francisco's lawyers, filled the rest of the space. At Christmas
its two hundred windows were illuminated with candles, and in summer its
courtyard, balconies, and great foyer were festooned with flowers. Anyone
with virtually any business in San Francisco could not avoid the block: here
were the offices of the Land Commission, the federal surveyor, the U.S. Corps
of Army Engineers, and James King of Williams's infamous *Daily Evening
Bulletin.* Here Adams & Company maintained a gold vault, dealers displayed
the latest in prints and jewels, an alchemist worked his mysterious wonders,
and the Hungarian Haraszthy put down his best clarets in a special cellar for
two years, while the firm of Kohler and Froehling stored 170,000 gallons of
wine and brandy a few feet away.

The Montgomery Block rose in answer to a clear need: six times in over
eighteen months San Francisco was devastated by fire. Property values plummeted. Fire insurance was at 6 percent. After the fifth conflagration a group

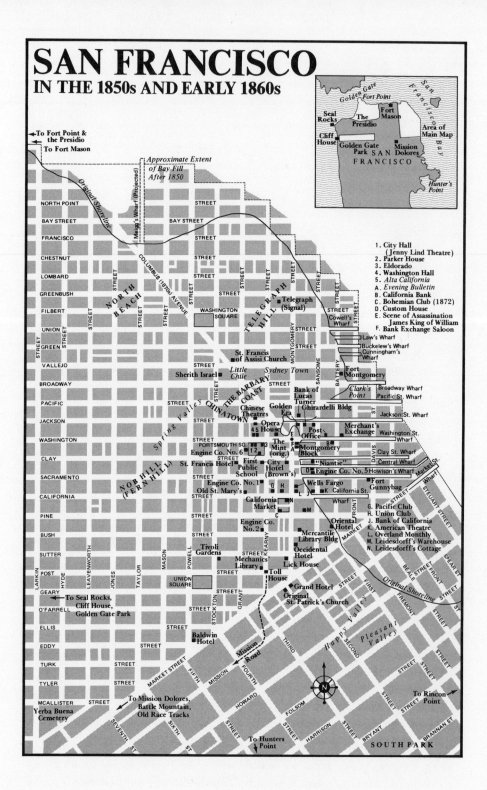

SAN FRANCISCO
IN THE 1850s AND EARLY 1860s

1. City Hall
   (Jenny Lind Theatre)
2. Parker House
3. Eldorado
4. Washington Hall
5. *Alta California*
A. *Evening Bulletin*
B. California Bank
C. Bohemian Club (1872)
D. Custom House
E. Scene of Assassination
   James King of William
F. Bank Exchange Saloon

G. Pacific Club
H. Union Club
J. Bank of California
K. American Theatre
L. Overland Monthly
M. Leidesdorff's Warehouse
N. Leidesdorff's Cottage

Golden Gate
Fort Point
Seal Rocks
The Presidio
Fort Mason
Cliff House
Golden Gate Park
Mission Dolores
SAN FRANCISCO
Hunter's Point
Area of Main Map

To Fort Point &
the Presidio
To Fort Mason

Original Shoreline

Approximate Extent
of Bay Fill
After 1850

Meig's Wharf (Projected)

NORTH POINT
BAY STREET
FRANCISCO
CHESTNUT
LOMBARD
GREENBUSH
FILBERT
UNION
GREEN
VALLEJO
BROADWAY
PACIFIC
JACKSON
WASHINGTON
CLAY
SACRAMENTO
CALIFORNIA
PINE
BUSH
SUTTER
POST
GEARY
O'FARRELL
ELLIS
EDDY
TURK
TYLER
MCALLISTER

NORTH BEACH
WASHINGTON SQUARE
COLUMBUS (1870) AVENUE
TELEGRAPH HILL
Telegraph (Signal)
Cowell's Wharf
Law's Wharf
Buckelew's Wharf
Cunningham's Wharf
Fort Montgomery
Broadway Wharf
Pacific St. Wharf
Jackson St. Wharf
Washington St. Wharf
Clay St. Wharf
Central Wharf
Howison's Wharf
Market St. Wharf

St. Francis of Assisi Church
Sherith Israel
Little Chile
Sydney Town
THE BARBARY COAST
CHINATOWN
Spring Valley
NOB HILL (FERN HILL)
Chinese Theatres
Golden Era
Opera 45 House
PORTSMOUTH SQ.
Engine Co. No. 6
The Mint (orig.)
St. Francis Hotel
First Public School
City Hotel (Brown's)
Engine Co. No. 1
Old St. Mary's
California Market
Engine Co. No.2
Tivoli Gardens
Mechanics Library
Toll House
UNION SQUARE
Baldwin Hotel
Mission Road

Bank of Lucas Turner
Ghirardelli Bldg
Post Office
Merchant's Exchange
Montgomery Block
"Niantic"
Engine Co. No. 5
Wells Fargo
Fort Gunnybag
California St.
Oriental Hotel
Mercantile Library Bldg
Occidental Hotel
Lick House
Grand Hotel
Original St. Patrick's Church

Clark's Point

To Seal Rocks,
Cliff House,
Golden Gate Park

Yerba Buena Cemetery

To Mission Dolores,
Battle Mountain,
Old Race Tracks

To Hunters Point

Happy Valley
Pleasant Valley
Original Shoreline
To Rincon Point

SOUTH PARK

LARKIN
HYDE
LEAVENWORTH
JONES
TAYLOR
MASON
POWELL
STOCKTON
GRANT
KEARNY
MARKET STREET
MISSION
HOWARD
FOLSOM
HARRISON
BRYANT
BRANNAN ST.
FIRST
SECOND
THIRD
FOURTH
FIFTH
SIXTH
SEVENTH
FREMONT
BEALE
MAIN
SPEAR
STEUART

N

of investors met to discuss erecting a structure that would make a statement to the future, one that would clearly say that they and their enterprises were in San Francisco to stay. At an initial meeting the group agreed on the solution, a fireproof building of the size and elegance to be a monument to the new state. A sixth fire, on June 22, 1852, intervened, and despite one estimate that the new block might cost as much as three million dollars, most investors persevered.

Credit for carrying the day and planning the block goes to the senior partner of HPB, General Halleck. He determined that the site must be on Montgomery Street and that the building must be of better construction than the customary "fireproof" brick, which proved to be anything but. Halleck felt strongly on this point, for too often the precious Spanish documentation he needed for settling a disputed land claim had gone up in smoke. John Parrott, a leading banker whom Halleck had known in Mexico, was putting up a building in record time using Chinese labor, and Halleck was impressed as he watched the structure rise, each mason working from sunrise to sunset (with an hour off) for $1 a day, half a pound of rice, and half a pound of fish. Halleck also liked the granite, imported from China, that Parrott had used. He too would use Chinese laborers, he decided—after all, bricklayers could command $25.00 a day and hod carriers $17.50, and they did not get the job done nearly so fast—but he would take far greater care with the structure itself. He particularly disliked the local bricks, a thousand of which cost $140 and which, being burnt at the San Quentin prison kiln, where fresh water was not always used in mixing the clay, were of poor quality. He would use only the best materials—granite certainly, and from California where possible—and would apply his own engineering skills to solving the primary problem of any brick or stone construction in San Francisco: how to build a fireproof structure on a foundation that would resist earthquakes and avoid sliding into the nearby bay.

First, the site: on Montgomery Street, "the Wall Street of the West," but precisely where? Apparently it was Trenor Park who suggested the precise location, a parcel of land owned by Fernando Wood, the New Yorker for whom Park was an agent. At first Billings was a bit doubtful, but Halleck liked the location and the price, $100,000, was right, or so Park assured them, and so the partners set out to raise the funds.[2]

Next, the money. Gordon Cummings, who was to be the architect for the block, said that $200,000 would be needed just to start. Park put in $33,000. Between them Halleck, Peachy, and Billings put up $150,000, drawing on every resource they had. Billings hurried back to the East Coast in the late spring of 1853 to raise $50,000 more. Local investors, led by François Pioche, head of the banking house of Pioche and Bayerque, put their authority (and money) behind the enterprise, and soon funds began to flow in. Billings was especially good at networking, talking with financiers, using his membership in the Pacific Club, frequently calling at Barry and Patten's bar, where—since no gambling was allowed—the better class of potential investors met over a bottle of wine, drawing all the while upon his considerable social charm to raise the cash. By July 1853, from 170 stockholders, HPB had $300,000 in hand.[3]

Now, the foundation. Halleck, taciturn and with few social graces, left the money-raising to his partners after putting in his own funds, and applied all his formidable skills to the structural problem. He was not Old Brains for nothing; he had not written scholarly monographs on a variety of engineering conundrums without having won the reputation ensuring that if he said the building could be made proof against all of San Francisco's outrageous fortunes, it could be. Some called him a fool, a few investors backed out, and once his plan was known, more referred to the scheme as Halleck's Folly. Peachy wasn't sure, but then he was a gambling man and ready to play the game as it came. Billings was sure, for he had studied Halleck as he had studied the law, and was confident that the "old engineer" (for so he was called, his scholarly demeanor obscuring his scant thirty-seven years) knew what he was doing. On handsome retainers now to several firms, Billings put most of his ready cash into the Montgomery Block as it took shape.

Excavation began on July 3, 1853. Buildings on the site were leveled, their pilings drawn, and, while the tide was out, exposing the mud flats, Chinese laborers and American mule drivers waded hip deep into the muck to gouge out a temporary basin. Working desparately against the tide's return great baskets of mud were carted away, and a hollow took shape along the tideland. Massive mats of redwood, carefully chosen because they worked best when damp, with a near-shoreside location to assure they would always remain so, were dragged into place by steam winches. With their butt ends facing Montgomery Street, the mats were hurriedly yet exactingly covered with a cross-layer of more redwoods. The tide returned and the mats held, with subsidence uniform at all points; when the tide again receded, the redwoods, now thoroughly ensconced in the mire, with the foundation twenty-two feet deep, were covered with two courses of ship planking, carefully dovetailed. Doubters watched as the planks were anchored with iron clamps, the whole bridged over through the night in the glare of naphtha lamps. Halleck had created a floating fortress.

At last, the building: onto Halleck's redwood mat went 1,747,800 bricks, each quality controlled; Halleck himself stood in the mire in front of the rising structure to superintend. The foundation was tapered toward the waterfront and honeycombed for lightness. It rose sixteen feet above street level, supporting great thick walls. More truss rods and steel beams, brought from the East Coast, assured that a stable structure would stand upon the foundation which, by virtue of its mats, honeycombing, and design, would shift slightly in the event of an earthquake; it would give without shearing. As to fire, a heavy fire wall was built across the back end of the building, which was to have no windows, so that fire might not sweep up from the waterfront, with an elegant courtyard providing the necessary light away from the street front. Should fire break out internally, iron doors backed in asbestos could be closed at once on each of the four floors and in the basement, restricting the fire to a single wing. The fire wall was concealed with four stories of ornamental glass, allowing for a lavish solarium and steambath house. The courtyard also meant that enough light and air reached interior rooms for them to command rents close to all but the best outer suites.

Next, the elegance: ornamental iron balconies, massive iron doors, hundreds of specially designed bits and pieces of decorative ware came from the East Coast, while from England flowed the new vitrified Portland cement, from Germany, France, and Belgium glass, from Italy marble for the fireplaces to be installed in every room. The brick walls were given three coatings of boiled linseed oil, brushed on before winter rains could begin their leaching, and the whole was covered with mortar. Stucco was applied, with a second coat floated on creating the polished look of porcelain. Two great wells were driven into the courtyard; the deeper supplied fresh water, the other salt water for the baths. Next to the wells stood a monument to modernity, the first private gas machine in the state, to provide light so that the premises might be conveniently used after dark since law offices often remained open until eleven o'clock at night. The effect was at once solid and elegant, a triumph of engineering, design—a bit Georgian, in the London style of the 1840s, yet to some eyes rather more Florentine bargello—and luxury, as gas lights and running water went into every one of the 150 offices, the fourteen street level stores, and the twenty-eight basement areas.

The total cost of the Montgomery Block cannot be known, for the records were destroyed in 1906, but folklore puts the figure at $2,000,000. Halleck once hinted that $1,000,000 was about right. Whatever the cost, this was the largest building the West was to see until 1875, when the Palace Hotel opened. It would stand firm against fire and earthquake, and would be brought down only by the decline of the area over which it reigned. In 1959, over 105 years after it was dedicated, Halleck's Folly would require two months of the wrecker's ball to be reduced to rubble, to the city's everlasting shame.

Finally, then, the name: in fact, the Washington Block, further to declare California's dedication to the great nation of which it was a part. Over the main facade Cummings placed a four-foot granite bust of Washington, with the president's name lettered out across the building's street front. (Rough-hewn heads of other worthies stared down from the cornice.) Only this aspect of Halleck's Folly did not take: though the bust remained in place for many years, the letters soon came down, for the location on Montgomery Street, and indeed Billings's own preference, argued that the building be called the Montgomery Block. So it was, except to the irreverent, who, whether because of the number of lawyers present or the plethora of black-tie affairs held in the lobby, or perhaps merely for reasons of alliteration, happily named it the Monkey Block.

Later, of course, came the opening. There had, in a sense, been several mini-openings, for Halleck created an office for himself on whatever was the top floor of the building as it rose, and during construction many came to see him and tour about. The more convivial Peachy and Billings added an elegant touch to the routine of rental, taking each tenant on the occasion of signing a lease through the entire block and then to dine at the Ivy Green, a fashionable restaurant. But the formal opening was on December 23, 1853, just six months after the first Chinese laborer had stepped out into the muck to hoist a basket upon his shoulder. Peachy was in charge, for he liked acts of theater. Paper lanterns dotted the block, fireworks discharged from the rooftop, a red carpet lined the entrance from curb to stairs, and a military guard flanked the carpet. Palms were set throughout the lobby, an orchestra played, and groaning tables laden with venison, high heaps of shellfish, row upon row of champagne bottles, and an iced meringue that modeled the Montgomery Block greeted the hundreds of guests. The governor, the chief justice, the Russian consul, officers of army and navy, and anyone who was conceivably anybody attended. Billings reported to his parents with his customary understatement that he had a grand time.

The Block remained the basis for Halleck's, Peachy's, and Billings's prosperity. From their offices they could see all that trafficked in the streets, within their corridors they could learn of most that was to be sold, over a glass at the remarkable Bank Exchange—which, despite its name, was an elegant marble-paved saloon where Wedgwood beer pumps revealed diaphanous figures allegedly from Greek mythology—they could close deals. Halleck's prime fascination had been in the challenge of the structure itself, and while he continued to do much work for the firm, he devoted more and more time to his scholarly inquiries, to a ranch he had purchased in Marin County, and to the social life his wife arranged for him so well. Peachy was happy to live well—an epigram of the 1800s had it that "Halleck went on to become chief of the federal armies, Billings became a railroad multi-millionaire, but Peachy lived"—and to pursue his political ambitions, virtually commuting on the river steamer to Sacramento. One could count on finding Billings somewhere in the building, if not in his office then in his favorite barber's chair or seated in the grand lobby talking with one of his many friends. There was, of course, a business manager, Colonel George W. Granniss, hired in 1856, who after the Civil War would take over the whole of the block and until 1894 run it for Halleck's widow, Elizabeth, even moving into Halleck's grand house on Second and Folsom Streets. Granniss faithfully gathered in the monthly rents: $1,000 for the cigar stand concession, $2,000 for the best suite, $200 for a basement cubby. People would pay what .they had to, for to succeed in San Francisco, it was said, one had to write on the stationery of the Montgomery Block.

Here was the California Billings liked most. Although he wrote often to his parents at home, missed them, and still thought of himself as a Vermonter, he had to confess that California had won him over. He liked the climate, the clear sky, the green distant hills, and the good health he enjoyed which, in the mid-1850s, was common though not constant. Letters from home were full of deaths, funerals, sickness, and more deaths: from canker rash or scarlet fever, measles that led to inflammation of the brain, from drink or drowning, from sleigh, carriage, and rail accidents, from consumption, seizure, brain fever (meningitis), madness, or suicide. It was so much better to feel the wind coming off the bay as he drove his fast horses through the streets of San Francisco, risking being thrown, as a friend said, all the way to Oakland. Billings liked to go up to the fourth floor of the Montgomery Block, which was effectively bachelor quarters, consisting of smaller offices with windows cunningly made to heighten the outside perspective of the building, and look down as though reviewing the scene from a grand balcony. He enjoyed taking a good Philippine cigar, tearing off its yellow ribbon, and sitting out on the open roof of the grandest building in the West—in part his building—to watch the sun paint great golden fireballs onto the waters of San Francisco's harbor. He liked the sense of movement about him, in the floors below, on the street, out in that harbor, yet stood apart from it. He was a happy man, content, surrounded by friends and admirers. Just thirty, Frederick Billings was more than a man of substance, a Vermonter who had made good in the instant city of San Francisco; he was, at the least, a minor social lion, a figure to be reckoned with, a participant in the key events of California's most turbulent decade, the 1850s.

 9

Justice in the
Instant City

D URING THE YEARS FREDERICK BILLINGS MADE San Francisco his home,
it was, in the words of the historian Gunther Barth, an "instant city."[1]
Though other such cities have existed in American history, none could combine both the sea and an interior mountain vastness. San Francisco was in a
process of constantly disorienting change throughout Billings's residence; in
forty years it would grow to a size that Boston required two and a half centuries to attain.

San Francisco was far more male, ethnically diverse, and youthful than
most of America. Materialistic—people had indeed gone there for gain—and
tolerant of high violence, it gave little thought to the long run. Those who
had lived in the more cultured cities of the East believed that San Francisco
would in time rival them, and, not being content to wait, they tended to a
boastfulness that often seemed quite ludicrous. The setting was magnificent,
and as settlers have done in all well-sited Western cities, the residents tended
to take credit for nature's grandeur as though they had created it themselves,
while also doing little to preserve it. Suicide rates were high, as was alcohol-
ism; heavy drinking was tolerated as a means of easing the reality of failure
and as a convenient mechanism for integrating the young into society. Though
the Sons of Temperance founded a branch in San Francisco in 1850, they were
never very successful, since saloons served good, hot, free meals to accompany
a set charge for liquor—two-bit, one-bit, and five-cent saloons proliferated—
and corner grocery stores had liquor available behind a back partition. Saloons
often were centers of prostitution and gambling as well, and by 1870, two
thousand flourished in the city.

Psychologically, an instant city differs from others. Because it is by definition rising in the unknown, less experience is available to city government—
Easterners who may have had such knowledge generally did not apply it as
they hurried off to the diggings—and relatively little laying on of seasoned
hands occurs. Rumor and fear play greater roles in such cities because of the

absence of long-standing authority figures. More tackiness, in building as in behavior, is acceptable, for one can presume the later correction of mistakes. Far less knowledge of an environment means far more initial damage to it. San Francisco of course fostered the cooperativeness associated with the frontier, founding early many associations, including those that represented ties to home, family, or the past (as with the New England Association or an association of settlers from Vermont, both of which Billings joined) but when clannishness mixed with gold fever and speculative land values, much suspicion also surfaced. Because the environment was unknown (and, from the perspective of the East, seemed adventuresome and even glamorous) there was far more tendency to declare that one had "seen the elephant" (the phrase for having had the ultimate experience), for turning the ordinary into the exceptional, and for plain exaggeration, especially in letters home. If faithful correspondents—and Frederick Billings almost never let a day pass without one letter East to a relative or friend—did not intentionally glamorize (and he did not), letters written with one eye on the clock to catch the closing of the steamer mail, nonetheless led to the conversion of many unverified rumors into fact.

San Francisco was both a mining town and an emporium. It was attached to the swings of changing technology, to the application of new techniques to hard rock mining, to building construction, to new forms of communication and modes of transportation. As an emporium, San Francisco experienced a spectacular flow of goods stimulated by commerce up and down the coast, across the Pacific, and inland to the mines (to which Nevada, and in particular the Comstock Lode, was soon annexed as a kind of suburb). Boom and bust cycles were virtually inevitable. Paper schemes for sudden cities, as with Bezer Simmons's Vernon, added to the cyclical aspect of commerce. In the 1850s, San Francisco was filled with middlemen: speculators in mine stock who stood between the mine and Eastern sources of investment. Those who speculated in real estate, as Billings did, were on the margin of the main speculation and were better able to pull out or modulate than those who dealt in mining stock or the merchants who depended upon them.

The society was a democratic one in most ways, though a growing nativism tempered that democracy. Frequent setbacks on the road to wealth kept society open, encouraging a wide range of friendships, for though wealth might denote status, it was also seen as much the product of Lady Luck as of hard work. The man in the gutter today might be bank director tomorrow. Gambling was a mania, and, since it turned on chance, Americans and immigrants were equalized in the gambling halls. Women faced a wide array of husbands and were more independent-minded since they had a genuine choice. A December 1859 photograph[2] of San Francisco's first high school graduates shows eleven young people, seven men and four women, looking out with confidence upon the world, a mixed class that one would not often have encountered at a similar level in Eastern society.

Very few San Franciscans gave thought to their city's welfare. Soon it might not be their city if they decided to move on, and even if they remained, it would change before their eyes. Billings therefore was quickly set apart, for well beyond most people he appeared to care about education, religion, and public preserves, about orphan asylums, libraries, and continuity. Billings did not do this because it brought in money—it often did not, preachers and teachers, then as now, having little of their own, and most merchants caring only whether HPB won cases for them. Instilled with a sense of noblesse oblige inseparable from his Vermont Whiggism, he did it because he considered it sound business policy and morally right. In part because of the long period of uncertainty concerning who held title to what land, San Francisco would become the center of the American title insurance business, and Billings was quick to see that title to land implied more than wealth: it meant a sense of continuity across the generations, a stake in the future and a tie to the past. Even before San Francisco might have been thought to have a history, Billings and others like him were founding societies for the preservation of the pioneer memory and subscribing to public monuments.

Instant cities are said to be heedless of history, having no time for it, but they also serve as the focus of a nostalgia that can be extraordinarily intense. Everyone is a sojourner, knocking the environment around with their elbows to make it comfortable to their needs, thinking of themselves as "really" from somewhere else. Often the most productive individuals are those subtly hostile to it, determined that they will force it to adjust to their perceptions of the good or moral life rather than themselves molding to its imperatives. Such individuals keep one mental bag packed throughout their stay, however long it may be. Frederick Billings did this, for why else would he so often speak of himself as "Frederick Billings of Woodstock"? California was home for a time, and he later considered that he had spent his most precious years there, but, even so, it was not his own ground. That remained, unmovable as a rock, in Vermont, far away from the dislocations and frenetic values of this instant city.

In his first months in San Francisco, Billings (and most others) found life physically miserable. Every morning the skies were swept of clouds, the crystalline air invigorating one for the new day, but by noon winds would whip in from the northwest, bringing a chill and often fog, and frequently kicking up dust from the dry streets that swirled about and deposited a coating on everything and everyone unfortunate enough to be out of doors. From November to February the rainy season made life wretched for those who lived in tents and difficult for those lucky enough to afford a room in a boarding house. Billings's good friend, Dr. George Turner, set the scene for a typical

November day: "A wet horse & a wet saddle & pounds of mud on the feet & legs, a wet india rubber coat dripping on all sides, slopping, slipping, comes on flouncing through the submerged clay"; muddy mats and dirty floors, wet wood and soggy kindling, dirty towels and muddy water for washing in, and clothes ruined from scraping the mud off them with a knife—this was the daily prospect.[3]

Life indoors was only marginally better. Hotel rooms were spartan, the walls so thin one was kept awake by all the sounds people make when alone or at least invisible. Cooking odors and far worse wafted their way upstairs, flying insects swarmed in season, and rodents crept into rooms to find warmth and crumbs of food. Even the best establishments might contain a saloon, a source of noise late into the night, and most likely there would be a "house for virtuous ladies"—a phrase entirely satirical—sufficiently nearby to remind one of the range of vices inherent in a predominantly male and floating population. The ague, or cholera, and an array of allergies were the natural order of things, and as early as May 1850 Billings was promising Dr. Turner that he would move out of San Francisco in search of better health. Every evening a friend would break a leg falling through or off a plank walk. One had to be alert and optimistic simply to get through the day.[4]

Still, Billings found this adventure incredibly exciting. As a young man he adhered at first to the local fashion, wearing hussar boots because of the mud and a Mexican sombrero, less because of the sun than the rain dripping from the buildings. He liked to walk the city, sometimes with friends though more often alone, or to take off almost without direction for a morning on his horse, to look out over the Pacific or along the sand dunes and up into the southern hills. He liked San Francisco at night, weird as it was: houses mostly of canvas, with inside lamps rendering them transparent, so that in the darkness some dwellings seemed to consist of solid light, the entire city a glimmering, shaking luminosity, as though the earth competed with the stars. Looking down upon the growing city from any high vantage, he could see extra-brilliant points of light—decoy lamps for the gambling houses—that would point the way to a game of monte (which Billings tried) or, less often, of roulette or blackjack (which he did not). The population of California in 1848 had been about 14,000; by the end of 1849 it was 100,000 and by 1852 would be 326,000. The majority of this growth was in or passed through San Francisco, entrepôt to the mines. Seldom has humanity overrun an area so quickly.

While Billings enjoyed the vibrant air of expectation that an instant city breathed, he also believed its people should be putting down roots more quickly than they were, especially where social services were concerned. His interest in schools, churches, hospitals, orphan asylums, and the like stemmed from a conviction that only through family life would San Francisco prosper. He was equally concerned with issues of law and order, with fire protection, with the orderly transfer of property (and the accompanying orderly keeping of records), and with sound finance. To the extent that an organized municipality could supply some of the needed social services, he was interested in city government, and, by extension, in state politics, though he considered politics a means to an end and of little pleasure in itself.

These were, perhaps, somewhat conventional views, at least for a young man of his background, but they were not widely held. As one examines the history of San Francisco from 1849 to about 1856, when the demand for protection against arson, hooliganism, and murder led to a virtual public uprising through a Committee of Vigilance, one encounters perhaps two hundred individuals who worked for the common good as Billings defined it. He was close to this group, and to the extent that his letters and such diaries as survived the many fires can inform us, at least half those individuals were his friends, riding with him, attending church and social meetings in his company, walking with him to North Beach, or dining at his club to discuss politics and the immediate problems of vigilante justice.

Historians are not agreed in their analyses of the causes of San Francisco's outbreaks of vigilante justice, nor can they concur on whether it truly had been necessary for residents to take the law into their own hands. True, in this heavily male society—there were, in California as a whole, twelve men to every woman in 1850—summary justice was far more likely than in a more family-oriented social environment. The rootless men who came there were expected to know how to use a gun and were quick to defend themselves with it, and though the incidence of crime was apparently rather low regarding theft, assault and murder with a deadly weapon clearly ranked high. Further, by the end of 1849 there were one hundred thousand Californians. Only 8 percent were native born: most had streamed in from elsewhere, bringing different concepts of justice, clamoring to shape a social system that would place one or another competing group on top—or not, as the more democratic-minded wished. California was, in the 1850s, a superb example of sociological jurisprudence at work; the law tended to become what the people wished to see it be. The atmosphere has been described as one of "legal amateurism," and in such circumstances the few legal professionals around were likely to feel distinctly uncomfortable.

On the whole the vigilantes did not operate where there was effective legal prosecution; they were not, many historians have argued, a mob. But discovery of gold had brought to San Francisco hundreds of young people who quickly formed into gangs. Properly speaking, the Sydney Ducks, with whom Riley and Billings had clashed over the Rincon reserve, were former convicts from Australia, not simply any Australians, though the term came to be applied to all, and most especially to those who were Irish. More important, there were the Hounds, sometimes called the Regulators, who were more determined to test the limits of such law as did exist. Some were discharged members of a regiment of New York volunteers who had arrived too late to see action in the Mexican War and appeared to be spoiling for a fight.

Billings strongly favored order and due process, and as a lawyer understood that the actions of the vigilantes were illegal even with the tacit support of a majority. But he also was inclined to leave a people alone in the pursuit of their customs, and he saw the resort to "justice by the people" as a legitimate expression of localism. After all, he opposed slavery as an institution, yet preferred that wing of the National Whig Party advocating gradual abolition, and was prepared to allow slave states to keep the institution of slavery intact provided no further spread of it occurred in new states. In time, however, experience changed his mind about the role of vigilante justice in a frontier community (and about slavery as well).[5]

Some precedent did exist for vigilante justice before the creation of the vigilance committees. In July 1849, during a patriotic parade, a number of Regulators smashed up several saloons, demanding liquor—it was Sunday—and assaulting Chilean families who were living in a tent encampment on the sand dunes. Some of San Francisco's most prominent and wealthiest citizens drew together a few of the city's more respected figures and arrested the Regulators, to try them before a citizen's court. The public's leader was Sam Brannan, a Mormon who had arrived in 1846; he had been the first to bring the news of the gold discovery down from Sutter's property to the Bay area. The court found the Regulator leader guilty of disturbing the peace and banished him and several other Regulators from San Francisco.

In February 1851, a prominent merchant was attacked and his store burglarized. After some months of agitation, some 200 San Franciscans organized a Committee of Vigilance and pledged that they would not allow criminals to escape punishment by virtue of "the quibbles of the law" or because of laxity and corruption on the part of police. Shortly, a Sydney Duck burglarized a shipping office in the heart of the city and taunted those who tried to stop him. He was captured and tried, with Brannan acting as judge, and when pronounced guilty was hanged in Portsmouth Square.

In July and August three more men were hanged, the last two before an audience of six thousand. Billings made a point of not going to what some considered festivities, though this may have been out of support for his friend the Reverend Albert Williams, for the two men who were hanged had been spirited away in the midst of a service Williams had been holding in the prison, and Williams was quite disturbed.[6] By the time the Committee of Vigilance felt that it could disband in June 1852, it had conducted four hangings and one public whipping, had deported fourteen individuals, and had handed over to the authorities fifteen more. (It had discharged forty-one.) In general the wider public had supported the committee in its self-appointed task, as had all but one of the city's newspapers.

Billings's approval of the Vigilance Committee of 1851 was shaped by his belief, shared by most men of his class in San Francisco, that the series of devastating fires that had swept through the city five times between May 1850 and June 1851 was the work of arsonists who, behind the cover of conflagrations, were bent on plunder. The city lacked any capacity to respond to widespread fire, and even the smallest blaze, given the congested conditions and almost universal use of wood and canvas for housing, could rage out of control in a few minutes. The fire of December 4, 1849 had wiped out much of the San Francisco that Billings knew. Fires of May and June 1850 caused damage estimated at $6,000,000—equivalent to about 20 percent of the total gold production that year—and another fire in September had left the rats in control of a good chunk of the downtown area. These four fires were followed by two more, in May and June 1851, and the blame for these, in particular, was laid at the doorstep of the Sydney Ducks.

The great fire of May 4, 1851, which came on the anniversary of the fire of May 1850, took several lives, consumed more than three-quarters of the city, and caused nearly $12,000,000 in damage. When booty was discovered in the surviving houses of some of the Ducks, the public felt it had a smoking gun. Thousands of people worked day and night to restore their properties, and within ten days 350 new structures were erected. But people remained deeply angry, for in many cases they had lost all their possessions, including treasured mementoes. The vigilante group had first begun as a voluntary night patrol to protect the reborn city from fresh fires, since it was evident that the authorities could not. On June 2 the cry of FIRE had raced through the city once more, and on June 10 an enlarged Committee of Vigilance had been formed with a somewhat broader intent than fire protection.[7]

Billings had been one of the many wiped out by the May fire. As he wrote to his parents, he saved nothing but a carpetbag of clothes that he had with him—for he was away on business in Sonoma when the fire struck—and some papers from the law firm's safe, which one of his employees had been able to carry up to the third floor of their building and, at the cost of burning his hands on the already-hot knob of the safe, thrown into the street. Billings lost most of his college papers, his diaries, his legal memoranda, daguerrotypes, lockets (including one containing a snip of Laura's hair), and the Bible Laura had passed to him as she lay dying. Papers used to help his clients were destroyed, as was his entire library, "equalled by only one in California." He had, he reckoned, lost $12,000 in furniture and appliances as well, and set his total loss at $20,000. With his parents he took the cheerful view: Halleck, who had been away, was safe, Frank was in good health and had found lodgings elsewhere, and, as for himself, he had his strength and was grateful for that. But he was angry, for himself and for Simmons, Hutchinson & Company, which had lost almost all its records in the midst of its efforts to settle its affairs.

Frederick's mother hurriedly dispatched six new shirts, fresh family pictures, lockets of hair, and other mementoes. Fortunately this new supply of family ties did not arrive quickly, for on Sunday, June 22, yet another fire swept the city. This fire burned eleven blocks, destroyed the offices of the *Alta California*, the leading newspaper, and a church, and left Frederick fully in favor of the Vigilance Committee. He lost little, for as he wrote his parents, he had nothing to lose. (In fact he lost a case of wine, which he did not confess to his teetotaler father.) This time it was Peachy who rescued the law firm's safe, dragging it into the street. This fire was also ascribed to an arsonist.[8]

What worried Billings was that vigilante committees, or groups calling themselves such, had sprung up elsewhere, and were not always bent upon eliminating hard-core crime. There was an overlay of nativism to the committees, a sense that foreigners were at the root of the trouble, which was paradoxical since nearly all the vigilantes were, in a sense, foreigners. Even those who felt permanently committed to California were foreigners when compared to the *Californios*, against whom they nonetheless often used the term. In short, issues of race and language were also present, and with others, Billings worried about the implications of such nativist definitions. He had been distinctly unhappy when Bezer Simmons had joined an early vigilante group in Sacramento, though on the whole it too was made up of worthy merchants and could be presumed (or so Bezer argued) to be a responsible body. But in 1851 Billings put these worries behind him.

Early in 1854 San Francisco faced a major depression. The first gold production had been based on relatively simple panning, but this placer gold was now gone, and expensive and bulky quartz-crushing machinery was needed to sustain the gold economy. Miners who lacked the ability to invest in such equipment were pouring into the streets of San Francisco; mines were no longer taking supplies from San Francisco's merchants; bankers began to call in their credit. The building boom of the previous year, when over 300 brick or stone buildings had gone up, 154 of them three or more stories high, collapsed and real estate values fell by 40 percent. Volume at the port was cut in half and business failures doubled. In the fall of 1854 Henry Meiggs, a highly respected businessman, politician, and philanthropist, fled the city leaving over $1,000,000 in debts.

There were nineteen banks and nine insurance companies in San Francisco early in 1855 and none subject to much control. The state constitution prohibited joint stock banks and the issue of paper money but little else, and anyone could set himself up as a banker, even if in his earlier life he had been a baker. In February 1855, just such a banker precipitated the state's most serious financial panic, and before the drama of the day was over Frederick Billings found himself caught up in a series of potentially damaging events.[9]

In 1849, Page, Bacon & Company had opened an express office in San Francisco. The partners were Daniel B. Page and Henry D. Bacon of St. Louis, Judge David Chambers and Henry Haight of San Francisco, and Francis W. Page of Sacramento. By 1852 the company was functioning primarily as a bank, with its offices in the Parrott Building. One of its principal competitors in the banking business was Adams & Company and, in transport, Wells Fargo & Company. Adams & Company was especially well known, with agencies in every town and many mining camps in the state. One of Adams's cashiers was James King of William, a twice-failed banker. The company recently withstood a serious run, having to pay out $428,000 in a single day, and apparently recovered, though the Meiggs affair left the city apprehensive about banking in general. Several people, Billings among them, were predicting a damaging financial collapse.

Page, Bacon & Company was a branch of a larger St. Louis house, and on January 12, 1855, Henry D. Bacon, the former baker, discovered that his commitments to financing construction work on the Ohio & Mississippi Railroad had gone sour. The St. Louis house asked the San Francisco branch to send all the gold it could obtain, and $1,000,000 in gold dust was hurried off by steamer just before the news arrived on February 17 that the St. Louis office had, in fact, been forced to close its doors. Though Daniel Page insisted that the two houses were distinct, a run began on Page, Bacon & Company, which paid out $600,000 on this first day. Without the gold dust just dispatched to St. Louis, the company could not meet its obligations and closed its doors on February 23. Five other firms, including Adams & Company, on which a run began on February 22, were forced to the wall, as San Franciscans withdrew over $3,500,000 from the six banks. Page, Bacon & Company and Adams & Company branches throughout the state were besieged and vaults were broken into; sometimes money was handed out before valid depositors could reach the scene. Mobs descended on the hotels where the bankers lived, and the Montgomery Block, where Adams & Company had offices, was a scene of "panic and chaos." Billings and Park were observed going in and out of the Adams offices, apparently working to stave off a total collapse.[10]

In this they failed, and charges of fraud soon were launched against Adams & Company and its general manager, Isaiah C. Woods. A young Englishman, Alfred A. Cohen, who had a good reputation for finance though no experience in banking, was asked to act as receiver. From this point the story became curiouser and curiouser.

The absentee partner, Alvin Adams, who lived in Boston, sued the other two partners, Woods and Daniel H. Haskell, to dissolve the partnership in order to keep the house closed, there being no state insolvency law. During the night of February 22, Cohen surreptitiously removed all the coin and gold dust from the bank's vaults, on such short notice that no one took the time to count the coin or weigh the dust. This trove was lodged with Palmer, Cook & Company, which was still solvent, reportedly as protection in settlement of pending suits. Cohen subsequently revealed that the officers of the company apparently had mysteriously disappeared. Adams and Haskell were somewhere in the East, and only Woods was actually present. Cohen soon left the city and Judge John S. Hager removed him as receiver and appointed the respected merchant Henry M. Naglee in his place.

During Cohen's absence rumors went round that he had taken much of the coin and gold dust and, further, had substituted inferior dust. Naglee won a suit against Cohen, who refused to turn over the assets he held, and in absentia the court ordered him imprisoned for six months. Two Irish fishermen found Adams & Company's last ledger and cash book in a canvas bag floating at the water's edge off the end of Montgomery Street, and demanded $30,000 for the books, which the authorities refused to pay. The books disappeared once again, finally to be discovered hidden under a mattress in the fishermen's room, but with the pages for February 21–22 torn out. Public opinion held that Cohen had tried to destroy evidence by throwing the records into the bay.

Billings was inclined to believe ill both of Adams & Company and of Cohen, and knew this collapse could mean the ruin of many of his own enterprises, possibly even the Montgomery Block, which depended so heavily on the general sense of well-being that arose in good times from the banking and legal fraternities. He was angry with James King, for Page, Bacon & Company had rallied and then collapsed for good under the vitriol that King had poured out against it in the *Bulletin*.

Then on March 15 Alfred Cohen, who had not gone to the East Coast, filed a complaint against Adams & Company, Frederick Billings, and Trenor W. Park. They had, he charged, extorted $10,000 from him. Further, Cohen declared, Halleck, Peachy, Billings, and Park were, and had long been, the attorneys and confidential legal advisers to Adams & Company, and he alleged that the firm had sought to take $250,000 out of the company. Billings and Park were said to be the principal advisers, and thus the blame for the collapse of Adams & Company might reasonably be laid to their doors. Additionally, Cohen claimed that Park had proposed the removal of the gold dust and the dissolution of the partnership, and the collusive procedure by which Adams sued Woods, effectively avoiding payment to creditors from personal funds. Finally, Cohen alleged, Halleck, Peachy, Billings, and Park had abused and intimidated James King, as chief clerk at Adams & Company on the night of February 22, into agreeing to pay them a retainer of $10,000 to see the company through the mess, and King had given them a note for this sum.

As this case worked its way through the press, and more slowly through the courts, circumstantial evidence started to pile up against Park, and Billings's name began to drop out. It soon became apparent that Park had definitely sought a retainer from King and pressed him hard; Park may also have used privileged information as part of the pressure. Billings had been present, together with the original lawyers to Adams & Company, and two or three other individuals, when Cohen was asked to become the receiver, since HPB's senior legal adviser, Oscar Shafter, had already drawn up the necessary documents, though without Cohen's name on them. At some point the company's legal counsel, apparently in Park's presence, recommended that Cohen remove the coin and gold dust to protect them against the mob that would undoubtedly descend, and Billings—who had gone after a judge to validate the appointment of Cohen as receiver—returned to find the transfer in progress under cover of darkness and rain. Witnesses were produced, some swearing that Park had browbeaten King, others that he had not, but none making any reference to Billings. His name was linked to the charge simply because Park that same month had said that in all matters relating to Adams & Company Billings was his partner. Billings publicly denounced this statement without denouncing Park, declaring that he was not an agent for Adams & Company; rather, Park was the sole agent, a point on which Halleck and Peachy supported him.

At this juncture, Park was challenged to a duel by Hamilton Bowie, a former city treasurer. Park implied that Bowie was at fault in the bank failings and pressed an indictment against Bowie for hiding the treasurer's records. These records would, he said, support a charge against Bowie's having embezzled $300,000 from the city the previous September. At Bowie's trial on March 15, Park did not show up, and though the sheriff attempted to serve a subpoena on him, he still failed to attend. The jury promptly brought in a verdict of not guilty, and Bowie at once sought satisfaction from Park, asserting that Park had known his charge was maliciously false. Park replied that he would not fight and Bowie publicly branded Park a scoundrel and a coward.

Relations within HPB were now seriously strained. Peachy thought it wrong to refuse a challenge and was outraged at Park. Billings was angry over having been named with him in Cohen's charges and embarrassed when Park was arrested while on board a ship that was to sail for the East, thus feeding the rumor that he was fleeing a weak case. Cohen added circumstantial evidence to his charges, with innuendo that tarred Billings as well as Park, and the law firm decided to place Park's defense in the hands of Oscar Shafter, who was able to get the charges quashed on complex technical grounds.

Cohen and Woods were now charged, in a different court, with having sold $400,000 in inferior gold dust to Page, Bacon & Company prior to the company's collapse. This extraordinary complaint, which held that the defendants had changed the color of inferior dust by chemical processes to resemble the superior product of the Southern Mines, brought public attention to fever pitch. While Cohen denied the charge, he admitted he had sold poor dust to the company long before he was its receiver, noting that if bakers and bankers could not tell good dust when they saw it, they were to blame, not he. He had not, he said, ever misrepresented the dust, even though he was overpaid for it. Shafter joined Edmund Randolph, lawyer to Page, Bacon & Company's creditors, in demanding the arrest of Woods and Cohen.

Cohen insisted that both Adams's attorneys of record, as well as Park, had asked him to act as receiver and advised him to remove the coin and gold dust, and that he did so legally, though in fact an hour or two before he was formally appointed receiver. He testified that Park had told him he intended to go to Sacramento to collect the company's assets and that Billings would do the same in Stockton and south. Cohen had opposed this, as did Woods when told of it, suspicious that Park wanted to get the money into his hands and then retain it for fees. To test Park, Cohen said, he had offered to retain him with a fee agreed to in advance, whereupon Park said that neither he nor Billings would be able to undertake the journeys. The next day Park had reversed himself, Cohen said, asking for a power of attorney to act for Adams & Company in Sacramento; apparently he did not mention Billings again. According to Cohen, Park then tried what was, in effect, an extortion: he said if he acted for those who wished to sue the company he could easily collect $50,000 in fees, but he would, out of friendship for Adams & Company, defend them against all comers for $40,000. Further, he hinted, he had two members of the state Supreme Court lined up to decide in favor of any cases he pled for the company. Though there was some confusion in Cohen's mind as to whether Park was soliciting the fee for himself alone, or for HPB, Billings was never present and was not mentioned again, and it seems reasonably clear on the basis of the surviving documentation that Billings was unaware of any details of Park's initiative, assuming such an initiative took place.

By this time public interest again was flagging, with even the press commenting that about enough had been heard of the matter. But a new sensation would soon revive the public's thirst for scandal. As Cohen walked from the court following the filing of his affidavits, his brother, Frederick A. Cohen, rushed up to Trenor Park as he was entering the Montgomery Block and, apparently without warning, struck him down. As Park fell, Cohen raised him up from his knees and struck him several more times about the face, breaking his nose.

In the end Park was not charged with perjury and no case was brought against any other member of HPB, which was permitted to act as the attorneys of record for Adams & Company. Frederick Cohen, in the face of one witness's insistance that Park had raised his cane to Cohen, was found guilty and fined $300, a small sum reflecting the doubts in the case. Park subsequently accused Cohen of being the person who had removed the critical pages from the Adams ledgers.

There was now a new set of charges, however, for James King said that Woods had instructed him to hold back a substantial sum of money during the run on the bank, and a former employee of Adams & Company, Henry Reed, charged Park and King with collusion to defraud the company. Park, Reed alleged, had betrayed the secrets of the house, having often been in the company's premises in the Montgomery Block, frequently as early as seven o'clock in the morning or as late as midnight, constantly moving about behind the counters and desks. A receipt was produced that appeared to prove that Park had, in fact, accepted a retainer from Woods—not King—for $10,000 on the day Adams & Company failed. Park implied that this was a forgery. By now the sensations attendant on the Adams & Company cases reached the point that the press had begun to satirize them with daily headlines behind which no story was to be found: "MORE AWFUL DISCLOSURES! MORE FRIGHTFUL ANNOUNCEMENTS!! STILL FURTHER TERRIFIC RASCALITY AND HORRIBLE TREACHERY!!!"

After "eminently able . . . and denunciary" argument on both sides, Judge Hager found that there was insufficient proof with respect to the charges against Park, despite the peculiar circumstances. While Park had, Hager implied, acted indelicately in the matter, no clear evidence existed that he had broken the law. He added, given the public speculation, that the conduct of Halleck, Peachy, or Billings was not in question.

There remained the charge that Cohen and Woods had fraudulently sold Page, Bacon & Company inferior gold dust, and on August 9 the company's creditors filed an additional count, to the effect that Cohen and Woods had effected sales through accomplices disguised as miners with whom they corresponded in code, substituting "barley" for "gold dust," thus having gone well beyond fraud to conspiracy. Two days later and a day before evidence against him was to be presented in court, Woods disappeared, coolly leaving behind a public statement to be printed in the *Alta* in which he said he was departing at once for the Eastern states to seek money to set right his obligations to his creditors. Instead, Woods was reported on a vessel bound for Australia. In fact, he appears to have gone to Peru.

On August 16, 1855, a formal notice of dissolution of partnership appeared in the San Francisco newspapers. Halleck, Peachy, and Billings would continue to practice law together. Park would open new offices in the Montgomery Block in partnership with Oscar Shafter, an arrangement, Shafter wrote to his father, intended to save Park "from a popular suspicion of having been discarded by the firm." Halleck, studious, proper, always careful, even more distant now, was surely happy to be done with Park, who seemed to have forgotten his duty to his clients. Peachy made it clear that he did not want Park in the firm, disgusted with a man who had publicly flouted the code of honor in the Bowie affair and had refused to seek revenge on Frederick Cohen for assaulting him. Billings was thoroughly sick of seeing his name attached, even slightly, to these affairs, and was particularly distressed at having it besmirched in the Eastern press. He must have seen Park as a divisive figure in HPB, though he never spoke out against Park in public (and, perhaps more significantly, did not speak out *for* him either).

Now King was attacking Palmer, Cook & Company for trying to obtain deposits of public funds, in part through the intervention of the powerful Democratic politician David C. Broderick. King warned readers of the *Bulletin* to withdraw their savings from Palmer, Cook & Company, and heaped calumny almost without restraint on the head of Broderick. In May 1856, King went too far, revealing in his paper that James P. Casey, a former Broderick supporter who had gone over to a rival Democratic faction, had spent time in Sing Sing prison before coming to California. Though Casey had asked King to suppress the story, arguing that King did not know the whole of it and that it was, in any case, behind him, King printed it nonetheless. That day Casey waylaid King and shot him down in the street.

The wounded King was carried into the Montgomery Block and made comfortable outside Billings's office. Casey was taken to the county jail. A mob gathered and it became apparent that the jail could easily be stormed. A vigilante group quickly formed and declared its intention to seize Casey and another miscreant, Charles Cora, who had shot and killed a marshal the previous winter, and who still remained in custody awaiting a second trial since the first resulted in a hung jury. Soon the entire city was in an uproar.

The remnant of the old committee of 1851 was assembled, deciding upon secrecy with each member taking a number for identification. William Coleman, president of the earlier committee, was asked to become its head once again, and a permanent headquarters was established in a building on Sacramento Street. The Vigilance Committee organized into military companies of infantry, cavalry, and artillery, assembled firearms from throughout the city, and placed sandbags around the building, which was soon dubbed Fort Gunnybags. The infantry companies began to drill, and Coleman let it be known that the headquarters had cells awaiting six prisoners.

On Friday, May 16, Governor J. Neely Johnson, aware that the situation was out of control, asked William Tecumseh Sherman to take command as major general of the California militia. Sherman, as a respected banker, would be accepted by the business community, or so Mayor James Van Ness hoped. Action was imperative, for with Sunday near, idle hands would turn to mischief. Sherman went to the district court, where he found about a hundred people, summoned by the mayor's writ, milling around waiting to see what action they would be asked to take. About two-thirds of those summoned had failed to show up, and among the men present, division prevailed, with the majority predisposed toward Casey and a minority leaning the other way but wishing to see the law observed.

In the second group Sherman found Billings and Peachy standing with Myron Norton, a district judge, Hall McAllister, attorney for the district, and James Dabney Thornton, known for his moderation. They accompanied the sheriff to the jail, where as a formal posse comitatus they elected as captain the absent Sherman, who had left for a quick supper before meeting the Sacramento boat on which the governor was expected. Billings, fearful that revolution was in the air, urged Sherman to hold firm to the law.

Sherman hurried to the jail and surveyed the scene. He concluded that the building could not be defended, but that unless King were to die, the Vigilance Committee probably would not attack the jail. He then went to meet Governor Johnson, and together they visited the sheriff, agreeing that defense in the face of a mob already twenty-five hundred strong and growing by the minute was out of the question. Johnson, Sherman, and others decided to parley with Coleman, who was an old friend of Johnson's. Coleman told the governor that the shooting of San Francisco's citizens must stop, and Johnson guaranteed that Casey would not be allowed to escape. If King died, Johnson promised, Casey would be indicted at once and the courts instructed to drop all other business to try him. If convicted, he would be executed. The vigilantes did not find these promises sufficient, however, and since the sheriff was known to be a friend of Casey's, they insisted they be permitted to post their own guard inside the jail.

Billings and Peachy were not members of the state militia, but as Sherman found that of the entire militia he could count on only seventy-five men, they volunteered to pass a long night on guard duty inside the jail, carrying oiled rifles in the crook of their arm and spelling each other. After one night of this they concluded that they were doing little good, for the mob outside the jail was growing by the hour, the vigilantes had their own guards inside the jail, and Sherman declared that his guards would not be the first to fire should there be an attack. Clearly what Billings or any other citizens could do under circumstances of virtual mob rule (for by morning the milling crowd had grown to five thousand) was very little. Indeed, nothing was done: on Sunday the Vigilance Committee ended its treaty, as Sherman put it, and left the two prisoners at the jail solely in the hands of the undermanned sheriff. A great mass, now ten thousand strong, descended upon the jail and demanded that

Casey and Cora be turned over to them, as the governor, mayor, Sherman, Billings, and others watched helplessly from the roof of a nearby hotel. Under protest the sheriff gave up his prisoners.[11]

Casey and Cora were given a trial of sorts, but on the afternoon of the second day the court was interrupted with the news that James King had died. It was decided that the prisoners would be hanged secretly on May 22, at the same time as King's funeral, in the hope—the committee later explained— that the crowd would be following the funeral procession. Militia and the sheriff's office were powerless to intervene.

As it happened, whether by accident or design, news of the impending execution soon leaked out, and the hangings immediately followed the funeral, so that the massive crowd for the first attended the second. Government in San Francisco was now in the hands of a well-disciplined, perhaps even high-minded, yet undoubtedly illegal clique. True, the Vigilance Committee con-sisted of some of the city's most honored men. But this time, unlike in 1851, there was a strong nativist tinge to public sentiment, party issues were des-perately entwined among the factions, and a large, unemployed mass of men, down from the mines with no place to go, were roaming the streets.

This was rebellion, revolution, civil war—the last being Sherman's pre-scient name for it—and two groups opposed to the Vigilance Committee now came on stage, one publicly and the other more privately. The first was the result of a meeting on May 27, with Van Ness in the chair, that led to the formal creation of a Law and Order Party to oppose the committee. Represen-tatives of this group (which included Archibald Peachy) called on the governor to suppress the "insurrection." Johnson, Sherman, and the secretary of state, D. F. Douglas, visited Major General John E. Wool, commander of the U.S. Arsenal at Benicia, and believed they had come away with a promise of the necessary arms should the governor decide to send the militia against the vigilantes. Then, after obtaining from San Francisco's sheriff a rather muddy statement to the effect that he needed the help of military force, Johnson ordered Sherman to call out the militia, issuing a proclamation declaring the City and County of San Francisco in a state of insurrection.

Sherman hoped to use moderation, believing that a show of force would be sufficient to end the troubles. Several persons, of whom the most telling was Henry Halleck, a member of the second group, the Conciliation Party, called on Sherman to urge that he embrace neither the Law and Order faction nor the Vigilance Committee, and that he find a means of ending the strife without bloodshed. Sherman hoped for the sheriff's resignation, which would cool matters, and was supportive of a public meeting assembled by the more respected members of the Law and Order group. Billings committed himself to the Conciliation Party. On June 4, Sherman called out the militia to protect peaceable assembly, and Johnson sent General Wool a note asking for arms.[12]

Bloodshed was averted by General Wool's reply to the governor: he would not supply arms. Johnson was thunderstruck, and Sherman felt deserted. A more formal requisition for federal arms and ammunition from the governor also met with refusal: either Wool had changed his mind or he had been misunderstood—and Sherman gave Johnson his resignation.

As Johnson, Sherman, and Wool stood in confrontation with each other in Benicia, members of the Conciliation Party arrived. The Law and Order group, denouncing Wool as a damned liar, were beside themselves with fury and impotence. The conciliatory leaders, which included Billings, Joseph B. Crockett, a well-placed lawyer, Judge J. D. Thornton, Bailey Peyton, former city attorney, and Frederick W. Macondray, now president of the Chamber of Commerce, said that they had not been given a sufficient hearing by the governor. Sherman told Johnson that Halleck, as a conciliator, ought to be his successor; he was the only figure with military standing who had not yet blotted his copy book with one of the factions. But Halleck would give no assurances on what steps he would or would not take if appointed, and Johnson thereupon asked Major General Volney E. Howard, who was sympathetic to the Law and Order group, to command the state militia. When Howard was able to get some arms from the arsenal at Benicia, the vigilantes seized them while they were being transported across the bay, depriving the militia of any force.

Billings was in a dilemma. Of 189 individuals identifiable by name and occupation as in either the Law and Order or the Conciliation factions, only seventeen were merchants. The remainder were military and legal figures, with a leaning toward the Democratic Party. Several of the Law and Order group were pro-slavery. Billings hesitated in this company, for though he was opposed to mob rule, and was close to Halleck, a conciliator, and Macondray, the breakaway Law and Order advocate, he was after all still a Whig and strongly pro-Northern on most political issues. Further, he felt that this entire affair had arisen because of King's provocations.

The Law and Order group now played into the hands of the Vigilance Committee. A vigilante body attempted to arrest the two men who had tried to transport arms to Howard; David S. Terry, a judge of the state Supreme Court who was present at the confrontation, knifed a vigilante. Judge Terry was arrested and tried. Obviously the Vigilance Committee could not hang Terry, so they prolonged the trial for five weeks while deciding what to do. When the man Terry had stabbed recovered, the heat was off the committee and they quietly released the judge. His action had, however, made a mockery of the high-minded statements of the Law and Order group, while the obvious ambivalence of the Vigilance Committee toward Terry's offense disillusioned its more extreme members.

By now cooler heads prevailed. The Vigilance Committee had saved face by twice more drawing blood, hanging two more men convicted of murder. General Howard had been unable to obtain arms but had shown his desire to do so. The Conciliators had been working steadily on both sides of the street, for they had friends in both major and some lesser camps. The San Francisco *Herald,* the only newspaper to oppose the vigilantes, had begun to make itself felt, despite having its advertising revenues cut below half. Billings and many others had concluded that the vigilante movement was interested in far more than justice. Certainly the properly constituted authorities had been rigid and frequently ineffective, and law-abiding San Franciscans needed better law enforcement, but Billings believed that defiance of Johnson and Sherman and the abuse of Terry were unacceptable. There was self-interest to consider too: most lawyers at the time were opposed to violent change, since they were the representatives of change brought through the legal process of government, and they collected fees along each step of the orderly way.

In August when tempers cooled, Billings was among the signers of a call for a mass meeting on political reform and was nominated as a member of the twenty-one–person executive committee of a proposed new People's Party. But he was not chosen for the executive committee; when his name was put forward to the assembled body, his and another nomination were shouted down, possibly because Billings was a lawyer—in the end no lawyer was admitted to the committee—though more likely because he had been a Conciliator. The new party would approve only the "pure," requiring an oath of allegiance and the promise that one had never been an office seeker.[13]

Years afterwards, with California only a memory, Frederick Billings would entertain dinner guests with stories of these stirring times. His tales described the outbreaks of 1849 and 1851, not of 1856, for in the earlier episodes he sided with the vigilantes. He spoke of how he and Colonel Granniss had passed sleepless nights at the height of the mob's near possession of city streets, guarding their premises against theft or fire, and Billings took some pleasure from seeing his guests' reactions when he told them he had once helped to organize a vigilante group and had seen an outlaw strung up from a nearby lamppost. This was the romance of youth recollected in the contentment of age, however, for while he may well have been an observer to such an event (though here too recollection was slightly colored, since none of the victims of the vigilantes was hanged from a lamppost—wharves, second-floor windows, store fronts, store signs, and gallows, yes, but not lampposts), his was largely the quiet role of a conciliator behind the scenes.

All his life Billings would play the conciliator, seeking compromise and the high ground when tempers flared. In the year after Billings's death the preeminent historian of California, Hubert Howe Bancroft, in his widely heralded and lengthy essays on "the builders of the Commonwealth," would praise Billings as a ruler of people, the equal of Napoleon, Garibaldi, and Peter the

Great. This was the hyperbole of high Victorian biography and obligatory for subscription history. But when Bancroft attested to Billings's possession of the "three primary principles of progressive force"—intellectual power, moral power, and the power of wealth—he may well have had in mind Billings's even-handed walk along the wild side of California's raucous early years. Bancroft transparently favored the vigilantes in his histories, often damning with faint praise those who opposed them. He made no mention of Billings in connection with the vigilante episode of 1856, though he did conclude that his "nature was the reverse of combative," and that he withdrew from political conflict. Billings had been, in Bancroft's argument, a good diplomat for an instant city.[14]

10

The University of California Is Born

ERASMUS DARWIN KEYES HAD SINGLED OUT Frederick Billings as a philanthropist and all San Francisco knew him as one, for Billings was prepared, first with his energy, later with his money, to support every cause he believed in. He helped organize the Mechanics' Institute, which included a library so that laboring men might read; he was a major figure in setting up the San Francisco Orphan Asylum Society; with Colonel David S. Turner he helped to found and partly fund a Protestant church for Chinese residents; he helped establish the Mercantile Library Association and to promote the San Francisco Bible Society (serving as secretary); he was president of the board for a Seminary for Young Ladies, and he worked on behalf of the state's first Agricultural Society. Billings soon acquired a reputation as a public speaker and was asked to dedicate the city's second and fourth schoolhouses and to speak at the opening of churches and civic amenities. Both his brother Franklin and Frederick served on the school board. He was interested in the performing arts and joined those who crowded into a school room on June 22, 1849 to hear San Francisco's first concert; he subsequently became the lawyer for the American Theater, which opened in October 1851, and later for the Union Theatre. As Milton Latham, then collector for the Port of San Francisco, wrote, no one was so "zealous, constant and unremitting in attention to the public good" as Billings.[1]

Of course, for a lawyer with Billings's interests, being zealous of the public good was also good business. He valued his reputation, disliked the notoriety embroilment in the affairs of Adams & Company brought, and, as a friend of Park's was to remark snidely, was "too modest" to do what he had to do to become vastly rich in the instant city. There is little reason to view his philanthropy cynically, for decades later, when anxious for neither fortune nor public office, he continued to accede to a wide range of appeals for his money, his time, or his oratorical skills. His wife would chide him that he never seemed to be able to say no to any good cause, and Latham concluded much the same, remarking that Billings responded to all the calls made upon him. To trace all these and his responses would be to reflect, through the eyes of one man, much of the story of California's social consciousness in the 1850s.

Billings remained constant to three public interests: the place of religion in public life, the importance of free, public education, and the creation of parks and preservation of scenic landscape.* He effectively practiced tithing, giving roughly 10 percent of his earnings to religious enterprises, while also supporting, with gifts that were large for the time in relation to his income, universities and schools, especially their libraries, and drumming up attention for parkland projects. After Billings married, he and his wife would continue this generosity.

Throughout his life Billings appears to have been without significant religious bias—he went to Episcopal, Congregational, and Presbyterian churches interchangeably, and when travelling occasionally attended Roman Catholic masses. To his wife Frederick wrote in 1873 that they should see good in the creeds they did not accept rather than accept ones they could see no good in, and he applauded Julia in her opposition to the Episcopal bishop of New York when he told his priests not to fraternize with other religions. Gifts went, as asked for and needed, to all denominations throughout his and Julia's lives.

On the Sunday after the Simmons-Billings party had come ashore in April 1849, they had gathered with the Reverend Albert Williams to say prayers and sing hymns in thanksgiving for their delivery to California. This first service was held in a schoolhouse at Portsmouth Square, and a pickup choir, with its female voices wafting to the street, soon attracted a crowd that girdled the school twelve persons deep, drawing people away from the gambling tables at the nearby Parker House. After preaching in the schoolhouse a few times, Williams organized the first Protestant society in the city. It had six members; this, the First Presbyterian Church, brought together Billings, nominally a Congregationalist, his friend Dr. Turner, John Geary's wife Margaret, and three others. The church rapidly grew and by summer upwards of two hundred worshippers were in attendance. Though the First Congregational Church was organized in July, Billings felt he could scarcely leave Williams's flock only two months after the initial meeting, and he remained a loyal attendant at the Presbyterian church until he returned permanently to the East Coast. A co-religionist, David Hewes, later wrote that Billings was, in those early months, "the soul and body" of the church, doing everything, including going along on Saturdays to clean out the dirty schoolroom used for Sunday services.[2]

Save for one intense schism, Williams and Billings remained fast friends for life. Daily prayer together aboard the *Oregon* and the memory that, as

---

* For a discussion of Billings's growing conservationist interests, see Chapter 25.

she lay dying, Laura had asked that Williams preach her funeral sermon bound them together. The first formal meeting to discuss building a church was held in Billings's law office, and thirty years later when Williams published his memoirs, Billings purchased several copies, declaring that nothing he had read had so brought back his pioneer days. Billings supported Williams when he founded a school, the San Francisco Institute, launched a short-lived magazine, *The Watchman*, and helped to open (together with Bezer Simmons, who became the first president) a branch of the Seamen's Friend Society. Billings paid for the cost of publishing Williams's farewell sermon in 1854, was treasurer of the church, chaired the committee that chose Williams's successors, and was chairman of the building committee in 1858, providing the largest single gift for a new church.[3]

Still, Billings was not without strong feelings on religious matters, however ecumenical his churchgoing, and when his parents asked what he thought of the Reverend Henry Ward Beecher, who had developed a substantial following in the East, Frederick replied sharply that he did not think him fit to go into a pulpit and preach. His judgment was based on Billings's perception that Beecher and Horace Bushnell had turned to mysticism and were mixing religion with politics in dangerous ways, with Beecher declaring that rifles would be better than the Bible in settling the dispute between free soilers and slaveowners in Kansas.[4]

As for Bushnell, Billings changed his mind once he met the man. His parents asked his opinion because the powerfully influential preacher, in poor health, was coming to California. As a Yale College graduate and Congregational clergyman, Bushnell actually stood close to Billings's own less articulate position, for Bushnell had drunk deeply from the works of Samuel Taylor Coleridge, just as Billings's teachers at the University of Vermont had, and he too felt that one ought to unite denominations rather than drive them further apart. Bushnell emphasized individual communion with God through Christ; he found orthodox theology too severe and championed a kind of romantic Christianity that appealed to a democratic age. When Bushnell reached California, Billings at first was aloof and critical, but a common interest in the new University of California soon brought them together, and a warm friendship grew. Billings read all of Bushnell's published work, and years later Bushnell was a regular guest in the Billings household in Woodstock, where Frederick and his wife Julia—who had read Bushnell before she met Frederick—and the venerable theologian discussed his most recent books. Both had accepted one of Bushnell's arguments in particular before their marriage: that churches ought to focus less on converting adults and more on giving the young an early, clear sense of the moral conduct implicit in Christian growth.

Of all his civic activities, Billings appears to have devoted the most attention to the founding of a university for California. Here the catalyst was his classmate from his days at Kimball Union Academy, the Reverend Samuel Hopkins Willey. Willey had come to K. U. A. from tiny Campton, in the mountains of New Hampshire, and he and Frederick had shared vacations together, the one too far away to go home, the other determined to remain at school and use the time for study. Promotion of a state university, even one without specific sectarian ties, was for Billings a natural extension of his belief, nurtured at the academy, in the educational influence of church attendance.

Willey arrived at Monterey on behalf of the American Missionary Society in February 1849 and remained for a year and a half. As the chaplain of the state constitutional convention in September, he came to know many influential men, and to all he preached his message of higher education. When the legislature was assembled at San Jose in December 1849, Willey attended, and while there he met Sherman Day, son of Jeremiah Day, a long-time president of Yale College. Day was eager for California to have a college of its own, and he and Willey brought into their discussions three other like-minded individuals: Billings, Reverend Chester S. Lyman, and Forrest Shepherd, both from Yale, the latter another alumnus of the *Oregon*.[5]

Willey did not wish to give the proposed college a denominational character. Still, Lyman, Day, and Shepherd attended Congregational churches in San Francisco or Sacramento, and Billings, though continuing to support the Presbyterian church, was regarded as a fellow Congregationalist at heart. While the new institution was to be nondenominational, most likely all were aware that both Methodists and Roman Catholics also had plans for higher education in California.[6]

On December 18, 1849, Willey, Day, Billings, Lyman, and Shepherd resolved that they would incorporate so that they might hold property for the foundation of "California University or College," and form part of a board of trustees. The group applied to the new legislature, which on April 20, 1850, passed the necessary law. Lyman had left for the East Coast the month before; in the meantime, an offer of land had been made by two non-Congregationalists, Captain Henry Naglee of San Francisco and Pedro Sansevaine of San Jose, on the condition that the institution be free of all religious distinctions. The trustees agreed, and issued a subscription paper for obtaining pledges for needed funds before an application could be made for a charter. To the original five trustees were now added three men especially close to Billings: John B. Geary, Thomas J. Nevins of New Hampshire, the superintendent of public schools in San Francisco, and Mariano G. Vallejo, the leading member of the influential *Californio* family. Meeting alternatively in Geary's and Billings's offices, the group prepared a charter, and Billings, as attorney for the majority of the donors, who pledged $30,750, presented the application.

The Supreme Court turned the application down: they required $20,000 in hand, in cash or its equivalent, rather than pledges. In this the court was wise, for several pledges (including one from John Sutter) were never met, though Billings tried to persuade the court that the land pledged to them was worth the necessary sum alone. The court replied that several of the land titles were in dispute and that the boundaries were unclear—which Billings should have known since this was his legal turf—and hoisted him and his fellow trustees with his own petard.

Billings was embarrassed and angry. He blamed himself; he had not done enough research, had not arranged to have property lines checked, and while he had obtained affidavits to testify to the value of the land, he did not have enough. He had let his fellow trustees down, or so he felt—it is not clear that additional affidavits would have made any difference to the court, and there was as yet no land commission that could have passed judgment on such additional boundary or title research as he might have done—and he was determined that he would not make the same mistake twice.

In the end Willey, Billings, and their fellow trustees got their university. Delay permitted time for the growth of a more grassroots movement, for agitation in the press, and for the setting of precedent favorable to the University of California. The U.S. Congress created the Land Commission in March 1851 to receive petitions from private land claimants and pass on the validity of titles. Any suspicion that the proposed college would serve a sectarian purpose was allayed with the addition of supporters from other denominations to the subscribers.

In May 1853 the Reverend Henry Durant, another graduate of Yale and the man destined to be the first president of the University of California, arrived. Durant wanted to open a female seminary in California, and he came, as he later said, "with college 'on the brain.' " A few days after his arrival, the Congregational Association of California and the Presbytery of San Francisco met in conclave in Nevada City; Durant attended. From the meeting came the decision to open an academy, which would be called Contra Costa College, in rented quarters in Oakland, across the bay from San Francisco. Durant was appointed superintendent, and he, Willey, Billings, and others immediately began a search for land for the college.

This time nothing went amiss. The Land Commission, which had convened in San Francisco early in 1852, had clarified many titles. A site was soon found in Oakland, funds were raised, and a building was begun. Land values had plummetted since the gold rush had crested, and in September 1854 the academy (now called the College at Oakland) opened in its own building, supported largely by fees. The following April a group of citizens successfully applied for a charter to designate the institution the College of California. Of the original trustees Willey, Billings, and Day remained, joined now by Durant and nine others. When the trustees met for the first time, on October 17, 1855, Billings was elected president of the board.*

Thereafter, even when away from California, Billings made the college a matter of great personal concern, paying particular attention to proving land titles and to the library. Several trustees, including Billings, had not forgotten their original intention to have the public support the college and, hoping to obtain a land grant, the trustees cast about for a suitable site. Oakland, it was thought, would not do for long, as it was growing rapidly, and, assuming a healthy economic recovery, land would be too expensive for the kind of expansion they anticipated. Thus Horace Bushnell, who was staying at the San José Mission, was appointed a consultant to help select the ideal location.

Bushnell loved California and the trustees admired him: they offered him the presidency of the new institution, which he neither accepted nor declined. For four months he looked about, joined at times by trustees who had local knowledge of probable land values, of possible railroad routes, or of sources for gravel and stone. Billings joined these expeditions, and he found that he quite enjoyed the ebullient Bushnell's company, perhaps because Bushnell was in good health for the first time in many months. Both men wanted the university to be the center of a great Western library, and this suggested a site near San Francisco; both were shocked by the persistent level of violence in the city, and wished a site away from San Francisco's "brutalizing vulgarity."[7]

On March 1, 1858, the trustees chose a new site in the lower hills of Berkeley, north of Oakland, with a view across San Francisco Bay. The spot was one first suggested by Captain Orrin Simmons, one of Bezer Simmons's brothers, who had settled on a Berkeley hillside in 1855. With Durant as president, the new College of California opened its doors in September 1860.

Frederick Billings would render the College of California two more major and many minor services. The first was to block an attempt by a group of Old School Presbyterians to establish a purely denominational college. Albert Williams, who had been in the East since 1854, returned to California in 1859 and proposed that a Presbyterian college be set up. Though Williams's successor, W. A. Anderson, and other members of the Presbyterian Synod of the Pacific spoke against the idea, a majority favored it, and the synod asked Billings to serve as a regent. Williams wrote to Billings to notify him of his appointment; after some soul searching, Billings went to see Williams and then, on October 13, released the text of both Williams's letter and his reply.

He was, he wrote, completely opposed to the separation by the Old School

---

* This almost certainly is the source of the frequent assertion in the Vermont press that Billings was "the first president of the University of California." At one point, he was offered the presidency, but declined it.

Presbyterians from the other Protestant denominations that had united to establish the College of California. To say that there was no common ground by which individuals of different religious denominations could work together for a liberal and Christian education and to create a great university was unacceptable, for the very idea of "a great university" had to be "large enough and catholic enough for all." California had an opportunity to show now how many forces might stand opposed to one another but how many might be united. Separate denominational institutions would be sickly and of no force, and any attempt to fragment support from what must become the cornerstone for a state university that would one day be second to none would be a confession that "each denomination loves itself better than the Truth—an admission so humiliating and so pregnant with evil that I sincerely hope the denomination to which I belong will not lead off in the sorry confession." He asked the synod to rescind its vote, predicting that the great majority of members of the Presbyterian Church would not endorse the synod's intent. In an editorial accompanying the text of Billings's letter, the *Alta California* declared his argument to be "terse, logical and sound."[8]

At this time state universities had not proven successful almost anywhere in the country; several closed their doors and most were not able to compete effectively with private institutions. Not until passage of the Land-Grant College Act, first introduced in the U.S. House of Representatives by Justin Smith Morrill of Vermont in 1857, would state universities acquire a firm footing. Billings was aware of this act, knew Morrill, and shared with the legislator a desire for publicly financed education at the tertiary level. Though the act was vetoed by President Buchanan, Billings fully expected it to be introduced again. In 1862 President Lincoln signed the Morrill Act, as it was generally known, providing federal aid in the form of land grants linked to railroad construction throughout the North and West. By this act California received 150,000 acres, and on the basis of this land the University of California was established in 1868.

Billings's sharp and public refutation of Williams's initiative created a rift between them that lasted nearly a decade. The minister replied sharply and the synod, with *ad hominum* rhetoric, refused to back down, so that the proposed Presbyterian institution opened with what many felt was unseemly haste, perhaps to prove a point, in the basement of a church. This City College of San Francisco would struggle until the late 1870s, when it was abandoned. Billings was not in touch with Williams again until the 1870s, when the two men became friends once more. (Williams made no reference to his breach with Billings in his autobiography, published in 1879). Billings concluded that he may well have done nothing in life more important or lasting than having been one of the founders of what would prove to be a truly great state university.

The second significant service Billings rendered the college was less dramatic: he suggested the name of the town in which the university would grow. A name was needed for the new college site in the hills above Oakland, since platting of the new town had begun and the sale of lots lay just ahead. On November 15, 1864, a committee was appointed to choose a name. Frederick Law Olmsted, landscape architect and at the time manager of John C. Frémont's former Mariposa estate near Bear Valley, offered several names, some of them including either Bushnell or Billings: Bushnell-wood, Billingsley, Billingsbrook, etc. By this time Billings was planning to leave California and would be in no position to defend himself—for so he viewed the matter, not wishing to have the new town named after him—but at least he did not have to face a street bearing his name, since the trustees decided to name all the streets in the unnamed community after American men of science and letters.

Perhaps in self-protection, perhaps because of some reading, perhaps from happy memories of his study at the University of Vermont, Billings came up with a name when he returned to California in the spring of 1866 to clean up some business matters. In May Billings visited the college and, while standing with some of the trustees at the base of the rock where the dedication had taken place in 1860, he looked out over the bay, filled with ships of all nations, across the city stretching beyond its wharves, to the Golden Gate. Some lines from an essay by George Berkeley, an eighteenth-century Anglican bishop of Cloyne, written in *On the Prospect of Planting Arts and Learning in America*, came to mind:

> Westward the course of empire takes its way;
> The four first acts already past,
> A fifth shall close the drama with the day;
> Time's noblest offspring is the last.*

He spoke the lines aloud, and someone asked, "Who said that?" Berkeley, he replied, and suggested to the group that the town might bear the bishop's name. Over lunch at Samuel Willey's home the trustees discussed the idea

---

* The stanza is invariably misquoted in its second line to read "The first four acts already past," and since nearly all accounts of Billings's comment follow this misquotation it may be that he, too, altered the lines. Horace Greeley, fellow Vermonter and editor of the New York *Tribune*, used the line earlier and had also misquoted it. The English pronunciation of the bishop's name was (and is) "Barkley"; however, the Irish pronunciation was "Berkeley," and Cloyne was an Irish bishopric. George Berkeley, though born in Ireland, was of English descent and most likely used the English pronunciation. In any event, the Irish pronunciation overwhelmed the English one, at least in America. Myth always plays a role in such moments: even allowing for some diminution of the rock over the years, the usual description of the trustees standing *on* it at the dedication requires poetic license.

and, at a meeting of the board on May 24, unanimously voted that the town should be named after Bishop Berkeley.[9]

Seven years later, for the commencement ceremonies of 1873, Billings arranged to have a portrait of Bishop Berkeley sent to the university. President Daniel Coit Gilman, who had left Yale's Sheffield Scientific School the year before to succeed Durant, recalled just the portrait he wanted: a work by the English artist John Smibert, then hanging in the Yale School of the Fine Arts. Billings offered to have the dean of the school, John Ferguson Weir, prepare "an artistic treatment" (rather than "a literal transcript") from the Smibert and paid him $500. At the time Billings wrote Gilman of the gift, he said that the name of Berkeley had come to him "as a sort of inspiration." Thereafter he maintained his interest in the university, sending gifts until his death. His widow, until her death in 1914, sent funds to the Pacific School of Religion, which was affiliated with the university, on the ground that, as Frederick had always wished, it was nonsectarian, and she endowed a chair in recognition of the friendship between Frederick and the trustees.[10]

Billings's greatest service to the University of California may have been in making clear its title to the land on which it was built. Having muffed it the first time, he got it right the second. But then in 1850 Billings was still struggling to learn the intricacies of California land law and had only very recently taken aboard Henry Halleck, his teacher on the subject, while by 1858 Halleck, Peachy, and Billings had long proven itself the outstanding California law firm where matters of land were involved.

11

The Land Problem

V ERIFYING TITLE TO AND BOUNDARIES FOR LAND is a complicated matter. For the lawyers involved, especially if victorious and particularly if
paid a percentage of the value of the land, it is also highly lucrative. This is
even more true in an expanding, litigious, capitalist society. Billings's law firm
usually was victorious and often was compensated in proportion to the economic value of that victory.

Nor were land claims in California, especially between 1849 and 1856,
contested under the best of circumstances. Indeed, there were few more complex land issues in America, for the problems of proving ownership were those
of a colonial environment: revolution, conquest, treaty, questionable authority, and even more questionable transition. To understand just why the firm
that Billings founded was so immensely profitable and in such steady demand,
one must know, at least in outline, the origins of these problems.[1]

The California land system was a mixture of Spanish, Mexican, and American law, the first often unclear, the second in dispute by right of conquest,
the third confused by what many lawyers could plausibly argue were incorrect
or overlapping exercises of interim authority prior to California's admittance
to the union as a state in September 1850. Spanish legal provisions relating
to mining and water rights, trespass, or the right of women to inherit often
still applied. But few lawyers in California could read Spanish, much less in-
terpret the niceties of Spanish law. HBP was an exception.

Records had been badly prepared and poorly kept by the Mexican authorities. The administrative center of California moved from Monterey to Los
Angeles in 1835 and back again the next year. Three governors were expelled
from California, one so abruptly much of his paperwork remained undone.
Boundaries were seldom marked, and even if they were, titles often referred
to only vague locations, "a clump of trees," a bleached steer's head placed on
the ground, a watercourse that was usually dry and that, over the years, under
the impact of flash floods would be altered out of recognition. Distances were

grossly inexact—the size of a league was measured by the distance a mounted horseman could race while dragging a heavy weight in a specified time—and were as likely to be spoken of in terms of time to travel them as in terms of physical measurement. Names were entered in documents in familiar ways, since everyone knew each other, so that lawyers searching records decades later could not be certain who among a dozen Juans a rightful heir might be. This was at least marginally acceptable to a people living in isolation, without pressures on the land, and with a high degree of illegal if effective self-government. It would change with the outbreak of war between Mexico and the United States on May 12, 1846.

On July 7, Commodore John D. Sloat landed a body of marines and seamen at Monterey, took control of the custom house, and raised the US flag. In his proclamation Sloat assured the "peaceful inhabitants" of California that they would continue to enjoy their rights and privileges. Did anyone who was not judged peaceful forfeit those rights (including land ownership) and privileges? Who was in authority in California? Commodore Robert F. Stockton replaced Sloat and issued a more belligerent proclamation, placing California under martial law. When General Stephen Watts Kearny reached California from Fort Leavenworth, Kansas, he insisted that the army was in charge. With this issue still unresolved, Stockton left California and gave up his post as governor to Captain John C. Frémont, who had personally concluded a peace treaty with the last resisting force in California. Whose agreements should be honored: Stockton's, Kearny's or Frémont's? Did a pardon by Frémont restore an individual to the rights enjoyed by "peaceful inhabitants"? Did Kearny's subsequent successful court-martial of Frémont for exceeding his authority answer this question? What of the fact that the president then pardoned Frémont and allowed him to resign from the army?

Some of these many tangled questions were not, in fact, real questions. But given the intense competition for land that broke out in 1849, the desire of many settlers to take land from the *Californios*, and the realization that California had dealt with land titles in its own way, it was understandable that clients and their lawyers would seek to muddy the waters by raising such conundrums. Nor were the inheritances of the Spanish and Mexican administrations and of the war—ended by the Treaty of Guadalupe Hidalgo, signed on February 2, 1848, by which the United States paid Mexico $15,000,000 for the land it had taken—the only sources of legal disputation.

As military governor Kearny had declared that laws not in conflict with the Constitution would remain in force, and he moved the capital from Los Angeles back to Monterey once again. There in May 1847, he was succeeded as governor by Colonel Richard B. Mason. When the Treaty of Guadalupe Hidalgo was ratified, on May 20, 1848, local residents considered military rule ended, but Mason continued to exercise authority while he awaited Congressional instructions about the organization of a civil territorial government. Such instructions never arrived, and Californians began to hold public meetings to create municipal governments. Congress could not reach a conclusion about statehood, for the number of slave and free states were in numerical balance, and the admission of California as one or the other would exacerbate the already intense controversy over slavery. Thus on June 3, 1849 the military governor, then General Bennet Riley, issued his proclamation calling for the selection of delegates to a general convention.

The situation was especially complex with respect to land holdings within the municipalities. Under Spanish law some of them had been pueblos, with particular political status regarding land ownership. Mission lands, especially in communities such as San Francisco, San Jose, and Los Angeles, had long been alienated. The temporary municipal governments organized in anticipation of early Congressional decisions continued to function, with the military governors permitting substantial local autonomy, but were their actions legal? The so-called Legislative Assembly of San Francisco declared that it would not recognize civil power as exercised by a military governor and deposed an *alcalde*. In Congress, Senator Thomas Hart Benton, a powerful figure (and, not incidentally, Frémont's father-in-law), espoused the theory that the Constitution had gone into force in California at the moment the Treaty of Guadalupe Hidalgo was ratified and that military governors were subordinate to the people, who had the right to legislate for themselves. He thus appeared to validate the decisions of San Francisco's assembly, which Governor Riley viewed as having usurped his authority.

The constitutional convention that assembled in Monterey on September 4, 1849 could not hope to deal with all the outstanding issues, but it did well. The delegates voted to form a state constitution in order to seek admission to the Union without the customary intermediate stage of territorial status—only Vermont, Kentucky, and Texas had previously done this. The convention voted that neither slavery nor involuntary servitude would be tolerated, provided for separate property rights for married women, set income from state lands aside for the support of public education, and sought to determine California's boundary. Each of these provisions would bear upon land titles.

The constitution was ratified on November 13, San Jose was selected as the capital, and on December 20 Peter H. Burnett, a pioneer from Oregon and a Democrat, was sworn in as governor. Riley resigned his powers, despite the objection of the legal-minded, who said that a civil government ought not to begin to exercise authority until Congress approved the constitution. The legislature then named two senators, William M. Gwin and John C. Frémont, one pro- and the other anti-slavery, and hurried them off to Washington, where in March 1850, they presented the state constitution to Congress.

The issue of statehood for California was fraught with perils for Congress, for its admission would destroy the delicate compromises worked out over the years to keep slave and free states in balance. Finally a new compromise offered by Senator Henry Clay and others—that the future territories of New Mexico and Utah should decide for themselves their status on slavery—broke the deadlock, and thus, on September 8, California was admitted to the Union.

Even before this, the federal government had moved to clarify land claims, since the peace treaty had guaranteed residents of California the "free enjoyment of their liberty, property and religion." Whatever the treaty may have said, however, it could not protect the *Californios* from the rapid influx of new Americans, some of whom simply seized land by force—after all, Mexico had lost the war, hadn't it?—or argued the taking of it from necessity. Most Mexican land grants had been huge, while the tradition of the free state settlers who had begun to arrive in 1848–1849 was for smaller parcels of land. Lack of water over much of California made the competition all the more intense. California was entering its period of the "squattocracy," and though the squatters did not acquire the social and economic power later associated with that term in Australia, they had a very real power much praised by Americans on the frontier: possession, which was, after all, nine tenths of the law. Thus the first secretary of the interior had felt compelled in 1849 to appoint an officer—William Carey Jones, another of Senator Benton's sons-in-law—to assess the validity of California land titles.

In May 1850, while Congress was debating the admission of California, Jones reported that the majority of Mexican land grants were sound, and that where evidence was lacking, the doctrine of continuous occupation often had established effective title. This report was not popular with the new settlers, of course, especially those who, perhaps not understanding the complexities of Spanish or Mexican land law, had taken up what to them appeared to be unoccupied land, only to learn that they were on it illegally. Further, Jones's conclusions were in conflict with those of Henry Halleck, who as Richard Mason's secretary of state issued a quite different report in March. Halleck had argued that the land grants from the Mexican period, especially in the earlier years of turmoil, were often spurious, for at the end of the Mexican period, when it was apparent that California might pass into American hands, the governor, Pio Pico, had made highly questionable grants to a number of his personal friends. The settler group preferred Halleck's report.

The Land Commission, established by Congress in March 1851, first met in San Francisco early in 1852. Though it uncovered several fraudulent titles, most were found valid, with issues of inheritance and precise boundaries being more troublesome. Publicity given to two particularly egregious claims led much of the public to believe that many claims were invalid, and when the commission concluded its work in 1856, without finding many so, several claimants appealed or turned to private litigation. Further, landholders who did not present their claims within two years were to forfeit their rights by law, and there were many claims dating back to the Spanish period for which records had long disappeared, necessitating costly and time-consuming searches in the archives of Mexico City.

The entire process provided inexhaustible openings for litigation, and specialized law firms were in great demand. Banks and other large business concerns that held land as collateral could not be bothered with pursuing each case separately. For an annual retainer of $30,000 or so, HPB would take on all comers, or none as the case might be. The old rancheros, not conversant with American law, needed maps from the surveyor general's archives, or testaments to long-term occupation of the land, and HPB could search the archives and obtain the testimonials. To pay legal fees, some of the *Californios* mortgaged their lands, and this practice led in time to further cases. Often, once title was clear, staggeringly complex issues of tax delinquency would surface. Cases might go on for years: one of Billings's earliest cases, that of the patent for the Simí Rancho, outlasted the law firm; the final decision was handed down almost a quarter century after he placed it on his books.

Even so, Billings did well from the Simí Rancho, and he was respected by the *Californios* he represented. Pablo de la Guerra, scion of the landed Ventura County family, was very pleased with HPB, for they had won in the early Simí litigation, and from his seat in Santa Barbara he worked as a silent "partner-resident" for the firm. In the north Billings grew especially close to Mariano G. Vallejo, whose huge Rancho Soscol in Solano County, as well as his Rancho Petaluma, where he lived in the foothills of the Sonoma Mountains, would pose major challenges to the Land Commission.[2]

As if all the complexities of California land law were not enough, with the discovery of gold on Sutter's lands at Coloma, they were given a new measure of intensity. Setting aside the question of access to water, one acre of grasslands might be rather like another, but one mountain stream, one placer bed, most definitely was not. Claim jumping was a constant threat: claims were often not properly registered or marked, and they frequently proved to have been made on land already under assignment. A large law firm that could send investigators to the claims as well as to the archives and law books would thrive: HPB was such a firm, and Billings and Peachy, both fond of travel, of the feel of the horse beneath them, were frequently on the move. Consequently, Halleck, Peachy, and Billings would handle the most important title dispute in California history: that of the New Almaden Mine, a textbook illustration of the problems the firm encountered during the fruitful partnership.[3]

Of all the California land cases, New Almaden was the most complicated. In the quantity of documentation produced to prosecute and defend, appeal and rebut, the case exceeded any previously brought before any district court in the United States. Jeremiah Black, the U.S. attorney general who presented the government's argument before the Supreme Court, said that it was "the heaviest case ever heard before a judicial tribunal." The transcript of only one of the trials ran to 3,584 pages in four volumes, and the manuscript draft for just one of HPB's arguments unfolds across blue-lined foolscap, page after page, 229 in all, largely in Billings's hand though sometimes in that of a scribe. Billings carried the brunt of the work, though Peachy also labored mightily in the courtroom; Halleck was personally involved and had to avoid a potential conflict of interest.

A great deal was at stake: the world's second richest quicksilver mine and the first quicksilver deposit discovered in North America—for a time almost the sole source of an ingredient essential to gold mining after the placer stage had passed. Until the discovery of the cyanide process in 1887, quicksilver— that is, mercury—from New Almaden made possible the development of the mining industry of California from 1853, and the whole of the Comstock Lode. New Almaden, which produced over $70,000,000 in quicksilver, was the single most valuable mine in California.

As early as 1824 a Mexican, Antonio Sunol, had tried to take silver from the mine and, failing, abandoned it. In 1845, on November 22, a Mexican army officer, Andrés Castillero,[4] also hopeful for silver, filed a claim for the land. In 1846–1847 Castillero sold part of his shares in the mine—now known to contain quicksilver and named New Almaden after the Old World's largest quicksilver mine, Almaden, in Spain—to an English firm, Barron, Forbes & Company, located in Tepic, Mexico, so that he would have enough capital to develop the site. Late in 1847 a small refinery was set up, permitting the first production of mercury in 1848. The near simultaneous discovery of placer gold at Sutter's Mill led to a soaring demand for quicksilver, since mercury was thrown onto the riffle bars of the pan or cradle, the sluice box or Long Tom, so that the small particles of floating gold would amalgamate. As quartz mining took over, quicksilver was even more important, for gold-bearing ore was ground and mixed with mercury and water, so that by gravity the gold would separate itself out under the attraction of the mercury. By 1854, quick- silver production had reached one million pounds a year.[5]

In Mexican law, title in minerals was held by the state; thus land grants in California did not automatically include mineral rights. Since the two rights were separable, one could claim a mine on land belonging to others. To perfect a mining claim as distinct from a land claim, one had to take three distinct steps, each with certain sequential technicalities. First, one reported the mine's location to the local representative of Mexico's Federal Division of Mines. Second, one visited the site with an official to have possession granted. Third, one reported to the Junto de Fomento y Administrativa de Mineria, which could grant benefits to stimulate development. Castillero felt he had taken these three steps. Because there was no one from the Division of Mines in California, he followed the prescribed procedure set down in his printed copy of Mexico's mining ordinances, filing two reports with the first *alcalde* at San Jose; the one for quicksilver was dated December 3, 1845. From the *alcalde*, accompanied by two witnesses, he obtained the necessary visit in lieu of a representative of the Division of Mines, and a statement of grant. Since Mexican law required that the mine be worked, Castillero promptly hired someone to build a furnace to begin smelting the ore. In April he left for Mexico to file his papers. A month later the United States declared war on Mexico.

Castillero met with Alexander Forbes, of Barron, Forbes & Company, in which Scotsman Eustace (or Eustaquio) Barron and his nephew William were partners. James Alexander Forbes (apparently unrelated), who was the British vice-consul at Monterey but lived in Santa Clara near the mine, had already informed his government of the claim. That summer Alexander Forbes acquired four shares from one of Castillero's partners. James Alexander Forbes was then placed in charge of the mine and in due course acquired four shares of his own. Robert Walkinshaw became supervisor in mid-1847. On February 2, 1848 the war with Mexico ended. Following a good bit of buying and trading, Barron, Forbes & Company had, by 1850, obtained all the shares, including Castillero's, and formed the New Almaden Company. All seemed straightforward at the time, though in fact these conditions, and the juxtaposition of events, were a recipe for litigation once California became a state. Ironically, the New Almaden Mine was the only significant mine in Alta California to have had all the essential steps of Mexican law carried out for it.

The mine was surveyed in 1848 by Chester Smith Lyman, a young Yale graduate, who laid out a rather odd parcel in order to meet two requirements: that mining rights should extend three thousand *varas* in each direction from the mouth of a mine, and that the mouth be as near the center as possible without encroaching on land claimed by individual owners for surface rights. He also surveyed two square leagues of land that Castillero had claimed in 1845. Lyman's survey did not extend the requisite *varas* to the north because of conflicting claims.

A further complication now arose. With seven partners, one of whom held one of the claims that forced the truncated survey to the north, Thomas O. Larkin, the former American consul at Monterey, formed a mining company and asserted a claim to mining rights in much the same general area as New Almaden.[6] When Forbes and Walkinshaw protested, Governor Mason, accompanied by William Tecumseh Sherman, visited both mining sites and decided that he would take no action, referring the matter to the civil courts. In the summer of 1849 Alexander Forbes displaced Walkinshaw as the supervisor at the mine, putting James Alexander Forbes in charge; Walkinshaw disputed this move and sought an injunction to keep the latter Forbes from operating the mine. In a doubtful decision, Walkinshaw was granted an order that removed Forbes, who appealed to the Superior Tribunal, a remnant of the Mexican judicial system, which put him back in charge. James Alexander Forbes did not prove to be a good businessman, however, and Henry Halleck was hired as the resident manager in May 1850. Walkinshaw then returned to the company.

Under Halleck the mine became an orderly big business. Even before the Land Act of March 1851 came into force, the prudent Halleck had drawn together the Castillero documentation. Several potential legal problems existed: was the claim valid under Mexican law? Did the newly appointed Land Commission have the authority to review this question since a mine was not land? Was the claim to three thousand *varas* or two square leagues valid? Were any or all of the rancho claims to the north valid? Was the survey accurate? Were the documents correct? Was all documentation properly executed by valid authorities? How did one prove the validity of those authorities? To Halleck, and to Peachy and Billings, as they took on the litigation, it seemed quite clear: the answer to the first question was Yes, and thus the other questions did not arise, save for some possible debate over measurements. But the United States was determined it should have this mine; as Attorney General Jeremiah Black would attest in 1860, "The United States have always met the claim with uncompromising hostility. From first to last they have shown nothing but the edge of the naked knife."[7]

On September 30, 1852, HPB filed a petition with the Land Commission, on behalf of the New Almaden Mining Company but in Castillero's name, for confirmation of the claim. It was largely Billings's work, as Halleck and Peachy were en route from Los Angeles. Legalists then and historians since have found the petition unusually clear. HPB had good reason to believe that it would be confirmed quickly. Nonetheless, they lobbied Senator Gwin heavily, with Halleck visiting him in Washington, and they tried to get a resolution of support out of the state legislature. Halleck reported that there was a worrisome problem: Robert J. Walker, a former Mississippi Senator, President Polk's secretary of the treasury, and governor of Kansas under President Buchanan, was bruiting it about Washington that he was the owner of the mines, and it was rumored that in exchange for supporting Buchanan's position on Kansas, Walker had been promised a favorable federal ruling when the matter came to court. Since issues of proof were crucial, Halleck tried to return to California via Mexico to see if he could get the original documents, but was turned back by the civil war there. Eventually, copies of additional documents were brought from Mexico by a former minister of exterior relations, who remained to testify to their authenticity. HPB meant business: they had involved the equivalent of the American secretary of state in support of their case.

HPB celebrated with champagne on the night of January 8, 1856, when the Land Commission found, two to one, in favor of the mining claim and of the three thousand *varas* from the mouth of the mine, though the claim to two square leagues was denied. Nonetheless, both sides to the dispute were bound to have a total victory, and in March HPB filed an appeal to obtain the two square leagues. The next month the United States appealed to obtain the mine. In the meantime, Billings was involved in the state courts in numerous legal actions against the heirs of conflicting and encroaching claimants, who did not, in the end, win on any of their actions.[8]

The mine was now worth $15,000,000. New Almaden was a dramatic sight: the refinery, the camps for the Mexican and later English miners, the loads of mercury-filled flasks ready for transport, the sheer magnitude of the enterprise the subject of much commentary. Halleck (using Gordon Cummings as architect once again) built a handsome mansion, the Casa Grande, where he lived when on his twice-monthly inspections to the mine and to which he and Billings brought many guests. On the East Coast there was enormous interest in the fabulous mine, and by way of his friend Robert S. Watson, Billings was invited to stoke interest higher by exhibiting specimens from New Almaden at the Crystal Palace in New York City during the World's Fair.[9]

The U.S. District Court for Northern California, to which the appeals were carried, was in the hands of Ogden Hoffman, of whom Billings thought most highly, in part because of his youth (Hoffman assumed the post when only twenty-nine). On the government side was assistant U.S. attorney general Edmund Randolph; Billings did not like him, and was inclined to think him more full of bluff and show than of the law (on this he was mistaken, for whatever Randolph knew of the law—and many thought he knew a great deal—he was an accomplished orator in the Southern style). To offset him, HPB chose to put up Peachy, also Southern and determined to defeat his fellow Virginian. In any case, no jury was involved, so oration was less important than careful questioning of witnesses.[10]

But the law's delay cost HPB dearly, for decisions were now being routinely appealed to the U.S. Supreme Court, in part because of the work of a dramatic new addition to the cast of characters: Edwin M. Stanton, one of the most formidable legal logicians of the day, a man who felt that virtually every California claim based on Mexican grants was fraudulent. Stanton was an enormously skilled rhetorician who wanted to "count the dead and bury them" while in California; if he went in for a fight, it was said, he went in for a funeral. Stanton arrived in March 1858, fresh from a widely hailed victory in the case of *McCormick* v. *Talcott, et al., survivors of John H. Manny*, a landmark in which he had destroyed Cyrus H. McCormick's effort to protect his former monopoly on reaping machines, winning against Reverdy Johnson of

Maryland, the most noted contemporary lawyer to plead before the U.S. Supreme Court.

Attorney General Black was so impressed that he sent Stanton to California to pursue the government's interests in the many disputed claims, and especially in a case that involved a claim to virtually the entire city of San Francisco: that of José Y. Limantour. The Limantour documentation appeared exacting: every piece of paper the law required was in place, properly signed and sealed, attested to, all apparently from the archives in Mexico City. However, an informant affirmed that many of the documents had been falsely dated, and Black was determined that no one would possess the whole of San Francisco or hold the U.S. government for ransom. While many of the land claimants and lawyers were nervous about Stanton's presence, HPB was, in fact, initially quite pleased, for they were utterly opposed to the Limantour claim, which if upheld would create problems for dozens of their clients. Mayor Cornelius K. Garrison and the city land commission were determined to defeat Limantour, offering HPB and another lawyer $10,000 to contest the claim.[11]

While historians still disagree on whether Stanton was right in all that he concluded, no one disagrees about Limantour. Stanton got Congress to pass bills making Mexican land papers compulsory and compelling their delivery to an archive (thus defeating several claims at one stroke) and punishing fabrication of claims (in preparation for Limantour). With subpoenas in hand and a Spanish dictionary on his desk he searched out every piece of Mexican paper he could find. "In the last few years a set of Mexicans has been plundering the United States," he wrote home, and he was "determined to throw a brick at them." (Limantour was in fact French, but Stanton's point was well taken.) He and his staff compiled four hundred folio volumes of records and then prepared his own documentation on every land grant either Spain or Mexico had made in California. Stanton then proved that the paper on which Limantour's grants were written had been fabricated well after the date they bore. He found Governor Manuel Micheltorena's account books and showed that they contained no entry for Limantour's payments. He presented detailed biographies of the men who had testified to the faithfulness of Limantour's documentation, proving that they were professional land witnesses. He demonstrated that Micheltorena had predated many official papers, either for a price or on behalf of old friends. Thoroughly exposed, Limantour slipped out of the country, no longer contesting his claim; his flight provided final proof of Stanton's research. Stanton then began to test every claim in which one of the perjuring witnesses had appeared. This included the New Almaden mine case.[12]

HPB knew what Stanton was doing; after all, he had his offices in the Montgomery Block and Billings and Stanton were both staying at the International Hotel. HPB was aware that some of their witnesses were professionals, though they had not known them as perjurers. Still, they were of a divided mind, for they also had not wanted to see the Limantour claim stand, and one of their witnesses, Pablo de la Guerra, had helped Stanton make his case against Limantour, which could only help them. They could live with the precedent Stanton had wrought but they did not believe it could be made to apply to New Almaden, for to their knowledge no forged documents existed. Now Halleck regretted not having been to Mexico, however, for they were working with documents provided by interested parties, and one or more of the pieces of paper on which they had relied might be proven false, casting into doubt genuine pieces. This is precisely what happened in July 1858 when a former employee of the New Almaden Company revealed that in 1848 Alexander Forbes had written to James Alexander Forbes asserting that the Castillero documentation had been obtained after the American occupation of California.

Forbes's note, thereafter known in the testimony as the "letter of March 28, 1848," may well have been false itself. The original was not available, the informant claiming it had been stolen from his carpetbag at his hotel. Nonetheless, Randolph and Stanton used the claim to good advantage. They also emphasized the irregularity of Castillero's not taking his claim to an official of the Mexican Division of Mines. Eleven witnesses testified they would not believe James Alexander Forbes under oath. Twenty-three witnesses testified they would not believe the informant under oath. Obviously, this was not getting either side anywhere, and HPB decided it had to visit Mexico to obtain more original, irrefutable documentation. Stanton was nonetheless so persuasive that the Circuit Court issued an injunction prohibiting any further working of the New Almaden Mine until the question of ownership was resolved. From 1858 until 1861 this very valuable asset remained inactive.[13]

HPB asked Judge Hoffman to be allowed to take depositions in Mexico. He denied this. HPB then tried to get a bill introduced into Congress to authorize discovery in Mexico. This failed. HPB then appealed directly to President Buchanan and to the secretary of state, Lewis Cass, while Billings wrote privately to his senatorial friends, to the Vermont delegation, and to other well-placed colleagues in Washington. Both sides blew up an epistolary storm in California, with letters to the editors of many newspapers becoming daily occurrences, and pamphlets appearing under various pseudonyms presenting the opposing sides in no uncertain terms.

On April 23 Black wrote the secretary of state a forceful letter of which HPB was quite unaware. Pointing out that HPB and other counsel in the Castillero case intended to go to Mexico City to obtain depositions and that they would be taking testimony in the presence of the secretary of the U.S. legation there or of consular officers elsewhere, Black instructed Cass that no officer of the United States could be permitted to cooperate, for they would be acting against the best interests of the government they were sworn to uphold. A private citizen had no rights in the matter and could not expect government officials to be neutral or supportive in such an enterprise. In short, the courtesies consuls normally would have extended to American law firms seeking depositions in a foreign country could not be extended to HPB. Further, Black said, depositions taken in Mexico were not valid; they must be taken in California in the presence of Judge Hoffman. Cass must, Black concluded, instruct all U.S. diplomatic representatives to abstain from giving the Castillero claimants any assistance.[14]

One can scarcely imagine a worse time to go to Mexico, especially for an orderly search for legal documentation and willing witnesses.[15] In 1859 the country was in the midst of a savage civil war, known to Mexican history as the War of the Reform. Angered by the excesses of the heavy centralist Santa Anna regime that ruled from Mexico City and determined to break the hold of the Roman Catholic Church on the Mexican peasantry, a number of young men of liberal or revolutionary bent had set in train the events that would, in the twentieth century, culminate in the great Mexican revolution. Two of these men, Melchor Ocampo, the son of a Michoacan *hacendado*, and Benito Juárez, a Zapotec Indian from Oaxaca, both of whom had served as governor of their states, had been exiled to New Orleans in 1852, and there they grew intellectually into the leaders of change. In 1856 the Reformers were successful in passing the Ley Lerdo, a law ordering all corporate bodies to give up their landholdings, especially the Church. A new, liberal constitution followed the next year, to remain essentially in place until 1917. The Ley Lerdo and the Reform Constitution had split Mexico, plunging it into war.

The Centralist-Moderates, as they were known, established their regime in Mexico City, while the Radical Liberal Federalists, led by Juárez, governed from Veracruz. The European powers recognized the Mexico City government while the United States leaned toward Juárez. Barron, Forbes & Company were identified with the Mexico City conservatives, in part because Britain backed the Centralists, in part because the company owned much land and, through its control of the textile industry around Tepic and its near monopoly over the port of San Blas, it naturally hewed to the propertied, and in part because many of Barron, Forbes's San Francisco clients and customers—who tended to be pro-Southern in the darkening American conflict over slavery—knew that the Mexican Liberals were, in theory at least, unlikely to be pro-Southern in the event of a northern civil war. Thus Billings went to Mexico not only in a time of chaos but branded with views he did not hold.

Accompanied by William E. Barron, Billings left for Mexico on January 5. Disembarking at Acapulco, the two men reached Mexico City in nine land days, arising each morning at two o'clock, and stopping at two in the afternoon to take advantage of the coolness of the nights. The stage journey was difficult and dangerous. Food was scarce, local *caudillos* could command a traveller's passport at any moment, and robbers had control over large stretches of the country. The ride from Cuernavaca was particularly tense, for two days earlier the stage had been stopped and robbed four times; the passengers on the last occasion were stripped of their shirts. (They were lucky: on one occasion stage passengers arrived in Mexico City covered only with newspapers.) Billings travelled with his papers in his hands, certain his baggage would be stolen; he was "prepared to arrive in Mexico City stark naked" provided those papers got through. Both men rode in the forward seat so they might see trouble as it came, with their pistols under the blankets they sat on.

The journey might well have cost Billings his life. At one point the stage outdistanced its escort, and at that very moment three robbers fell upon the travellers, one putting a gun to Billings's head. This was the closest he came to a premature death, for just in time the escort hove into sight and the brigands fled.

In Mexico City Billings set to work collecting documentation, meeting with Barron's lawyers, and seeking out witnesses. He persuaded a number of highly placed individuals who were informed about the New Almaden title to return to San Francisco with him, chartering a special steamer to stand by at San Blas, to which they would travel by stage. Elderly, rich, frightened by the revolution, these men constantly had to be encouraged and reassured lest they drop out. If they would all come with him, Billings reported, "I shall save the case of the mine certain."

An intended short visit was prolonged by Eustace Barron's illness and by the government's disarray. One could not count on finding any official in his office, broken appointments littered the engagement book, mail was lost to robbers unless sent by special courier. Important news from San Francisco reached Billings by messenger from Acapulco only because it was under seal of the British consul and addressed to the British Minister, and even then robbers kept the courier tied to a tree for several hours before releasing him with his package. Costs were mounting, for so much was unpredictable: the special steamer was to arrive at San Blas on March 10 and wait for Billings and his group, though on February 15 he wrote that he had no expectation of reaching the port in less than thirty days. The group expected to be robbed, so the steamer was to have a supply of new clothes on board in case they showed up naked.

By mid-March the situation had grown much worse. Juárez was firmly in control at Veracruz, the government troops had lost control over the San Blas route, and William Barron decided he ought not to travel back to San Francisco with Billings, who now felt virtually imprisoned in Mexico City. He spent his time rearranging archives and ferreting out more bits and pieces of paper. He read the work of José Maria de Lafragua, the minister for foreign relations in 1846 who had already testified in San Francisco in 1855 on behalf of the claimants, and went to see Lafragua's publisher, hoping there might be data in draft copies of his manuscripts that had been omitted from the published form. He wrote long letters home and he waited. And then he had a good thought.

With more time on his hands than he had expected, Billings decided to go to the College of Mining and interview the faculty. There he found four scholars who remembered Castillero's discovery quite well. He was also able to describe in detail the precise steps through which the submission would have passed, the way in which the faculty had reacted to Castillero's claim, and why the Junto de Mineria had failed to provide Castillero with the development aid he had asked for. Most important, by examining the original documents at the college as well as in the archives, and by building up contextual and circumstantial evidence about conditions in the capital at the time Castillero had filed his claim—evidence he bolstered with extracts taken from contemporary newspapers drawn from the archives—Billings was able to provide a convincing answer to why one of Castillero's key documents was misdated by a month. For already those attacking the claim were making a good bit of the fact that an important document bore the date of April 24, 1846, which meant that a slightly earlier document of the old Larkin (and now Fossat) claimants had precedence, while the Castillero claimants had always insisted that the document properly should have been dated March 24. Billings was able to demonstrate that the junta that met in March had discussed the Castillero claim and thus knew of it; that a new and ignorant clerk had made other errors in dating other documents that came to the junta at the time, and that the error was surely one of transcription since the scribe had a subsequent enclosure dated April 24 in front of him at the time he was working with the document of March 24 and, as the mind sometimes will, mistakenly applied the later date. Billings felt his research had materially bolstered the Castillero case, and he was increasingly eager to get on his way.[16]

Billings's senior client, the elder Eustace Barron, died on April 11. He had not lived to see his name vindicated "from the foul charges of the base conspirators against New Almaden"; Billings felt doubly obliged to clear Barron's memory and confront his accusers on his behalf. As soon as he had put the funeral behind him, he was determined to find a way out of the Mexican impasse, and began casting about for some means of getting on to San Blas, perhaps with a reduced number of his distinguished gentlemen. Then, suddenly, the siege was lifted, perhaps only temporarily, and the road was open, though not likely for long. Billings rode out to see the battlefield for himself and returned to hire carriages.

One plan Billings had brought to Mexico City was to persuade the government to issue a statement, over an official seal, stating its formal endorsement of the validity of the New Almaden title and attesting to the legitimacy of the documentation. He even hoped that the Mexican government, owing much to the Barrons, would protest Stanton's proceedings as a violation of international law and of treaty relations with the United States. It is difficult to see precisely how Mexico could have sustained such a protest, and one may doubt that Billings ever really hoped for such a document, perhaps reasoning that were he to ask for it, he would at least obtain his simpler wish, the document attesting to the validity of title. This paper was, in fact, nearly completed, or so he reported—no copy of it has come to light, reasonably enough, since it would have been in draft only—when the United States, contrary to the position taken by most European nations, suddenly moved to recognize Juárez's government, putting an end to relations with the regime in Mexico City. Billings could hope for no succor from Juárez, and he now pinned his fading hopes on the word that an old friend, Robert M. McLane, a former congressman from Maryland and U.S. commissioner to China and Japan, had arrived in Veracruz as the new American Minister. Billings hurried a letter off to him, asking him to intervene in the New Almaden matter. He received no reply, and it was with mixed emotions that he learned, on April 6, that the United States had formally recognized the Juárez government.

From San Francisco HPB had made an offer to President Buchanan and Secretary of State Cass that they send to Mexico City "any disinterested man of their own selection" to examine the matter, at HPB's expense. HPB agreed in advance to admit this person's report as evidence in the case. Black opposed this, for he had made clear his view that any depositions taken or research conducted in Mexico was invalid; witnesses had to be personally present in California. Further, any new document coming from Mexico had to be authenticated by the Mexican Great Seal. He knew this was contrary to Mexican law but said this was the claimants'—and Mexico's—problem. Under the circumstances and by his understanding of Mexican law, Billings knew he would never attain anything to satisfy Black. New Almaden was "an outrageously hard case," one of mental as well as physical siege, and Billings now had little hope of success until there was a new attorney-general. The government's conditions were, he thought, utterly unreasonable.

On April 25 Billings and nine witnesses left Mexico City for San Blas; his intended three-week visit had lasted over three months. He was armed with special passports for safe conduct, and he hoped to be able to rely on orders President Miguel Miramón had issued nearly six weeks before to local authorities, some of whom no longer were in place, to provide escorts for the party. With the essential papers hidden in the stuffing of their coach's seats, the group took twelve days to reach the sea, six hundred miles away. All arrived in San Francisco safely on May 14, ready to bear witness to the Castillero claim and the legitimacy of Barron, Forbes & Company's ownership.

His Mexican mission had been an adventure for Billings, and it clearly stimulated him, for it combined his love of travel with his love of research. While en route and in Mexico he had lost only one day to illness. He enjoyed the hardships, rather liked the image of himself—with a brace of pistols, a hammock for the nights on the road, and soap for bathing in mountain streams— as a risk-taker. He knew death was genuinely possible, for though he had not studied the intricacies of the local political situation, he had been told of the many murders along the trails. He realized that the centralist army was still held a semblance of control along his escape route, but that he would not be able to tell whom to trust, since roaming bands of deserters, still in remnants of uniform, often fell upon the unwary. He made his will before leaving California. Now he returned to retrieve it.

By now Peachy in San Francisco, along with Barron, Forbes's lawyers in Washington, knew of Black's campaign against the admission of any evidence brought from Mexico. Billings was on the West Coast with nine more-or-less eager witnesses and depositions from eleven others, and Reverdy Johnson and John A. Rockwell were in Washington trying to coax an elderly, very tired, and sometimes confused secretary of state into exerting what they believed to be the rights of his office over the "bad law" of the attorney general. At seventy-five, and on poor terms with President Buchanan, Lewis Cass was the only northern conservative Unionist in the Cabinet carrying the battle to those he saw as defeatists. The lawyers contemplated approaching him through an intermediary but thought better of it and decided on a frontal assault. On October 4 they sent what proved to be an ill-advised letter.

Summarizing the long history of the case and the pro-Castillero position, Rockwell and Johnson fell into language that was intemperate and snide: "The subject is one so plain, not only to any professional man, but to any one of ordinary intelligence, that we should despair of making it clearer," they wrote with Black obviously in mind. The attorney general's imputation that the Castillero witnesses were to a man ready to commit perjury was "utterly reckless and gratuitous," "random" judgments passed upon individuals Black did not even know, deeply biased as he was against men who included the leader of the Mexican bar, former cabinet officers, distinguished academics, and foreign ambassadors. They hinted at collusion between Black and Cass, they cited Macaulay on the English revolution, they implied that heads ought to roll as King Charles's had done, and they concluded that fairness demanded that all proof be presented on both sides and the matter be left to the courts.

This letter clearly did no good. The startling news of John Brown's raid on Harper's Ferry intervened, and late in the month a desperately harried Cass replied in a cool twenty-line communication that he refused to receive the letter or place it in his files, returning it with the suggestion that learned counsel submit an application "free of objectionable remarks." They did so, making precisely the same argument with their *ad hominum* observations removed, on November 29. At this point a crucial pamphlet disappeared and Cass had to write that he could not find documentation previously supplied by Rockwell and Johnson. Even President Buchanan took the same line, observing that the letters he presumably had received from them (he could not really remember, he said) had been misplaced. The only very thin silver lining in all this was that on December 6 a statement appeared in the Mexican *Diario Oficial del Supreme Gobierno* supporting HPB's line of argument, this being as close as Billings was ever going to get to the Great Seal Black was demanding. Even this tiny triumph hardly mattered; later that month the Juárez government signed the McLane-Ocampo Treaty which put the United States (and the Black-Cass-McLane group) thoroughly in the driver's seat on matters Mexican.[17]

In Sacramento Peachy had been working steadily to get the California state legislature to pass a resolution instructing the Congressional delegation to use all "best endeavors" to validate the claim through persuasion and legislation. He also wanted the state legislature to lift the injunction on the mine. Billings spent a week on this in Sacramento, while Gregory Yale went to Washington to lobby the California delegation. William Barron was practically in residence in Washington, rolling logs. On March 21, 1860 the California state legislature voted, twenty-two to seven, that the U.S. Senate ought to discontinue action on the claim and lift the injunction. As soon as Black heard this he reported to President Buchanan that the injunction had been granted by a court and only a court could dissolve it; a federal judicial opinion was always more valuable than one of a state legislature; the legislature was concerned about the rise in the price of quicksilver and the loss of jobs brought on by the mine's closing, and while legitimate issues, these concerns had nothing to do with the legal matter. Neither the president nor the U.S. Senate took any action.[18]

In October 1860 the case at last came before the U.S. District Court for Northern California, Judge Ogden Hoffman Jr. presiding. Because the district court was functioning as a circuit court, however, prevailing procedure called for the Circuit Court judge, H. Hall McAllister (not to be confused with his son Hall McAllister, at one time an attorney for the New Almaden mine), to sit with the district judge. The case was, as the press noted, the most important civil action ever to have been tried in California, and the attorneys for both sides displayed unusual skill in marshalling evidence and arguments, drawing a large audience. For the Castillero claimants there was Reverdy Johnson, who had come from Washington, and Judah P. Benjamin, a senator from Louisiana and expert on Spanish and French land law,[19] assisted by HPB. Arrayed against them were Edmund Randolph, assisted by the attorney for the Fossat claimants, A. P. Crittenden, and by Edwin Stanton from a distance. Johnson was to be paid $35,000, Benjamin $25,000 if successful.

Peachy opened for HPB, Randolph followed, Benjamin came next. Each spoke at exceptional length and with more than the customary Victorian oratory, appealing to the courtroom audience—there was no jury—fully as much as to the two judges. Randolph's statement took a week and Benjamin's response lasted for six days, the long arguments giving the attorneys time to revise tactics en route. Benjamin seemed to some almost in a frenzy of language, and the press hung on every word. For a moment he appeared to forget who his allies were, correcting one of Halleck's translations, invoking testimony from John C. Frémont, who was not in good odor on land matters at the time, and impugning Hoffman's youth. Perhaps Benjamin felt it would be difficult to get sympathy for British interests pressing a debatable Mexican claim in an American court. He knew about the "letter of March 28, 1848," on which Black and Stanton had rested much of their case, and he took this head on, arguing as Billings had advised that true or false the challenged documents were not essential to the case. In this way he prepared for Randolph's sharp personal attack on Forbes, which he knew would come. In turn Benjamin attacked Attorney General Black personally, as an instrument of an overbearing central government, and at one point he questioned the integrity of the court, implying that McAllister had already made up his mind.[20]

Benjamin's strategy may well have boomeranged. His attack on Black and the central government, coming from a Louisiana senator when the Union was nearing a crisis point, cannot have appealed to the pro-Union faction in the audience. For his remarks about the court Judge McAllister rebuked him, and Johnson, who spoke again after Randolph, had to use valuable time to defend his colleague and mollify the judges. Still, Randolph may have played into Johnson's hands, for he then tried to exploit Benjamin's peroration even after the lawyers had shown they understood he had gone too far, and attacked Benjamin personally and not at all well, giving the eloquent Southerner a chance for a final, stinging reply. Randolph may well also have gone on for too long: his closing argument fills three hundred pages and wearies even the dutiful historian.

When the decision was handed down in January 1861 everyone lost. Judge Hoffman found the Castillero claim and two square leagues of land valid; Judge McAllister upheld the claim to the mine but not to the two square leagues, and he cut down the three thousand *varas* as well. As his was the controlling vote, his decision stood. Some of Billings's friends promptly implied that McAllister had been bribed, but Billings would have none of this, declaring the judge to be utterly honest, though wrong.

Billings and Peachy were appalled. They had lost further ground, despite spending $200,000 on the Mexican expedition and the presentation of witnesses. Halleck's reputation had been damaged, since it was clear from Stanton's work, which Randolph cited, that he had not done so fine a job as an archivist as supposed, he was shown to have translated inaccurately, and his own publication had been used against him. By now general opinion on the East Coast appeared to see the New Almaden claim as fraudulent, and even so good a friend as Robert Watson was writing to Billings to tell him that HPB appeared to have taken a line "sinker and bobber." Except in the business community (and not widely so even then), there seemed to be little sympathy for a foreign company that not only sought full control over New Almaden but had also gained control over New Idria, another significant quicksilver mine nearby, and had moved aggressively to control by combination Redington, a new mine north of San Francisco, so that only two relatively small quicksilver mines would stand between Barron and a North American monopoly.[21]

HPB had no choice but to appeal. The government still wanted everything, and the Fossat claimants also appealed. For the Supreme Court the New Almaden Company retained Reverdy Johnson, back on his own turf again. They added John A. Rockwell, who was thoroughly familiar with the case and had also written an important work on Spanish and Mexican law. Initially, Benjamin was a member of the team, but with the outbreak of the Civil War he left for the South, where Confederate President Jefferson Davis named him attorney general and, later in the year, head of the Confederate War Department. His place was taken by Charles O'Conor, a highly regarded member of the New York bar. For the government there was Black, no longer attorney general, and Stanton. HPB was not part of the picture when the case was argued in January 1863 since the firm had been dissolved. Nor was Randolph, who had died. Peachy was part of the team, however, and he again would make the opening statement, and Billings was on retainer.[22]

On March 10 the Supreme Court ruled, five to three, against the Castillero claimants, this time completely: no land and no mine, either. In one dissent, Justice James M. Wayne accepted the decision of Judge Hoffman in the District Court. The ninth justice, Roger B. Taney, did not take part, though it was reported that had he done so he would have concurred with Wayne. For the majority, Justice Nathan Clifford ruled that Castillero had failed in strict compliance with the law requiring a denouncement to an officer of the Mexican Bureau of Mines. The court did not rule on the question of fraud. Billings wondered how the court could have assimilated all the arguments in only thirty days, for the record had grown by another thousand pages, and he concluded that the justices—including two freshmen—had deliberately taken the easy way out and not examined the fraud charges at all. He was consoled only a little by the judges' observation that the documentation in the case—

Billings's doing—had never been equalled in extent or clarity of presentation. Surely not: the New Almaden proceedings now filled ten massive volumes.

Still, Billings was not quite out of the New Almaden fray. The mine, from which the injunction had been lifted in 1861, was operating again, and the Fossat claimants, through the Quicksilver Mining Company, a new enterprise whose president, Samuel F. Butterworth, was a front for the pestiferous Robert J. Walker, now sought to have the mine seized to prevent depletion of the asset. This company offered to operate the mine for the government as an incentive to seizure, and through the attorney general, now Edward Bates, and a friend of President Lincoln, Leonard Swett, Lincoln was persuaded to issue an order to the U.S. marshal for Northern California to take possession of the New Almaden mine. Interestingly the order, dated May 8, embraced three thousand *varas* in all directions from the mouth of the mine, even though this measurement had not been sustained. Swett was to administer the property.[23]

As Lincoln's order had been kept secret, it was not known in California until Swett arrived in San Francisco in July 1863. With the order in hand and accompanied by Swett, the marshal went to the mine, where they were turned away by Sherman Day, the mine's chief engineer, and a body of armed men; someone—most likely Billings—had sent a warning ahead. On July 9 Frederick Low, Republican candidate for governor, sent a telegram to the secretary of the treasury, Salmon P. Chase, to warn him that forcible seizure of the mine would be "terrible." The secessionists, led by his Democratic opponent, John G. Downey, would, he said, use such a blatant intervention to stage a general uprising. After all, the federal government had already taken over Black Point and the Presidio, ejecting squatters, and the situation was tense enough without this added intervention. Low urged Chase to have the order withdrawn at once. The commanding officer of the U.S. Army in California also urged Henry Halleck, now general-in-chief of the federal army, to defer action on the order lest "great excitement" create problems at a most delicate moment in the efforts to keep California sound for the Union. Lincoln replied the next day, telling Swett and Low to consult together and not make a riot.

Billings and the banker John Parrott, who held an interest in the mine, also sent telegrams to Halleck, and to Billings the general honestly if incautiously replied that the order to seize the mine had been "surreptitiously obtained." When this telegram was leaked, Attorney General Bates demanded in Cabinet that Lincoln dismiss Halleck. The President told the Cabinet that Halleck has been "hasty and indiscreet" but nothing worse; Stanton, now secretary of war, raged against anyone who would not back his position, which perhaps was somewhat compromised by the fact that he was in the employ of the Quicksilver Mining Company at the time. Lincoln wanted no riot in his Cabinet any more than he wished one in California and he remained calm, waiting for the news of how California had taken the word of the attempted seizure.[24]

When that word was made public on July 10, a storm broke about Lincoln's head. He had referred to a long-dormant act which, if universally applied, could be used to justify the eviction of what the government could now view as squatters from many mines. Rumors spread that the government might sue for the value of all the quicksilver taken from the mine. Swett tried to draw a distinction between mines based on fraudulent titles and certified as such by court action and all other mines, but his argument proved that he did not know the facts of the case even as well as the putative man in the California street, and the Barron company was able to win points with a scathing reply in the San Francisco *Bulletin* which also revealed that Swett held an interest in the company about to take over the mine under the order.

Indeed, Swett proved particularly inept. He admitted to a friend that the Quicksilver Mining Company planned to pay him $10,000 if he could pull the seizure off, and after his ill-advised march on the mine he had gone to see Billings to ask if Barron, Forbes & Company would make some arrangement with him directly. He may have meant nothing venal by this query, but it gave Billings the opportunity to reply that the mine owners were no more likely to recognize the right of the government to interfere with them through agents than by soldiers and that no concessions would be made to Swett of any kind.

Now the San Francisco *Bulletin* named Stanton, Black, and Walker, who was revealed to be one of the Fossat claimants, as not only being behind the attempt to "steal the mine" but as all deeply self-interested financially. At last public attention focused on Walker, who managed to remain behind the scenes to this point. Watson had been warning Billings against him since 1854, asserting that Walker was working the halls of the Senate and of the Supreme Court and that Billings must do something to counteract him. A former Jacksonian Democrat and noted *éminence grise*, Walker had bought in to the Fossat claim as early as 1850, and in 1857, as governor of Kansas Territory, had declared that climate, not morality, would ultimately delimit the expansion of slavery, an argument displeasing to abolitionists and Southerners alike. He was taking every opportunity he could to espouse a centrist, Northern-expansionist view, and was using the rhetoric of Union to cloak his personal stake in the Fossat claim, or so Billings believed. Most likely sincere, Walker nonetheless seemed not to understand how delicate the union issue was in California, and he had played into the hands of those who advocated an independent California.

Then, on August 26, 1863, the New Almaden Company surprised nearly everyone by selling out to the Quicksilver Mining Company, letting a property said to be worth upwards of $25,000,000 go for $1,750,000. Perhaps the owners knew that the decision in the Fossat case was going to go against them, perhaps they had concluded that they simply would never win, quite possibly they believed that in the current state of Anglo-American relations, in which Britain was widely viewed as pro-Southern, they would be unlikely to be able to return to the mine. Certainly they believed they had been forced into a fire sale, for in 1872 Barron, Forbes & Company would present a claim for $14,000,000 to the American-British Claims Commission, sitting in Geneva to deal with reparations arising from Britain's allowing the Confederacy to construct privateers in Britain ports, stating that they had been forced to sell at a nominal price in the face of persistent government harassment. The claim was denied.

To this day there can be no final resolution of the New Almaden case, though whatever an historian might conclude retrospectively, what counts is possession, and the Quicksilver Mining Company had it. Most historians who have examined the records conclude that the company engaged in much bribery. Of course, bribes may be paid on a matter legally valid for fear that courts, or government officers, will not see the truth. Biographers sympathetic to Stanton or Black conclude, perhaps without a full examination of the record, that they were correct in maintaining that the Castillero claim was fraudulent. Again, it may not have been, but Stanton and Black may well have believed it so nonetheless. Walker may have spent money freely to engineer a conclusion favorable to the Fossat claim, but then the Barrons spent money freely to engineer the reverse, and despite charges flung in both directions of bribery and coercion, nothing has ever been proven. Surely if either side had clear proof of such a charge, they would have used it, for one more law suit in such a tangled affair could hardly have mattered. Historians favorable to the North and Union are inclined to take the government side; scholars who lean more toward states' rights favor the Castillero position. Issues of race, of attitudes toward Mexico and Mexicans, and toward the expansion of the United States obtruded then and may still influence judgments now. On the larger issue, historians who tend to support the "settler view" on the land laws and the work of the Land Commission are bound to be sympathetic to Stanton, Black, and the government, while those who have concluded that in good measure the old Mexican claims were valid, and that most sales to settlers were also valid, are more inclined toward Castillero. In short, judgments on the case tend to be colored by larger issues, now as they were then.

Although New Almaden was lost, the HPB lawyers were paid very handsomely for their work, yet not so handsomely as Black, who is believed to have received $180,000 as his fee in the final case before the Supreme Court. HPB used all the weapons of their profession: persuasion of politicians, the production of friendly witnesses, placing the best face on ambiguous evidence, manipulative organization of data, imputation of doubt upon those who spoke for the opposition. The Forbeses were unscrupulous, and HPB passed over this in silence because there was no expectation in the law that one should undermine a client. However scurrilous charges became, no one ever mentioned Billings as having engaged in any questionable activities (there was a hint—very faint but there—that Peachy used more than his silver tongue to persuade one or two of the California state legislators to support the pro-Castillero resolution sent to Washington). In all Billings's hundreds of letters to his parents and siblings, to his partners and witnesses, to his friends and political acquaintances in Washington—at least those letters that have survived—not a suggestion of doubt appears on his part about the legality of the Castillero claim. Of course, as one of the lawyers in the case he had to maintain a publicly optimistic view, but surely in his private moments, when depressed and anxious, he would have let something pass to a co-conspirator, to Peachy or Halleck or William Barron, if he was worried about some of the documentation. He never did. Whatever conclusion one may draw, one cannot doubt that Billings firmly believed his case. That is what lawyers are meant to do, if possible, and if they do not believe it, they are paid to put the best face on matters for their clients nonetheless. Billings believed it, and when the case was lost he was bitterly disappointed, though by 1863 certainly not surprised. Probably the government was so determined, certainly in the midst of the Civil War, to secure the mine or at the very least take it out of the hands of a foreign company, that the narrow decision was virtually inevitable.[25]

The long saga of New Almaden had seemed to bring out the worst in people. After all, the richest quicksilver mine in the world, as it was by 1860, was at stake. Vast sums of money changed hands, much of it in legal fees, much no doubt illegally. Lincoln's friends, especially Leonard Swett, had used him badly; Peachy may have purchased supporters; the Forbeses and the Barrons employed every pressure available to them in Mexico; Walker had connived shamelessly; Cass and Buchanan had acted weakly; and Black had changed his stripes very quickly after leaving the government, as Gregory Yale was quick to point out when he published his own book in 1867, *Mining Claims and Water Rights*.[26] When the mine was not producing quicksilver it was still producing fees. Billings had some moneys to show for it, and now more huge volumes stamped with the name of Castillero sitting on his shelves, representative of what many in the profession said was the finest research ever done until then by an American lawyer.

12

The Best Chance

S ECOND ONLY TO NEW ALMADEN IN MAGNITUDE, yet exceeding it in
influence on Frederick Billings's life, was the exceptionally complex matter of Las Mariposas, or the Frémont Grant, perhaps the best known land
grant in California history. Las Mariposas threw Billings into intimate contact
with one of the period's most romantic and controversial figures, John Charles
Frémont. Billings both made and risked a small fortune over Las Mariposas,
further broadened his legal connections and reputation, meeting for the first
time many of the figures with whom he would do business as a railroad magnate years later, and as a result of the frustrations dogging him throughout
the case, decided upon his return to the East Coast that he no longer wished
to practice law. His connection with Las Mariposas also quickened his interest
in conservation.

John C. Frémont had not wanted Las Mariposas to begin with. Busily
engaged in making a revolution, war, or coup—that is, in helping to take
California for the nation in 1847—he had asked his friend Thomas O. Larkin,
the American consul in Monterey, to purchase some land for him at the old
San José Mission, which had been secularized under Mexican law. However,
in February Larkin bought Frémont trouble instead, purchasing from Juan
Bautista Alvarado ten square leagues of land in wild, mountainous country far
to the east of San Jose. This land had been granted by Manuel Micheltorena,
the Mexican governor who rewarded his friends with somewhat casually defined largesse, when Alvarado had asked for pasturage. The land was not pasture, it was ill defined as to location, and as hostile Indians controlled it, one
could do nothing on it. Alvarado was happy to be rid of it.[1]

Rancho Las Mariposas, Spanish for butterfly, ran for fourteen miles along
the Mother Lode.* The grant comprised 44,380 acres, though its boundaries

---

* Most books refer to Las Mariposas as the Mother Lode, but this term was not coined until
1867–1868, after which it becomes appropriate. The usual term in the 1850s was the Johnson
Lode.

were indeterminate—"floating," in the parlance of the time. Gold was discovered on the property in the spring of 1849. By October a steam-driven quartz mill, perhaps the first in California, was in full operation and squatters were in full protest. Frémont, therefore, moved to protect his grant. He had a questionable deed of sale, for when Micheltorena had granted the land to Alvarado, he attached the condition that it could not be sold, transferred, or mortgaged. Frémont seems to have been aware that Alvarado had never settled on the land: in 1849 he had hired a man to occupy the land for him, as Mexican law required. Still, no permanent occupation was achieved, and on January 21, 1852, Frémont filed a fresh claim under American law.

The first gold found on the property had been along the creek beds, and Frémont hired a group of Sonorans to work the stream for him. It was clear that if he wanted to make his fortune Frémont must find the mother lode from which the placer mines had come. Soon found, assayed samples proved exceptionally rich. Frémont knew he must pin his floating grant down, and he decided to locate his ten square leagues of land in such a way as to take up the bottom lands along Mariposa Creek, the Mariposa and Agua Fria diggings where placer mining already was in progress, and land toward the high mountains where the quartz veins led.

Soon Frémont's Sonorans were washing out gold too rapidly to remain silent, and as the word spread, a horde of prospectors descended on Las Mariposas. Frémont knew the law; even if valid his grant had not conferred title to mineral rights. Further, rightly or wrongly, the public generally viewed placer deposits as the property of anyone who worked them. The Sonorans clashed with the newcomers and, near Christmas, went home.[2] Frémont had no time to attend to his property—if indeed it was his—for he had just been elected, together with William M. Gwin, one of California's first two U.S. senators. One was to serve a short term and, after drawing lots, Frémont was chosen. He left for Washington on New Year's night, 1850, after putting a caretaker in charge of his distant *rancho*.

Frémont was dashing, impetuous, and famous: indeed, he may well have been the most famous man in California in the 1850s, and he certainly was the only person immediately identified simply as "the Colonel." Born in Savannah in 1813 and raised in Charleston, he had eloped in 1841 with the sixteen-year-old daughter of the formidable Thomas Hart Benton, senator from Missouri. He became expert in astronomical, topographical, and geographic observation by the time he was twenty-four. With Kit Carson as his guide he explored the Wind River Range in Wyoming and, with his wife Jessie's help, published a highly popular account of his deed. In the winter of 1844 he crossed the Sierra through deep snows, showing up dramatically at Sutter's Fort. His third expedition, in 1845, opened up a new trail across Nevada; at Hawk's Peak, near Monterey, he ran up the American flag in what was still

Mexican territory, and in May 1846, when unhappy American settlers rose against the Mexican government in the Sacramento Valley, he inspired the Bear Flag Revolt. Commodore Stockton appointed him civil governor of California, and, as we have seen, General Kearny arrested him for mutiny and insubordination. A court-martial trial found him guilty, after which Frémont resigned from the service. In 1848–1849 he led another expedition, this one financed by Senator Benton, in search of a Pacific railroad route. He was, if not a legend in his own time, at least a very good story, the "Pathfinder," recipient of scientific medals, lavish praise, and harsh denunciation.[3]

Frémont not only had to win a quick confirmation of his claim, since squatters were taking away large sums daily, but he had to win the right to bar intruders, so he put William Carey Jones, author of the influential land report that reached conclusions opposing those of Halleck's more substantial tome, in defense of the title, hoping to get it to the Supreme Court as quickly as possible. Jones, not at all incidentally, was married to another of Senator Benton's daughters. The strategy worked: the Mariposa claim was docketed first before the Land Commission, and on January 10, 1853, the commissioners confirmed the title. The case then went to the District Court where Jones argued it once again. There Judge Hoffman reversed the Land Commission on the ground that the conditions of the grant had not been met. Jones appealed to the Supreme Court, where Frémont's title was upheld on February 8, 1856, with two dissents. This left the issue of mineral rights. Frémont's lawyers quickly had the tract resurveyed and two days after the court's decision the tract was rushed to patent, with President Franklin Pierce handing the document to Frémont personally ten days later.[4]

The contrast with New Almaden is striking. Frémont's floating grant was vague, and in any case the Colonization Law of 1824 had prohibited such grants. Alvarado had never occupied or improved the land; the manner in which the claim was "swung"—rather like a frying pan, someone noted—was irregular. Mexican law did not permit assignees unilaterally to locate a grant, and thousands of miners suddenly found that improvements they thought were theirs now sat on Frémont's land. Yet Frémont had been confirmed in his property. The decision was unpopular and smacked of favoritism.

By now Frémont, who had been living in Washington, was thoroughly distracted from Las Mariposas, for he would soon be a presidential candidate. Sectional tension was at an all-time high. The old Whig Party had disintegrated and the Democrats were in disarray. Some Democrats approached Frémont to see whether he might not be their standard bearer, and he rebuffed them; then the Republicans made their approach. Late in 1855 Joseph Palmer, head of Palmer, Cook & Company, called on a number of the new party's stalwarts to argue that Frémont would be an ideal nominee. Nathaniel Banks, a powerful congressman from Massachusetts, cleared the way, and when Frémont's close friend Francis P. Blair, a highly regarded Washington journalist, and his son Frank P. Blair, a powerful figure in St. Louis, came aboard, the bandwagon was rolling. In June 1856, the first Republican National Convention met in Philadelphia and nominated Frémont and, as his running mate, William L. Dayton, a former U.S. senator from New Jersey.

The election of 1856 went badly for Frémont, exposing him in a way that national political prominence tends to do, and not for the better. Chosen because he was thought to be a national hero, he was charged with having grossly exaggerated his role in California at the expense of Stockton and Kearny, with recklessly taking his men into great danger, and with being a Papist. Above all, his commitment to an anti-slavery position cost him both the moderates and the immediate abolitionists. He carried eleven states. California gave him only 18% of the popular vote.

Shortly after the election Frémont returned to California to look after his properties, which included lots in San Francisco and elsewhere. In February 1857, he and his family moved to Bear Valley for Frémont's only sustained attempt at hands-on management of his mines; they would remain until the spring of 1859. Jessie gave the cottage at Bear Valley a fresh coat of whitewash and dubbed it the White House, while her husband wrestled with the need to improve the mines. Frémont had many creditors on the mine, with respect to other properties, and abroad. At this point Halleck, Peachy, and Billings enter the story.

With his new patent, Frémont had asked the state to measure his seventy square miles along both sides of the Merced River in one long strip. The state refused, stating that the grant must be compact. Frémont nonetheless secretly defined his "float" so as to include valuable and working mines physically in the possession of others along the Mother Lode. On the pretext of surveying for a ditch to bring water to his mines, Frémont completed his new survey, the boundaries of which he hoped to keep secret.

In April 1857, Frémont leased the Mount Ophir mine, which the survey had confirmed as his, to his Bear Valley caretaker, Biddle Boggs. The mine already was in production, for the Merced Mining Company had taken it up in 1851 under a quit-claim, and the owners of the company insisted they were on government land, two miles away from Frémont's property. They estimated their investment in improvements on their property at $800,000 and viewed Frémont's assertion of ownership and Boggs's presence as provocations. While the Merced Company did not contest Frémont's ownership of the land, its lawyers pointed out that his grant had been for grazing and agricultural purposes only, and they were determined to test Frémont's claim to subsurface rights.

HPB represented the Merced Mining Company while Rufus Lockwood and seven other lawyers defended Boggs and thus Frémont. Lockwood was a strange, self-taught lawyer of undoubted brilliance who was known to have been on retainer to Palmer, Cook & Company, and many miners deeply resented him. They also thought of the company as the fountainhead of Black Republicanism—that is, of Frémont's presidential candidacy—in California, and in a county where he had received only 6 percent of the vote, this bevy of big city lawyers was bitterly unpopular. The juryless trial began on May 4 in the new Mariposa County Courthouse, built two years before on land Frémont had transferred to the county to meet his tax assessments, and the second floor courtroom was packed with interested parties, the miners themselves.[5]

Frémont was present and tension ran high; most of the miners were carrying guns, and the judge and attorneys knew that they risked being shot if they did not appear to be fair. On behalf of the Merced Mining Company HPB contended that the Frémont grant did not confirm to its description, that it had been surveyed secretly and illegally, and that various technical requirements had not been met. Further, Frémont had never formally extended his claim to include the Mount Ophir property. Finally, the Alvarado grant had been for grazing only and had conveyed no mineral rights. On each of these points, with the weight of the Supreme Court decision on the land claim behind them, Frémont's lawyers were able to respond effectively. On July 2, 1857, the judge found in favor of Frémont on all points, though the decision in the gallery went to the Merced Mining Company and HPB.

There was, of course, an appeal to the California Supreme Court, during which Frémont began construction of a new quartz mill near Bear Valley, hard by the Josephine and Pine Tree mines, which were being worked by the Merced Company. In 1858 Associate Justice Peter H. Burnett, with David S. Terry concurring and Associate Justice Stephen J. Field dissenting, overturned the decision. Rumors at once went out that all three justices had accepted bribes; Gregory Yale was thought to have been responsible for having persuaded Burnett to his decision, while Frémont was said to have given Field $25,000.[6]

At last, in November 1859, the state Supreme Court granted a rehearing and Frémont was confirmed in all he claimed. Court personnel had changed: Joseph G. Baldwin, who had been one of Frémont's attorneys, had replaced Burnett on the bench and, though he did not participate, this left the decision to Stephen Field, who was now the chief justice, and W. W. Cope, who had replaced Terry. Field's decision was regarded as a significant statement in defense of the rights of large property owners, and he would go on to the U.S. Supreme Court where he would win a reputation as a leading conservative and advocate of big business.

For some time after leaving HPB, Trenor Park had been working on Frémont's behalf among the money men of San Francisco, and had persuaded Frémont that he needed an infusion of capital. With the final legal decision in hand, Park now pressed Frémont on the need to expand the operation, especially at the Princeton Mine, and put the need at a million dollars. In 1857 Park had acquired a $65,000 mortgage on part of the property in lieu of legal fees, and by 1859 he put his interest at $250,000. Mark Brumagim, a Marysville financier active in San Francisco, supplied crucial operating capital, while John Parrott and A. A. Selover also held interests. In 1857 Frederick Billings acquired a share by purchase from his friend Hall McAllister, and by 1860 Park and Billings each held a one-eighth interest. Other investors held a third one-eighth, and Frémont remained in majority control with five-eights.[7]

On June 20, 1860, Park took over as manager of the entire estate. In lieu of salary while managing the mines, Park would receive an additional one-sixteenth interest. From 1860 Park was, for all practical purposes, running the mines, and as part owner he was in a powerful position, though technically Frémont, Billings, and other owners had authority over him in his role as manager. Park now had a very rich cow to milk.

One of the persistent problems at Las Mariposas was that those making the decisions were not single-minded about their obligations. Frémont wished to be rich but lacked knowledge or steadfastness. Park was hoping to be elected a U.S. senator from California, using his connections to Frémont and his father-in-law, Hiland Hall, who had returned to Vermont, where he was elected governor in 1858 and again in 1859. In 1860, Park did not get the Senate seat, but he continued to be hopeful, and in 1864 he would be angling for a vice-presidential nomination in the event the Republicans chose Salmon P. Chase, Lincoln's secretary of the treasury, who had the backing of a group of party dissidents. Billings felt that Park was not paying attention to the cow, having taken as much from it as he thought likely, and that Las Mariposas would never produce the $100,000,000 so many predicted for it.

The first task, Billings argued to Park, was to realize some income, and then to make the necessary improvements so that the mines could become more productive. He advised Frémont that he could tax the placer miners for the privilege of continuing to pan the estate's streams, and Park set the sum at $4 a month, which he promptly began collecting. Billings also urged that more water be brought to the mines and that a large new stamp mill be built. He was less sanguine about the construction of a railroad from the Pine Tree mine down the precipitous slope to the river, though Frémont and Park were enthusiastic. Billings feared that Park would puff up the prospect of success, so in the summer of 1860 he sought out his own direct contact at Bear Valley, asking a supervisor, James Clark, to report privately to him. In the fall, Billings put Sherman Day, whom he trusted from his experience at New Almaden, temporarily in charge at the Princeton Mine, and Day made it clear to anyone who asked that Billings, not Park, was in authority.[8]

By the fall of 1860 the mines were producing upwards of $100,000 a year, but were so encumbered by debt service and taxes, little was left beyond wages.

The court cases had cost $200,000, and the owners owed nearly half a million dollars for development costs. As a miner's proverb had it, "it takes a mine to run a mine," and Frémont did not have the tidy and consistent money producer to break open this nut. Refinancing abroad was the answer, or so the owners concluded.

In December 1860, Billings agreed to accompany Frémont to Europe, acting as his lawyer in all matters, and travelling independently in Britain and Germany while Frémont tackled the French, where he and his wife had excellent contacts. They drew up an agreement providing Billings with another one-sixteenth interest in Las Mariposas as his fee. Park and Billings had a magnificent and very expensive portfolio of photographs prepared by the California photographer Carleton E. Watkins. They commissioned reports from French, German, and Scots mining engineers. They also drafted a handsome prospectus, which they took to England with them for printing and binding. The prospectus promised a yield of a hundred tons daily and a gross income of $40,000 to $50,000 a month, or a net of $16,000, to any consortium that would purchase a part interest in Las Mariposas.[9]

Though he liked to travel and had never been to Europe, Billings was reluctant to leave. He did not like Frémont very much. He thought him a braggart, possibly even a fake. He considered Frémont vain, bent on self-promotion to a greater degree than even a self-promoting age ought to tolerate. He knew him to be a bad businessman, inattentive to detail, uninterested in the long run, eager to make a killing so as to return to the good life, the bright lights of Washington, and political power. But Billings wanted the Mariposa mines to succeed, for he thought they were his own best chance to become a multimillionaire, and he wanted Frémont's wife Jessie to be well cared for.

Jessie Benton Frémont was contradictory, dynamic, and very intelligent. Many men found her fascinating, and Billings was among them. He felt she was the making of Frémont, despite a marriage in which Frémont was neurotically dependent on his wife, and while Billings knew that she exaggerated and had a convenient "forgettery" when her husband's interests were at stake, he thought she was essentially honest. He was, clearly, just a bit smitten with the vivacious Jessie, and he kept an eye out for her best interests. Billings was furious when he believed Frémont was disloyal to her.

Billings liked to visit with the Frémonts and with the guests that they gathered about them during the short period in late 1859 to early 1861 when they lived at Black Point. Here Billings talked politics and conservation with Frémont, who professed to have an interest in the redwoods while owning a company that was chopping them down. Conversation with Jessie was far more interesting, and especially so at Black Point, which she loved above all her many homes. She was poised and commanding, attracting to her drawing room the best writers, the most entertaining conversationalists, the wittiest of wits and sharpest of journalists the city could offer. There Billings encountered the writer Francis Bret Harte and the satirist J. Ross Browne, and Herman Melville and Samuel Clemens, both to become famous later. There was Carleton Watkins, the photographer, and Albert Bierstadt the painter. Billings enjoyed talking with the two Frémont boys and the shy daughter, Lily, and though he was not in the Frémont home as frequently as he had been in the Pattersons', he felt at ease there. He was sorry to give up these visits and his growing prominence as a pro-union speaker to go to Europe with the Colonel, for the nation appeared to be on the verge of war.

13

"Saving California for the Union"

THE HISTORIAN HUBERT HOWE BANCROFT had observed that Frederick Billings did not enjoy public controversy, and this was clearly true. Though his instincts were political—he was a natural conciliator: between his father and his mother, with and for his brothers and sisters, between his hot-headed Virginian partner and those with whom he was tempted to duel—he had no taste for the rough and tumble of party politics. As his good friend Alfred Barstow wrote, Billings was "disposed to be exclusive." Still, if he did find himself locked in combat, Billings would not, Barstow added, "cease from the struggle until he saw the end of it and saw it satisfactorily."[1] Thus while he struggled to see his way clear on the impending crisis over slavery, once war had begun he was unwavering in his support for the cause of Union.

In Vermont and during his first half-decade in California Billings held naturally to the Whig persuasion. For him, Whiggism embraced a sense of order, a commitment to hard work, devotion to family, and a patriotism that did not require the denigration of others. He favored a purposeful intervention by the federal government to subsidize internal improvements and regulate banking. He believed that the values of Vermont were, or ought to be, the values of the nation. Though at times uncomfortable with his fellow Whigs at his club, and especially with those who seemed to glorify industrialization as an end in itself, he had avoided direct challenge to them. He wanted, he wrote to his parents, a home of his own, which must be "the most agreeable place in the world"; this meant, he concluded, that he ought not to allow himself any political ambitions, for politics inevitably brought controversy.[2]

But the growing national dilemma over slavery would not permit Billings to opt out of some sort of political commitment indefinitely. In 1852, when a Whig candidate for the California Supreme Court died, a member of the Whig State Central Committee in charge of nominating the new candidate told Billings he could have the nomination. Friends were also urging him to run for the state senate. While he declined to pursue either office, Billings was aware that an appearance of indifference to the needs of the party whose principles he espoused was not consistent with his widely known public-spirited activities on behalf of other causes, and he accepted membership the next year on the state Whig committee, serving for two terms. In 1855 Billings hoped to be chosen as Whig candidate for Congress, and went to Sacramento where he made a sincere effort, only to lose on the twenty-third ballot thirty-six to eleven. In 1856, still on the state committee, he did not attend the state convention, pleading ill health though in truth he was simply disillusioned with the party. Indeed, he thought he would have to vote for the Democratic nominee, James Buchanan, in the coming national election, for he could not support Frémont, and he regarded a vote for Millard Fillmore, the Whig who had become the national Know-Nothing nominee, as thrown away. Further, the calls of the firm might draw him into the Democratic Party, for at last Peachy looked as though he were on the verge of success, possibly to be chosen as a candidate for the U.S. Senate.[3]

Almost up to the outbreak of the Civil War Billings continued to sit on the fence. On the one hand he disliked men like Charles Sumner, the outspoken abolitionist senator from Massachusetts; on the other he was outraged when Sumner was violently assaulted on the Senate floor by Representative Preston Brooks of South Carolina: "this shooting, pounding bowie-knife-kind of spirit has got to *go down!*," he wrote.[4] He understood that abolitionists were "anti-slavery," as he was, but he did not assume that all advocated the same policies. Though the HPB partners had been members of the Colonization Society, which proposed to relocate free blacks in Africa, he realized this was not a realistic solution to the race issue. He thought the only resolution to the bloody strife in Kansas, where Free Soilers and pro-slavery elements battled for the state, was to abolish all government there, hold a new election under the supervision of the U.S. Army, and let the residents decide for themselves the question of slavery. While he opposed slavery for California, he also believed it to be constitutional and defensible where it was the will of the electorate. This position, endorsing popular sovereignty, in effect was a vote for nationalizing slavery while restricting the institution to the slave states. However, after the Dred Scott decision in March 1857, popular sovereignty as a solution to the problem of slavery expansion was, in fact, moot, for Chief Justice Roger B. Taney and the majority of the Supreme Court had effectively ruled that free soil did not mean free men.

Billings felt uncomfortable with the rising tide of the Republican Party. As chairman of the Republican State Central Committee from 1856, Trenor Park unsuccessfully tried to persuade Billings to support the new party. Park was blatantly using promises of patronage to bring the undecided into the tent, and Billings disliked this intensely. He moved toward the Republican Party only when he thought it might nominate Senator William Henry Seward of New York, even though Seward was identified with anti-slavery radicalism, because he believed that Seward would stand up to the South without bringing on war. Equally important, if the South did attempt secession, he believed that Seward would resist. To Billings the issue now was preservation of the Union by any means necessary.[5]

Until 1860 slavery was not the major issue in California politics, however, and in the parlous condition of state party realignments after the statewide defeat of the Whigs in 1855, Billings was an attractive candidate to those seeking a centrist. At some point early in 1857 a group of powerful Republican financiers and political figures proposed to back him for the U.S. Senate, believing that John B. Weller's seat was vulnerable. Billings's reaction to the possibility reveals why he could never expect to succeed in elective politics in California. He went to Sacramento with John Thomas Doyle, a well-placed and scholarly lawyer friend, at the invitation of those who said they could send him to the Senate. Receiving the committee in his cabin aboard the Sacramento evening boat, he asked what they proposed to do. They told him how many votes they could command, and from whom, and how easily he could receive the Republican Party nomination. According to Doyle, who wrote of the meeting later, Billings inquired, "Well, what else?" After some backing and filling, one of the delegation at last said that Billings must pay a certain sum of money, and advised him on how to arrange the payment so that, if he did not get what he had paid for—election—he could recover the money. "Is that your whole proposition?" Billings asked. "Have you stated it completely?" Yes. With that he said that he did not think it worth his time to go ashore, for he would have nothing to do with such an arrangement, and returned to San Francisco.[6]

This was a principled stand but also a realistic one. By taking it (and in the retelling of the story later by friends) Billings retained his integrity. He also avoided battle for what was, he surely knew, a losing cause. A Democratic victory was clearly in the cards, and while an exceptionally clever and unscrupulous person might have had some chance of victory through a split Democratic vote, this was exceedingly unlikely.

Talk, always mutely present, now began of founding a California Republic, of letting the East go its own way to war or not. Some advocates of an independent nation on the West Coast, who invisioned adding Oregon, Washington, and California's colony, Nevada, thought that three American nations would be better than one, and that the federal government, whether in the hands of pro- or anti-slavery leaders, had never paid proper attention to California's needs. This was not true, but it played well to those who felt neglected or wanted not to have to decide between slavery and anti-slavery. By 1859 there was a further internal threat of disunion, for a bill was introduced in the state Senate to separate northern and southern California into two units, the latter to be a territory named after the Colorado River. Thoughts of two Californias or of a separate California Republic alarmed Billings fully as much as speculation about possible Southern secession.[7]

The proposal for two Californias had complex origins. The state's revenue system relied on a property tax by which miners escaped most payments while the large ranchers, especially in the southern part of the state, carried a disproportionate burden. Assemblyman Andrés Pico demanded tax reform on behalf of southern Californians in nearly every legislative session, and in 1859 took the bold step of proposing that the six southern counties in the state form a separate territory in preparation for independent statehood. The state legislature passed Pico's measure, and suddenly the prospect of disunion seemed very real to Billings, who opposed any separatist movement both on principle and on highly practical grounds: he supported the almost universal California demand for a transcontinental railroad, but wanted it to enter the state in the north, to the benefit of San Francisco, not in the south, as the separatists advocated, and he believed slave expansionists were behind the attempts to create a separate state government in Los Angeles.

Party alignments were also in chaotic disarray as a result of the bitterly fought campaign of 1859, which led to the death of David Broderick. Drawn into a duel by David S. Terry, Chivalry stalwart and justice of the California Supreme Court, Broderick was shot and died on September 16. Many people believed that a pro-Southern cabal had deliberately set up the duel, for Senator Gwin was known to have said that Broderick ought to be either clubbed or killed. The circumstances of the duel were peculiar, and rumor quickly had it that Broderick had been murdered. One of Broderick's friends, Edward D. Baker, a noted public speaker, delivered an oration in Portsmouth Square to the largest crowd ever assembled in the state. At length Baker told the throng that Broderick had died because he opposed the extension of slavery, a crowd-pleasing position to take in the fall of 1859.

The Democratic Party was now irretrievably split, and at the state convention in February 1860, there was talk of independence for California. The Republicans, now united, sent a delegation to their national convention in Chicago, where Abraham Lincoln was nominated. Billings was among the many Californians who hoped Seward would be chosen, in part because he had been a consistent champion for California's statehood in 1850, in part because they believed he would be a stronger president more likely to maneuver the nation away from war.[8]

The Charleston Democratic convention broke into two factions, meeting in Baltimore and Richmond. The first named John C. Breckinridge of Kentucky their presidential candidate, and, in an effort to hold the West, chose John J. Lane of Oregon for the vice-presidency. The second, after a further secession of southern delegates, nominated Stephen A. Douglas as President.

For a moment Frederick Billings was a Democrat, for he thought Lincoln too weak to hold the Union together and believed Douglas just might. In California the pro-Breckinridge faction issued a long address, signed by sixty-five public figures, and the pro-Douglas group responded with an address of their own, bearing one hundred and fifty names. Among the signers were Halleck and Billings. California's senators thereupon made it known that they intended to support Breckinridge. Billings stood with Douglas in 1860, in some measure because he had not encouraged a resolution offered by one of the California delegates calling for an independent Pacific Republic. But in the national election in November, Lincoln won as a minority president, with 1,866,352 popular votes, while Douglas and Breckinridge split 2,224,938 between them.[9]

Historians generally credit two men above others with the oratorical skills, so important to the nineteenth century, that "saved California for the Union."[10] One was Edward D. Baker. The other was a young Unitarian minister who had recently arrived from the East Coast, the Reverend Thomas Starr King. Billings almost immediately became King's ally and travelled with him, sharing the podium throughout the northern part of the state, and in 1861, in a letter home, King declared of himself, two other preachers, and Billings, "We have saved California."[11]

Baker's chief contribution came on October 26, 1860, just before the presidential election. He had gone to Oregon in February and had been elected senator that September when the legislature was seeking someone to oppose a candidate chosen by ex-Governor Lane, now on the national Breckinridge ticket. On his way to the East Coast, Baker stopped in San Francisco, where his admirers (including Halleck) persuaded him to address the public. Baker's oration has been hailed as the most eloquent ever delivered in California; he was credited with swinging San Francisco, Sacramento, and many of the mining towns to Lincoln in the state election, in which Lincoln received 38,699 votes as against 38,032 for Douglas and 33,970 for Breckinridge. By the time the vote was counted in California, news of Lincoln's victory already had arrived and word soon followed that a confederate congress was to meet in Montgomery, Alabama.

Though various books credit Baker, Starr, or Billings with "saving California for the Union," this statement is frequently misunderstood. The risk now was not that California would join the seceding states; rather, the worry of unionists was that the idea of a Pacific Republic would grow. Such a movement would, of course, play into the hands of the Confederacy. Billings was utterly opposed, for though he had spoken often of the honorable past, present wealth, and future power of California, and had compared it to the independent nations of Europe, he wanted to see that power applied for the benefit of the United States. Thus he and Starr King toured the state together, the Unitarian minister speaking out on the theme of national unity and Billings following up with a talk calculated to deflate the independence movement.[12]

Starr King had arrived in San Francisco on April 28, 1860. At thirty-five, a year younger than Billings, he was well read in the classics, deeply versed in theological disputation, charming, quick, and resembled Billings physically. Starr King won an immediate friend. He was not a typical Boston Unitarian, for he happily debated matters other ministers regarded as closed, and the two men often discussed their respective beliefs late into the night. King had just published an attractively written book of essays, rather in the manner of Thoreau, on the mountains of New Hampshire. Called *The White Hills*,[13] the book explored the legends, poetry, and landscape of New England, and Billings like it very much. King made it known that he intended to write a book about the Yosemite Valley as well, and Billings helped organize a trip for the minister. They had much in common, including frequent—and growing—bouts with illness that put each into bed with nervous exhaustion for two or three days at a time.[14]

Billings had begun his attacks on disunionists when invited to address the annual meeting of the California State Agricultural Society in 1858. The invitation had come in April, the address was in August, and he had worked and reworked this talk to an unaccustomed degree, for he realized he was speaking on an issue that had not crystallized. Billings previously had addressed the society and been well received, but on that occasion limited himself to extolling the virtues of agriculture and predicting the greatness of California. Standing side by side with Starr King in 1860 and 1861, Billings now reshaped his 1856 and 1858 agricultural talks to meet the broader need.[15]

Then on April 25, 1861 the pony express brought the news that thirteen days before, General G. T. Beauregard had ordered the bombardment of Fort Sumter in Charleston's harbor, and after forty hours the fort had been surrendered to the confederacy. Republicans, Douglas Democrats, and some of Breckinridge's followers now rallied for the Union. In San Francisco a vast public meeting was held on May 11 with 25,000 people attending. The Democrats, Senator Latham, and James A. McDougall (who two months before had succeeded Gwin as California's second senator) together with Thomas Starr King exhorted everyone present to support the Union cause. With the nation's flag flying from nearly all the city's churches, and with the vast throng milling about on the city's streets, the mass meeting passed ringing resolutions in support of national unity. Not until July, at the first battle of Bull Run (Manassas), however, did the North—and California—fully understand that the war was likely to be long and bloody, and that the "secess" in California just might succeed.

The Civil War brought the eventual end of HPB. Peachy had become increasingly difficult to contain, and his outspoken pro-Southern statements, once a useful balance to Halleck and Billings, were now a liability. Halleck knew that if there were war, he would return to the U.S. Army. Four days before news of the firing on Fort Sumter, the partners inserted a notice in the *Alta California* announcing a grand auction of the firm's entire library of three thousand volumes, the largest law collection in the West. Peachy, who had married a politically ambitious Irish woman in 1859, withdrew to Sacramento, where he hoped to influence opinion favorably to the South. Halleck told Governor Downey of his intention to resign his commission in the state militia and left for his new command in the army, sent off from the Folsom Street Wharf to the echo of a thirteen-gun salute by the Pioneer California Guards. Granniss remained, to keep a single-room office open in the name of HPB while awaiting developments and Billings's return from Europe, where he had gone in January to raise loans for developing Las Mariposas.[16]

California's contribution to the northern war effort was of three kinds. The state supplied infantry and cavalry to pacify Indians (thus releasing troops to the front), to guard the overland mail route (which some, including Billings, feared would be subverted by the reputedly pro-Southern Wells Fargo), and to remove Confederate positions in New Mexico and west Texas. California's mines contributed precious metals to the nation's war chest, as gold dust was carried by sea and over land back to the East. And more than any other state, California supported the Sanitary Commission. When not on the rostrum speaking against separatism and disunion, Billings's contributions to the war effort were made largely through this commission.

The Sanitary Commission was the creation of the Reverend Henry W. Bellows, a Unitarian minister from New York. It was one of the Civil War's success stories, for despite frequent factionalism within it, it raised substantial sums of money. As a civilian auxiliary to the medical bureau of the War Department, the commission supplied funds, goods, and workers to tend the wounded of the North and of Southern troops left by retreating armies, ministering to the comfort, morale, and shattered limbs of thousands of men. The commission held systematic fund drives, generally called "sanitary fairs," across the country, culminating in a great fair in New York City in December 1863. Billings contributed quite large sums to the California fund drives, and when in New York in 1863 worked on behalf of that fair. Starr King and he played effectively upon the slight embarrassment felt by many Californians who remained safely distant from the fields of battle, and King commended Billings to Bellows as "the man I love more than any other in California."[17]

California contributed by far the most to the Sanitary Commission, and Billings stood second only to King in espousing its needs. He discovered that he possessed a new and remarkable oratorical power, refined by his study of King's own methods and his passion for the cause. As Alfred Barstow wrote later, Billings was aware that he had a dangerous gift, for "if he called upon them to do it, [the people] were sufficiently frenzied to go out and commit murder." On another occasion a former student of Billings from his Vermont years, George B. Reed, reported that Billings was the only person who had been able to quiet a noisy political gathering, by exclaiming as he stood up to speak, "I am from the land of Long John Stark and Ethan Allen," referring to two of the founders of Vermont. Jessie Frémont, on whom rousing political oratory was never lost, urged Billings to run for the U.S. Senate, and though he may have suspected that this was manipulative flattery to keep him working on behalf of Las Mariposas, he also now allowed himself to believe that she might be right.[18]

The first mass meeting held in San Francisco on behalf of the Sanitary Commission on September 14, 1862 brought in $6,600 in gold;[19] by the end of the year Californians had contributed nearly $370,000. When Bellows appealed specifically to California in October 1863 in preparation for a mass meeting in New York to raise half the total needed, the state responded handsomely. By the end of the war California had provided 25 percent of the total operating funds of the Sanitary Commission. By then, however, King had died, succumbing to diphtheria and exhaustion on March 4, 1864. Billings, on the East Coast at the time, was devastated, for King had reinvigorated his religious and political convictions.

With hindsight the historian may conclude that neither Billings, King, Baker, nor anyone else had, in fact, saved California for the Union, for the majority of Californians never forgot that their interests lay with the North. Events determined California's position, and the Unionist newspapers such as the *Alta California* and the Sacramento *Union* clearly contributed more. Nor had Gwin's proposed Pacific Republic ever had much following behind it, useful as the idea was to keep Washington aware of California's expectations. What mattered, however, was what people believed, and a genuine fear existed in the northern part of the state that the southern counties, which had tended toward a pro-slavery position before the war, might well go it alone. After all, the state legislature had forwarded the separatist resolution to the U.S. Congress, which in due course and without war would have acted, though now northern Congressmen were bound to bury the resolution. Billings was shrewd enough to realize this, and never embraced the claim made on his behalf by others that he or anyone else had "saved California for the Union." He did know, however, that his efforts had helped in a major way the raising of the state's quite considerable contribution to the Sanitary Commission, and he believed that speeches like his had kept a preoccupied Washington alert to the need to make concessions and gestures toward California's unionists.

 14

Arms Purchaser

FRÉMONT AND BILLINGS SAILED for the East Coast in January 1861 on a journey that threatened to go sour even before they were through the Golden Gate. Shortly before their departure, Jessie was thrown from her carriage and badly injured, so she remained behind. With a ticket in hand, Frémont secretly offered it to another woman, Mrs. Margaret Corbett, without confiding in Billings, and throughout the voyage and while on the East Coast Frederick became increasingly agitated, reporting to Park of his growing suspicions that there also was a second agent along, charged with carrying out much the same task as Billings. This is distinctly possible, for Frémont had done this sort of thing before. The situation was "disgusting," Billings thought, and he felt particularly bad for Jessie.[1]

When Frémont and Billings arrived in New York, the colonel scurried off to Philadelphia, and Billings was even more distressed, for this time he was certain that Frémont was visiting a lady with whom he had had an earlier liaison. He was amazed to discover that Frémont proposed to bring Mrs. Corbett and a child to Europe with him, and Billings believed Frémont was urging him to take an earlier vessel so that he would not discover they were travelling together. Frémont had used up his time in dalliance, or so Billings felt, so that he would have to attend to the situation himself. This would delay his own departure at a crucial time, since war was a distinct possibility and it was essential they make their first overtures in Europe quickly.

Frémont and Billings met briefly with President-elect Lincoln (who, it was rumored, wanted Frémont to be minister to France, though no offer was made), and sought such financial allies as were left from the campaign of 1856. In New York Billings called on William Aspinwall, Isaac Sherman, and August Belmont, international financiers, and entertained his mother and sisters, who came down from Woodstock. He also went to Boston, where he invited his father to stay with him, and made arrangements for his brother Oliver to accompany him on part of the European trip as his secretary.

Shortly before Frederick and Ollie left for Europe, now behind rather than ahead of Frémont, Frederick apparently decided to clear the air, for he expressed his concerns to the Colonel. Frémont responded that he was disappointed in him: "I thought more highly of you than I suppose that you could be shocked by any of the small rascalities." Though not married, Billings did not consider infidelity a small rascality, and he was outraged. Still, he was hopeful for the business enterprise itself, writing to Park in an unusual burst of enthusiasm that if he and Frémont succeeded in Europe they would make themselves "easy for life."[2]

One wonders at Billings's optimism, for he and Frémont scarcely could have been going to Europe at a worse time. Eight Southern states had seceded and Southern forces had seized Federal forts and arsenals at ten locations. Judah Benjamin had written Billings shortly before Christmas that an independent Southern confederacy was certain. He had also offered to look out for Billings's Mariposa interests while Billings was in Europe and provide letters of introduction, which Billings decided to have sent on. They did not come, of course, for in April, less than a month after Frémont and Billings arrived in Europe, the feared war broke out, and Benjamin was a member of the Confederate cabinet.[3]

Further, California's image in Britain had suffered a sea change. In the early 1850s California was regarded as a desirable place for Britons to immigrate to and as a land of investment and opportunity. Beginning with the vigilante crisis of 1856, however, California acquired the reputation of a lawless state where human life and property were not safe. Even in peacetime the British would have required hard persuading, and there would be no peace.[4]

Frémont was to focus on Paris, Billings was to concentrate on England, and A. A. Selover, who had a stake in Las Mariposas and crossed the Atlantic with Billings, was to take over from Frémont when the colonel was appointed minister, should the rumors prove correct. In Britain Billings drew on every contact he had—there were not many—and on those that Frémont, or the American legation (where the secretary took a liking to him) could provide, to create an ever widening circle of acquaintances. His best contacts were fellow Vermonter Henry Stevens, a distant cousin, and Stevens's younger brother, Benjamin Franklin Stevens. Billings was able to begin what he did so well: networking. Using these contacts, he distributed the handsome prospectus that he had had printed in London, and thrust upon friend and financier alike one of the two stereopticons he had with him so that they might see for themselves his many Mariposa views.[5]

For a month Billings was optimistic, finding the financial community "supportive" and the *Times* of London, the thunderer that shaped so much public opinion, favorable toward Las Mariposas. So too were the several members of Parliament with whom he dined, he reported to Park, and since London was "the center of the world" he surely would succeed. He was shrewd enough to soon see that the British were merely being courteous, however, and concluded that they neither trusted Lincoln nor believed that he could avert war. When the news reached London of the firing on Fort Sumter—news that did not arrive until nearly two weeks after the event because the Atlantic cable had been broken—he met with a good bit of ridicule as well, since no one could understand how the fort could have been bombarded for thirty-four hours without the loss of a single life. Whether there was to be a real war or a comic opera farce, the British public appeared to Billings essentially pro-Southern.

Having found London unresponsive, Billings then journeyed to Paris, where Selover was at work. Frémont was now moving back and forth between the two capitals, with Las Mariposas having slipped toward the back of his mind; he had written to Lincoln to offer his services and was preparing to depart for Washington. Selover had succeeded in getting Emperor Napoleon III to examine a set of Watkins's photographs, and wanted Billings to help him. Frémont had not received the ministry and now France would require much cultivation on behalf of Las Mariposas.

The French seemed open and friendly, and Billings fell in love with Paris, which he pronounced the most beautiful city in the world, and with the Louvre, where he dragged a somewhat reluctant Oliver. Compared to Paris, he wrote, London was "a sort of coal hole with all the dwellers heavy with beer." Money, however, was even tighter in France than in Britain, and when the Bank of England moved its rate of discount from 6 to 5 percent, he returned to London, where he had a fruitless meeting with the deputy governor of the bank.

By June Billings was convinced he was not going to succeed. He had met with the secretary of the Board of Trade, Sir James Emerson Tennent, and dined at Sir James's home, where he was regaled with tales of Tennent's governorship of Ceylon. He had taken breakfast with the noted social arbiter Monckton Milnes on several occasions, but Milnes directed the conversation to what the United States might learn from the Crimean War or picked Billings's brain for legal advice. He had spent a weekend with the banker and philanthropist, American-born George Peabody, who warned him that the British took a long time to get to know. He had become friends with Sir Henry Holland, Queen Victoria's personal physician, and had frequented the Royal Geographical Society, meeting with Admiral William Smythe and the Franco-American adventurer Paul Du Chaillu, who was recently back from his explorations on the Gabon River. He had talked with James McHenry, Irish-born but Philadelphia-bred, who had established a business in Liverpool in 1853, a man eager to discuss investments of all kinds. He had been taken under the wing of the American wife of the Belgian minister to London, Sylvain Van de

Weyer, who knew everyone, having been on post since 1831. Mrs. Van de Weyer was a daughter of Joshua Bates, a senior partner in Barings, and she and her husband lived in a grand house in Portland Place, where she "gave advice to stray Americans." None of this social life, enjoyable as it was, was helping Las Mariposas, and Billings was becoming more and more restive, for he wanted to do something for the war effort.[6]

On the whole, however, these were not the right people for the task at hand, though Billings, new to British society, did not sense this until too late. Stevens, in particular, was a mixed blessing: while many liked him, an equal many did not, and to be included at one of the parties he hosted at his Camden Town house was to be cut out of another party somewhere else. The British called Stevens the Yankee Peddler without a shop, and the circles to which he had access were generally pro-Northern, so that by going through him Billings had little chance to reach the pro-Southern aristocracy who, he soon learned, needed persuading of the rightness of the Union cause.[7]

Billings concluded that his place was back in California; he was accomplishing nothing on behalf of Las Mariposas and felt he was out of his depth. He would, Billings decided, return home, after a quick run to Scotland and Ireland. In Glasgow he persuaded Henry D. Rogers, a distinguished mining chemist, to examine Las Mariposas' records with a view to endorsing the value of the Princeton Mine, and tried, through the Chamber of Commerce, to see whether anyone in Scotland was interested in investing.[8]

In Dublin Billings received word that Frémont had been appointed to the command of the Department of the West, with headquarters in St. Louis, and that the Mariposa mission was converted to an arms- and munitions-buying expedition on behalf of his army. Billings felt certain he would be successful this time. Much of his effort already had been given to speaking out for the North, for social conversation was about the war, not about mines. He had paid to have a series of articles from the *New York Times* reprinted as a pamphlet in order to provide statistics on the relative productivity of free soil and slave states.[9] Through Henry Stevens he also had arranged for the printing and distribution of the farewell address Charles Francis Adams had delivered to his Massachusetts constituents before Adams left Congress to take up his diplomatic post in London, for Billings felt this, and a similar speech by Edward Everett, a former American minister to the Court of St. James's, which he included in the pamphlet, were admirable statements of the North's case. Now he would do something more substantive.[10]

Billings felt certain that British private enterprise would sell to the highest bidder. There were, he had learned, many ways to subvert Britain's neutrality act, which was viewed in the North as pro-Southern. The North was buying saltpeter to forestall the South; it was shipped on letters of credit on Brown & Company of New York City in small parcels to Holland and then on to the

States. The saltpeter came from Calcutta and was, in effect, being transhipped, so British laws were being broken. The North was also buying up lead, since most of the American supply came from Missouri and was going to the South. The Northern merchants had borrowed a leaf from the Crimean War, importing large quantities of red paint which could be restored to its metallic state with the loss of only two pounds out of twenty. Billings wrestled, though not much, with the question of whether all of this was illegal and concluded that while British exporters might be breaking the law, American importers were not. He did know, however, that the British were talking of tightening their legislation, and if purchases were to be made they would have to be done so quickly.

Frémont realized that the North had insufficient arms, and though he had neither authority nor money, he examined the latest in rifles, field-guns, and ammunition while in Britain and France, so that he was able to tell Billings and the Stevens brothers precisely what he wanted. In London Minister Adams, and in Belgium the new minister, Henry S. Sanford, offered their support. Frémont had contracted for $75,000 worth of cannon and shells in England and had ordered ten thousand rifles in France; Adams agreed to pay for them. (Seward subsequently approved the draft and authorized an additional $175,000.) Billings hurried from London to Frankfurt, by way of Paris, with Sanford instructing him to double the arms order if possible, using drafts on Adams, who made Billings an official bearer of dispatches so that he might travel freely on government business. His entire mission had changed from promoting Las Mariposas to arms-purchasing and even diplomacy.[11]

Billings settled into a small apartment on the Avenue Montaigne and pursued his two goals, pressing Las Mariposas on the comte de la Garde and Baron Rothschild while negotiating delivery of the promised rifles which, Sanford warned, Frémont had not in fact pinned down. To Park, Billings confessed that Las Mariposas was dead: "if I had a million [friends in France] and the estate was all gold, I don't think I could work up anybody to the scratch." By the end of the summer, Billings had completed his purchases and the arms were on the way.

Billings and the Stevens brothers achieved a good bit before George F. L. Schuyler, the first official commissioned by the Northern government and dispatched expressly for the purpose of buying arms, arrived in London on August 14, 1861. By then Confederate arms procurers had been on the scene for three months and suppliers were raising the prices well above anything Frémont or Billings had anticipated. Further, the individual states had placed orders with gun shops in Birmingham and London that were far larger than anything Billings was authorized to handle. While Frémont had ordered before leaving, he left it to the two men to make payment, accept delivery, and ship purchases to America, with Henry's brother Simon to receive them in

New York and send them on to Frémont in St. Louis. Billings and the Stevenses were successful in shipping 204 cases of artillery shells from England, and from France 450 revolvers, 1,500 rifles, and 102,000 cartouches with two million percussion caps. Not yet shipped by the time Billings left, though securely into the pipeline, were five hundred more rifles, eight six-pounders with shells, and an indeterminate number of horses.[12]

As it turned out, Billings was fortunate that he was not involved after August. Failed orders, prices that shifted on the winds of the news or the currency of the discreet bribe, and what he believed to be a distinct tendency on the part of the British and French to favor the Confederates had inflated prices beyond his capacity to pay, and once Adams knew that Schuyler was on the way, he would not, indeed could not, approve any further drafts for Stevens and Billings.

As receiver of the initial shipment of arms, Simon Stevens had seen an opportunity for a killing. He learned that five thousand Hall carbines would be purchased for $12.50 apiece. Stevens called on J. Pierpont Morgan, then only twenty-four, who knew of Frémont's purchases through his father, J. S. Morgan, a partner of George Peabody. Morgan had just handled a final draft from Billings for £1,000, and he believed Simon Stevens thoroughly reputable through his connection with Billings and Henry Stevens. He loaned him $20,000, with the rifles as collateral. Shortly it became known that Secretary of War Simon Cameron had already arranged to purchase the rifles at $3.50 each, on the ground that they were outmoded. Simon took possession of them and had a New York gunsmith rifle most of them for $.75 each before he sent half the order on to St. Louis. Almost at once a select House Committee appointed to look into government contracts got wind of the deal and in September took testimony on it. Simon Stevens was censured, the press had a field day writing about merchants of death, and though Frémont apparently did not know that he had initiated the sordid affair, he too came in for angry public comment. Billings was tired of the entire unseemly arms business and wanted no more to do with it.[13]

Billings left for the States on August 24, 1861, to confer with Frémont and, he hoped, to meet with the president. Sanford and Charles Wilson, first secretary of the British legation, supplied Billings with letters of introduction to Lincoln, urging the president and secretary of state to confer with him about public attitudes in Europe toward the North. Billings probably never had an opportunity to present these letters to Lincoln, for by the time he was in Washington Frémont had issued an ill-timed and unilateral emancipation proclamation, by which he declared free all slaves coming into his lines from Confederate-held territory, creating consternation among all but the most dedicated abolitionists. Billings was so closely identified with Frémont, he may have thought it unwise to use the letters, at least until Frémont had been rehabilitated.[14]

Billings had hoped to get back to California quickly, but this was not to be. Arriving in Boston on September 5, he went home to Woodstock, only to receive an importunate telegram from Frémont to meet him at Willard's Hotel in Washington. Billings discussed with Frémont the financial situation at Las Mariposas, proposing an exchange of bonds for stock. After Frémont returned to his command, Billings set out to escort Lizzie Patterson and her children to Vermont, when yet another telegram summoned him, this time to St. Louis. Sending the Pattersons on ahead, he hurried westwards, to find that Frémont wanted his advice on how he was running his department and how best to defend himself against his attackers, rather than on business matters. Frémont also wanted him to return to Europe to purchase more arms, and Billings at first opposed this, since there was a competent government agent already there. Billings felt Frémont had lost the public: the Hall carbine affair, his aloofness, his emancipation proclamation, by which he had moved ahead of the Cabinet in Washington, and the plain fact that he had won no military victories meant that probably he would be removed.[15]

Still, the weary Billings returned to Washington to make one more effort on Frémont's behalf, and in doing so knowingly gave up any real chance of appointive office, at least until another throw of the dice. One charge against Frémont was that of favoritism toward Californians, so Billings rather opportunistically once again referred to himself in all situations as "F. Billings, Vermont." He pleaded that Frémont must either be sustained and upheld or be "slaughtered." He believed in Frémont as a fighting man, whatever he thought of him in other ways. All was to no avail, however, for on October 24 Frémont was relieved of his command. Billings knew he would be held accountable for the company he kept. He did not know that Frémont, while still seeking supplies for St. Louis, had proposed that Billings should be sent to Canada to purchase horses at a price of $30 a head more than he could obtain them for in nearby Quincy, Illinois, and though Billings had never been consulted on this proposal, the War Department might think he was part of an unseemly speculation.[16]

There was no point in going to Britain for Las Mariposas again. On November 8 a U.S. sloop of war had stopped a British mail packet, the *Trent*, in the Bahama Channel, and forcibly removed James M. Mason of Virginia, the Confederate States' special commissioner to Great Britain, and John Slidell of Louisiana, destined for an identical position in France. British public opinion was outraged at what was viewed as a boldfaced kidnapping of diplomatic personnel and a violation of British neutrality by an armed vessel. The British fleet was put in readiness and preparations set in train to send troops to Canada in case of an outbreak of war with the Lincoln government, to which a demand for an apology and the return of Mason and Slidell were sent. The draft instructions to the British ambassador in Washington, Lord Lyons, were softened at the last minute by Albert the Prince Consort, but it remained a cold challenge to Lincoln. An American who had long lived in London wrote to Seward that "999 men out of a thousand would declare for immediate war."[17]

Accordingly Billings sailed for Havre on December 9 with instructions to proceed to Bremen, carrying letters of introduction to German bankers and financiers to make a final effort on behalf of Las Mariposas. He met with the American consul and Minister Adams, who was visiting, and called on J. Ross Browne, who was in Germany collecting material for a book. Briefed, Billings went on to Hamburg, Heidelberg, Mannheim, and elsewhere, met with Baron Erlanger and several financiers, and received the unhappy report that they would do nothing to help him.

Billings put most of his hope in the elector of Hesse-Kassel, who allegedly "had millions he didn't know what to do with" and had been thoroughly "indoctrinated with the Mariposas." The German financiers Signitz & Adelberg had written to Signitz' brother in New York, who in turn had been endorsed by the powerful banker August Belmont, and they reported that the elector wished to invest in Las Mariposas for the benefit of his son-in-law, a German count. Billings was told that the German mining inspector, Ulrich von Bieber, already approved negotiations and was looking forward to a visit to the mines, and that Signitz, acting as broker, had obtained the endorsement of other leading German bankers. Thus Billings remained in good spirits until he reached Frankfurt.

The German *affaire* proved to be "a most egregious humbug." Signitz lived four hours from Frankfurt, a snowy rough journey, and was neither a broker nor someone in contact with moneyed men. He was, in fact, simply a small merchant pulled up by his New York–based brother to impress Belmont with his own connections. Having spent the money to go to Germany, Billings sat down to write to potential investors in Leipzig, Hamburg, and Geneva, to be told that no one wished any business in California, convinced as they were that the Pacific Coast might not be part of the Union. Billings learned the truth that all fund raisers must face: the notion that men of vast wealth do not know what they want to do with their money is a myth. He was also thoroughly sick of Europe: "Every step I take in the Old World makes me love the New the More," he wrote. Abruptly, upon hearing negative tidings from Geneva, he returned to the States.

Billings was deeply worried about Las Mariposas, for he had a goodly sum sunk in it, and Park's reports were increasingly gloomy. The mines could swallow the Montgomery Block up and leave him a broken man. There had, in fact, been offers—the French, for example, had said they would loan $1,000,000 for three years at 5 percent per annum, with the property to be turned over to them for management and, at the end of the three years, one-half of the estate to pass to them—which, if accepted, would ruin them all. As no succor was to be found in Europe, he would try Washington one more time. There, too, he failed, and in late February Frederick and his mother settled in at the Astor House in New York City. Any solution to the Mariposa problem would have to be found there.[18]

This visit to New York changed Billings's life. On Sunday, March 2, 1862, Frémont showed up at the hotel and asked Billings to stay behind to discuss Las Mariposas with him while Sophia Billings went on to church alone. The next day, perhaps feeling a little guilty, Frederick agreed to accompany his mother on a visit to an old friend, Eleazar Parmly. At his house Frederick met Parmly's unmarried daughter Julia—a chance encounter, as she was at home only because it was raining and she wanted to wash her hair. His diary for March 3 contains the single entry: "Memorable above all other days as the day when I first saw Julia Parmly (no other fact can be recorded.)"[19]

Cliché or not, it would be love at first sight for both Frederick and Julia, and arms purchases, Las Mariposas, and Frémont were utterly forgotten.

15

Julia

FREDERICK BILLING'S PARENTS AND FRIENDS had been worried that he was almost past the age when romantic love might fell him. They knew he would not take a wife for business or purely social reasons; he had written to his mother, when she pressed him uncommonly hard about a candidate, that he could marry only for love.[1] He liked the company of women, he professed to want the kind of family life he enjoyed when visiting with the "band of brothers" from California, and, except for Laura's marriage, appears to have been genuinely pleased when each of his brothers and sisters left the family circle to marry. At thirty-seven he was on the verge of permanent bachelorhood. Numerous pictures survive of Frederick Billings. His mother thought of him as one of her "two handsome children." Women who tried to matchmake for him always assumed his wife would have to be intelligent, witty, and attractive, since they found him so. He was, at least in 1861, and according to a certificate of safe passage issued to him in lieu of a passport, five feet, seven inches tall—not particularly small for the time—with a full forehead, brown eyes, a rather large mouth, brown well-trimmed hair, a ruddy clean-shaven complexion, and an "ordinary nose." Despite the nose, he was decidedly a catch.[2]

Frederick was always popular with the ladies. He seemed to come even more alive in their presence, for his bantering, happy speech, his quick turn of phrase and pun, his well-stocked mind and solicitous manner won over women. The wives of good friends were always telling him he was their second favorite man. Perhaps his relationship with his mother, to whom he brought a sense of high spirits in the presence of a somewhat dour father; perhaps the need to cheer his sisters when times were relatively grim and there was no one else to do it; perhaps the simple fact that he lived in an environment and a time when men were in demand—after all, the bloodletting of the Civil War eliminated an entire generation of males in both the Union and the Confederacy—perhaps all this made him naturally flirtatious in a way that women at once recognized as unthreatening. Whatever chemistry was at work, the ladies—young and old, happily married and on the lookout—loved him, forgave him, wrote to him, sighed over his picture (or said they did), defended him to their friends, and were happy in his company and pleased by his attention. He was, after all, a handsome, rather jaunty man, intelligent and worldly wise, and very successful.

Frederick clearly was of two minds. He believed that marriage meant large families, hard work to provide for wife and children, and a happy and close home life. He was fairly obviously smitten with Lizzie Patterson and, in a different way, with Jessie Frémont. He had confessed to his mother that he could not understand why Captain Patterson was away so much: if he ever got married, Frederick wrote, he would "stick to his wife like wax." Yet he enjoyed his freedom, the chance to dash off to San Mateo or the Napa Valley on the spur of the moment, and the company of men in the clubs and libraries of San Francisco, and when he did marry, though he was utterly faithful, he was frequently away. Every time names came up, he thought them not quite right.[3]

Paradoxically, Billings's notable pleasure in children may also have contributed to his reluctance to marry. He had a way with children and they were attracted to him, yet he had seen so many die, and so many parents thrown into grief that broke them, that he hesitated to risk this himself. The frequency of child-death—at birth, during the first year, two or more at a time as cholera or some other disease ran its course—had often put mothers into a "glorious frame of mind" that was, he felt, next to madness. Though they endured by seeking spiritual consolation, he was not certain that he could be as strong, should he have to live through the death of a beloved child. No one was exempt from such pain: his mother, his sister Lizzie, the Watsons, the Pattersons, all lost children prematurely. His sister-in-law Mary Parmly Ward would have seven and lose three. No one quite knew why this must be so—public understanding of germ transmission would not be widespread until the 1890s, and neither boiled instruments nor clean beds were viewed as important to childbirth—and everyone knew that for the woman the experience of giving birth could be excruciatingly painful, with labor often a dreadful ordeal that could last through a long day into the night, deliveries performed almost always at home with the aid of a midwife, ergot administered sometimes far too heavily to ease delivery, and anesthetics limited.[4]

The family had a phrase for those times when, sunk in introspection, Frederick seemed worried because he had not found the right person: he had a touch of autumn leaves.[5] Twice before he met Julia Parmly his reading of those leaves seems to have turned a moderately serious interest away. There was, first, Agnes Bates, one of the children of the Reverend Joshua Bates, the president of Middlebury College. Agnes was the youngest of the Bates daughters, and Frederick apparently met her when in Woodstock for his sister Lizzie's wedding in 1853. Agnes pretty clearly fell in love with him, but Frederick did not reciprocate and in 1857 told his parents so quite directly, declaring that Agnes was a good friend and nothing more.[6]

Sophia Billings had another candidate in mind in any case, and in this she was supported by one of Frederick's long-time Woodstock friends, Eben Whitmore. Sophia knew the Eleazar Parmly family in New York City quite well and visited with them often. Eleazar had boarded with the Levi Mowers in Montreal at the same time as Sophia and had occasionally escorted her out to tea. Frederick corresponded with Mary Parmly, Eleazar's second daughter, born in 1831, and appeared to have been interested in her. Whitmore tried to help Frederick conduct his courtship from a distance, even drafting a letter of proposal, but nothing came of this, and Mary became the wife of Charles Ward, a wealthy and sophisticated banker who shared her interest in European travel.[7]

Frederick may well have forgotten that Eben Whitmore had casually mentioned Julia Parmly to him in 1855, when she was a bridesmaid at her brother Ehrick's wedding. The auburn-haired young woman had "looked very beautiful," Eben wrote, and "held her head up like a wild deer in hot pursuit." She, too, might well marry soon, he noted. But Frederick had not shown any interest, and when Mother Parmly was taken ill two days before Mary's wedding and died soon after, Julia settled in to look after her father in the dutiful manner of the times.

Eleazar Parmly had a flourishing society dental practice on Bond Street in New York City. At the time of Frederick's admission to the family circle, Parmly probably was the best known figure in the United States in dental surgery, having played a central (and mistaken) role in the "great amalgam war," concerning the desirable qualities for fillings in teeth, and having founded the *American Journal of Dental Science*. He was also a lay preacher in the Disciples of Christ, a forceful public speaker, an ardent advocate of prison reform, and a prolific amateur poet, producing long verses for virtually any occasion, including somewhat improbably titled stanzas to the dental profession. At the time of his death in 1874, Eleazar Parmly was "the wealthiest dentist this country has ever known." He was also, wrote a medical historian in 1923, the person who had done the most for the profession of dentistry in his time, perhaps "the greatest dentist in history."[8]

Julia's mother Anna Maria had been an orphan. The daughter of Captain William Smith of Bermuda and Herriot Valk of Charleston, South Carolina, she had lost her father six months before she was born, when the vessel he commanded was wrecked on the bar outward bound from Charleston harbor. Her mother died within the year, and Anna Maria was taken in by a Charleston couple, John and Peggy Ehrick, who moved to New York City, where she grew up. She married Eleazar Parmly in 1827, and the widowed Mrs. Ehrick lived with the Parmlys until her death in 1832. Twice bereft of parents, Anna Maria placed a great emphasis on family ties that she passed to her daughter, for when it came time to think about leaving her father, sisters, uncles, and cousins behind in New York City, Julia would sense a great void opening up before her, which only Frederick as her constant companion might fill.[9]

Certainly Julia had contemplated marriage before she met Frederick. In 1862 she was twenty-seven, a bit older than most young women who married for the first time. There had been admirers before—Anton Wenzler, a painter, and the writer George W. Curtis—but she always treated potential suitors coolly, discussing books with them and keeping up a lively discourse that obviously fascinated men who wanted to marry an intellectual equal. Her friend Mary Caswell told her that she knew her fate would be to marry "some creature compact of fire and spirit, who would come down like a falling star."

Enter Frederick Billings. Apparently the night after she met Frederick, Julia told a suitor, John Sherwood—who had just given a ball in her honor that evening at which he had proposed to her—that she had met someone else, though Frederick had not yet proposed himself. Somehow she knew he would, and she planned to accept.

Pictures of Julia show a slim, well-turned-out young woman with clear eyes and an attractive expression, pensive in the fashion of the time, though clearly full of life. She almost always was in good health (except for a worrying spell in 1872) even though she took far too much on herself, and she often seemed to feel inadequate. "If you had only exerted your influence over me before my character was set, I might have been tolerably methodical," she would later write to Frederick when he was away on one of his many trips. She called herself "incompetent," she worried that she had offended her husband for not looking pretty enough, she blamed herself when their thirteen-year-old son lost his Latin books. She was, in fact, quite methodical and very able.[10]

Both Julia and Frederick were decisive people. Frederick was smitten at once, and for years afterwards would record in his diary the anniversary of the day he first met Julia. He laid siege, vigorously and persuasively, and would brook no opposition, postponing his intended return to California and calling almost daily. They talked and walked, exchanging notes, progressing from "Mr. Billings" to "My Beloved Fred" in less than three weeks. He was "half sick and half in love," he wrote his father. Julia decided that she must do nothing hasty and ought not to marry him, but on the night she was determined to tell Frederick this, "He came and when he had left I remembered I had forgotten to tell him of my decision."

On March 31, Frederick and Julia were married in the Parmly home on West 38th Street. Julia remembered that she was in white silk and tulle veil; Frederick could not recall what he wore and was "the most agitated bridegroom . . . ever," so beside himself he could hardly speak, according to one observer. The next day Billings sent a telegram to San Francisco to announce to his world that he and Julia were married, and they left for a brief honeymoon in Philadelphia (where they read Longfellow's "Courtship of Miles Standish" to each other) and returned to New York to pack for the trip to California. They set everything right with all those "whose noses were out of joint" for not being invited to the wedding with a grand reception on the evening of April 9 at which the Philharmonic played and five hundred friends and wellwishers filed by the happy couple. On April 11, 1862, Frederick and Julia sailed for California on the *Northern Light*, seen off by a tiny band of family and friends.[11]

With marriage to Julia Parmly, Frederick acquired not only an entire new set of relatives and friends, but also access to the world of New York society, of international as well as domestic art. Eleazar Parmly was a rich man—quite possibly as rich as Frederick, for he owned over a hundred properties in New York, New Jersey, and Ohio, a handsome townhouse, patents on several significant inventions for oral surgery, and good investments. When he died he left a fortune of $7,000,000.[12] Eleazar knew the Vanderbilts, the Roosevelts, the Astors. Julia had visited Washington Irving at Sunnyside; Catherine Maria Sedgwick, the creator of the domestic novel, often dined in the Parmly home. John Brown, the martyr of Harper's Ferry, had taken a meal at their table after his attack on Pottawatomie. There also were many cousins—and cousins counted for much in the days of arranged marriages, with much visiting about, constant letters asking people to be attentive to the needs of this or that one, spinster ladies staying for extended periods.

One of Julia's interests was religion, through which she provided an additional circle of church friends such as Julia Ward Howe, to whom she was drawn by distant kinship and the teachings of the Disciples of Christ. The Disciples of Christ were dedicated to eliminating the divisiveness in Protestant sectarianism and were on the way to becoming the sixth largest religious denomination in the United States by the end of the century. Begun among reforming Baptists in western Pennsylvania, eastern Ohio, and western Virginia, and led by Thomas and Alexander Campbell (and thus sometimes called Campbellites), the Disciples believed in the Bible as the sole spiritual authority. Though Julia read many sermons, methodically clipping those she liked and preparing scrapbooks, she would, twice a day, turn to the Bible for sustenance.

In a time when religion was often a contentious matter, Frederick and Julia appear to have experienced little strain over the subject. However, Julia read to moral purpose, not for entertainment, and while she liked to hear Frederick laugh aloud as he sat with a volume of Josh Billings, his humorous namesake, in his hands, she was made of sterner stuff. She refused to read Harriet Beecher

Stowe's book about Lord Byron, a scandal when it was published in 1869 for openly hinting at the poet's incestuous relationship with his sister. It was not the author's place "to reveal such a horror," Julia thought. After the birth of their last child, Julia turned increasingly to evangelical religion, admitting Dwight L. Moody to her scrapbook and her drawing room, and supporting the Northfield Seminary and the Mount Hermon School when he founded them in 1870 and 1881 respectively.[13]

There were other interests, however, for she was an intelligent and inquiring woman. For Frederick and her relatives Julia often was an arbiter of taste, appearing to defer to her husband yet usually getting her way. He had begun to acquire his first works of art through Henry Stevens when in London in 1861, bringing back some prints—a Murillo, a Raphael, a Rosa Bonheur, a Benjamin West—and through her father's contacts in the world of art, Julia introduced Frederick to the purchasing of originals on canvas. He would buy his own Albert Bierstadt, perhaps to bring a bit of California into the Woodstock home, perhaps because he remembered in old age a remarkable evening at Jessie Frémont's when she unveiled a Bierstadt with the artist present. He would acquire paintings by Asher Durand, John Kensett, William Keith, and Thomas Cole. Julia would establish the ambience of home, and so far as one can tell from the correspondence, when they differed she was almost always right. In 1870 a decorator proposed a theatrical plan for her sitting room, crimson and gold paper, a rose and salmon ceiling, cornice lines and corners "allegorically treated," and she expressed her horror at the whole thing. Frederick would enlarge and enhance the Marsh house in Woodstock and would be credited with it, but it was in fact a statement fully as much of Julia's taste as of his.[14]

Frederick had said he would want "many children" if he married; he and Julia would, in fact, have seven. For Julia, interest in her children would be constant in her marriage. Frederick would be away from his children and home a great deal, especially when the family was in Woodstock and he in New York City. Over the years he would often miss birthdays, Julia's as well as the children's, and he was not at hand to cope with childhood illnesses. Julia would dutifully write him thank-you notes for presents and tell him of their sleigh rides and their hope that he would come home soon.

One may wonder about these absences. After a home visit there usually would be a letter exchange in which Frederick would express his sorrow that he had to be away, or that he had been cross, distracted, or irritable, and Julia would reply that he had been no such thing, or that she was certain the children (or servants) had not noticed it, or that his irritation was quite understandable and that she should have had the children more properly dressed or kept them quieter. While Frederick doted on his children, at least from afar, he does not appear to have enjoyed their company much, contrary to his experience in California with the children of friends. Of course, Frederick was an older father—fifty-two when his last son, Richard, was born and forty when he and Julia had their first child—and he must have worried a bit about the age gap even before the wedding, for Robert Watson had written to tell him he must not fear the difference in age between himself and Julia.[15]

Over the years, Julia would provide most of the parenting. This was customary at the time, and certainly so in their circle. Frederick would room at the Brevoort Hotel in New York, even when Julia brought the children down to stay at her father's house, until he bought a townhouse on Madison Avenue in 1881. He seldom accompanied the family to the Parmly summer house, Bingham Place, a large and rather ornate affair with a rolling lawn in Oceanic, New Jersey. Julia would write, adding notes from relatives or friends, giving news of the children, their illnesses and bruises, reporting on the servants—generally in cheerful terms, though there is a cook with a temper whose gravies are greasy, and Julia takes four letters to fire her—and the weather, passing along neighborhood gossip (Mr. and Mrs. This-and-So were to begin divorce proceedings but "have made their peace over a quart of oysters"), and commenting on Frederick's own rather more cursory missives. Parenting, and being the wife Frederick wished, needed, and loved, filled much of Julia's day.[16]

Frederick was determined that the first child be a girl, and he wanted to name it Julia, while she opted for Laura. Notwithstanding, the first child was a boy, Parmly, born in San Francisco in February, 1863, followed in 1864 by Laura (the name on which the proud parents agreed in the end). In 1866 Frederick Jr. ("Fritz") was born, then in August 1869 Mary Montagu, named for Julia's older sister, followed by Elizabeth in 1871, Ehrick in 1872, and Richard in 1875.

Frederick always thought the girls were quite special. Laura, a precocious child with excellent spelling, grammar, and penmanship far beyond her age, would not marry until she was in her late thirties, and she and her husband, Frederic Schiller Lee, a professor at Columbia University, would lead a varied, full life. Mary Montagu, a bright, attractive, somewhat frail young woman of thoughtful bent, often travelled with her mother until she married John French, a highly successful lawyer, in 1907. She was close to her parents, editing her father's letters from Mexico for publication in the 1930s and, until her mother's death in 1914, working with her on family history. Elizabeth, the third daughter, would remain unmarried, living at home in Woodstock, helping with the grandchildren when they came to visit, writing frequently on wildflowers and gardening and, like her grandfather Eleazar, composing poetry, which was occasionally published.

Julia and Frederick Billings would lose two of their sons, Parmly and Ehrick, in the space of seventeen months, the first in 1888 and the second in 1889. Frederick was devastated by their deaths. A phrase he used about his old friend Charles L. Dana, that "the springs of his life were all drying up," applied to him from the time Parmly died,[17] and when Ehrick followed, Frederick would, within two months, have an incapacitating stroke from which he never recovered. He dearly loved his daughters and enjoyed their company, but in keeping with the expectations of Victorian fathers, it was for his sons he had the highest hopes. Much is revealed about him, and his marriage, from his approach to their rearing.

Parmly, the first-born, and his parents never quite emerged from a cloud of mutual incomprehension. He was put into a neck and chin brace as a small child, enduring this discomfort and humiliation because Frederick and Julia were convinced by a doctor that the shape of his chin was not right. By the time he was six, he would fall into "consequential moods," acting badly whenever his father was away, which was frequently. Still, he had a good sense of self-deprecating humor, and a great deal of charm.

One senses that Parmly tried hard to win the approval of his parents. He wrote to his father, in formal prose, that he would do all that he could to make life pleasant for his mother—this when he was nine years old—but his high spirits too often burst forth. He wanted, and got, a pony and was most proud of it; he invented a squirrel trap of which he was equally proud; and he pressed his father very hard to allow him a pair of ice skates, even getting a note from Frederick's physician to testify that skating would not hurt his neck or back. The correspondence between father and son is a revealing, sad testimony to how generations reach out and do not quite touch. When away at college Parmly would write charming, apologetic letters in the manner of a student, rather as Frederick had done to his own parents, but when he accounted for $114 his father had sent him, Frederick lashed out about $12 still unmentioned. Parmly was not above playing on his father's sense of guilt about his absences, invoking his bad back to win the gift of a boat while away at camp. As to the skates, he would get them, though only after Frederick, as an overly cautious and middle-aged first-time father, had agonized over the decision until much of the pleasure had gone out of the gift.

By the time Parmly was eighteen he was drinking and playing poker to the point that a despairing Frederick concluded that it was "easier to bury a child than to see him live to go astray." A year at Williams College went badly, and Parmly transferred to Amherst. In the summer of 1883 Frederick and Julia, deeply worried, engaged a tutor to take Parmly to Europe to "improve his manners and attitude." On board ship Parmly spent his time playing cards with a young woman who claimed to be a widow, to the horror of the tutor, though happily he met two young ladies of the "right type" and discovered a sudden interest in the Paris opera in their company. While Parmly learned a bit of French, he did not attend to the sights as his tutor wished—perhaps understandably, since the tutor's notion of a good time appears to have been to go to the local French Protestant Church and the prison. Still, across the summer a bit of culture rubbed off, and by the time the travellers reached Ireland, they had achieved a sufficiently happy relationship for the tutor to tell Parmly that he hardly needed to kiss the Blarney Stone.

Soon after graduation Parmly headed west to try his hand at the "privations of a frontier life," according to his sister Elizabeth, and for "health, fortune and the good life," by his own word. His first business venture went badly, however: he purchased horses in Washington Territory, intending to drive them through to Jamestown, North Dakota to sell. The horses broke loose and it took Parmly, who was frail and working almost alone, ten days to round them up, by which time it was too late to make delivery. He appealed to his father that he would be bankrupt unless he could have a loan of $1,700, advanced at 5 percent interest, which carried him through the year. In 1884 he filed for a preemption claim, but since he had not built a cabin on the property, he was challenged by a claim jumper, to whom he yielded after spending a night at the claim in a tiny shack that he and two others ran up in a day.

Parmly then entered into a partnership with his cousin Edward Goldsmith Bailey. Bailey had hoped to buy a ranch near Miles City but had not been able to raise the funds; after working as a land agent and engineer for the Northern Pacific Railroad, he joined with Parmly in the growing town of Billings—named after Frederick because he was president of the Northern Pacific at the moment the surveyors reached the tent city—and together they established a successful supply business and, in 1886, a bank. Parmly was still eager to prove himself to his father, and spent much of his time in local politics, once being asked to be candidate for mayor, though he declined to run. By 1887 Bailey and Billings was on the verge of dissolution because Parmly lacked the "quiet persistence" necessary for success. Rather, he seemed to be spending his time with the ladies, and having his photograph taken in woolly chaps as though he were a cowboy.[18]

On May 7, 1888, on his way east to visit his parents, having just returned from a trip to Alaska, Parmly became ill in Chicago. Friends alerted Frederick and Julia, who caught the first available train, but they arrived too late, Parmly having died from "congestion of the kidneys," or uremic poisoning. One finds today, in a box in the Mansion attic, all of Parmly's school books methodically stored away, most likely by Julia's hand. There is a metallic chemistry exercise book, passing as his, yet clearly loaned by a fraternity brother, mute testimony to the borrowed life of this attractive, rather wistful romantic who reminded Frederick of his own brother Richard.

Ehrick died seventeen months later, on October 7, 1889, ten days short of his seventeenth birthday. He had been ill for some time and sickly since birth, and it was widely believed in the family that he had been much weakened by

Sophia Farwell Wetherbee Billings (1796–1870), Frederick Billings's mother.

Oel Billings (1788–1871), Frederick's father. Both portraits of his parents, by the locally famed artist Benjamin Franklin Mason, were painted in 1869 and now hang in the Mansion.

Frederick's sister Laura, c.1840, perhaps the first American woman to die in the gold rush.

Laura's husband Bezer Simmons (1810–1850). This portrait, attributed to Robert M. Platt, was prepared in 1849, shortly before Bezer, Laura, and Frederick sailed for California.

Evening in the Tropics, oil on canvas by the American artist Frederic E. Church (1881), purchased by Billings as a reminder of the difficult Chagres River crossing of Panama in 1849.

During the height of the gold rush, crews abandoned their vessels for the gold fields until the harbor in San Francisco was clogged with ships,

San Francisco's magnificent setting was scarcely matched by its uncontrolled and ugly urban environment. (From William Taylor, *California Life Illustrated* [New York: Published for the Author by Carlton & Porter, 1860].)

some of which served as hotels. This two-part view was taken early in 1851 from the roof of the Union Hotel on Portsmouth Square.

Fire was the greatest enemy to a city built almost entirely of wood. On May 4, 1851, Frederick Billings lost nearly all his possessions when a conflagration began in Clay Street and consumed over three-quarters of the city, causing nearly $12,000,000 in damage. (From *Gleason's Pictorial Drawing-Room Companion,* I [July 12, 1851].)

Panorama of San Francisco, taken in 1862 and attributed to the early photographer Carleton E. Watkins. This is the view to which Frederick brought his bride Julia in the spring of 1862.

In December 1853 the law firm of Halleck, Peachy, and Billings completed the Montgomery Block, shown here in an 1853 lithograph by Britton & Ray. It survived the great earthquake of 1906 but was reduced to rubble by a wrecker's ball in 1959.

Billings (left) and Trenor W. Park (1823–1882) in 1856.

Henry Wager Halleck (1815–1872) (seated) and Archibald Cary Peachy (1820–1883).

Jessie Benton Frémont (1824–1902).

Promotional photograph by Carleton Watkins showing the miner's town at New Almaden as seen from Cross Hill in 1863.

Billings sent the portfolio of Watkins's photographic expedition to the Yosemite valley to Professor Louis Agassiz of Harvard. These photographs helped convince Congress that the Yosemite valley should be set aside as a national preserve.

Julia Parmly (1835–1914), in an 1865 oil portrait by Anton H. Wenzler.

Billings posed for the famed photographer Mathew Brady in 1865 upon returning permanently to the East Coast.

Sixteen teams break the prairie at the Dalrymple Farm, twenty miles west of Fargo, Dakota Territory, in this 1878 photograph by F. Jay Haynes.

Construction of the line was delayed at the Big Cut west of Bismarck, Dakota Territory, by shifting surfaces, high winds, and an insufficient work force.

Grim little towns, harbingers of progress, grew up along the rail line, their tiny depots the focus of expectations. Haynes portrayed this sense of waiting at the depot in Bismarck in June 1877.

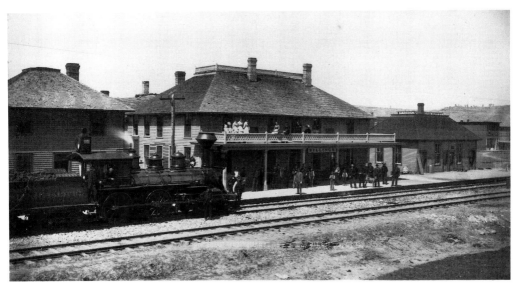

From the balcony of the Headquarters Hotel, photographed by Haynes in 1883, Frederick addressed the townspeople of Billings, the tent settlement named for him that would become the largest city in Montana.

Billings's son Parmly (1863–1888), shown in the inset in the year of his death, and nephew Edward Goldsmith Bailey (1857–1940) opened a private bank in Billings with Frederick's backing in 1886.

Henry Villard (1835–1900).

On September 8, 1883, at Gold Creek, Montana, the officials of the Northern Pacific Railroad drove the ritual "last spike," signifying that the line was complete from St. Paul to Portland. F. Jay Haynes was there to record the first train.

Promotional photograph by F. Jay Haynes of the Grand Canyon of the Yellowstone.

A wood engraving used on the 1884 Northern Pacific timetable to advertise the splendors of the new dining cars. (From Edward W. Nolan, *Northern Pacific Views: The Railroad Photography of F. J. Haynes, 1876–1905* [Helena: Montana Historical Society Press, 1983].)

The geysers and terraces of Yellowstone National Park were a major attraction, though they were ill-protected, as this Haynes photograph at Mammoth Hot Springs, showing a man standing on the travertine terraces, attests.

George Perkins Marsh (1801–1882) in 1875.

The Marsh estate (background) in Woodstock in the late 1860s.

The Powers Farm on Church Hill, Woodstock, *c.*1860–1865, representative of the denudation of the Vermont forests.

The Billings Mansion, shown here in a watercolor view made for Henry Swan Dana's book *History of Woodstock* in 1886.

South Park Street, Woodstock, *c.*1870, showing the house given by Frederick Billings to his parents.

Billings family portrait, including all the children, by E. R. Gates of Woodstock in 1884.

Woodstock from the slopes of Mount Tom, near the time of Billings's death.

Samuel E. Kilner (1858–1928)—"the faithful Kilner"—shown here in 1892.

George Aitken (1852–1910), shown here c.1890.

Julia Parmly Billings, her eldest daughter Laura (far right), her second daughter Mary Montagu, and her youngest daughter Elizabeth (far left), shown here in 1896. Mary and her husband John French had three children, one of whom, Mary, married Laurance Rockefeller in 1934.

being dragged by the arm up Mount Tom, behind the estate, by his nurse, though in fact he died of heart disease begun by a childhood attack of scarlet fever. Ehrick was the best student of the children, standing first in his class at the Berkeley School in New York for four consecutive years until his failing health forced him to withdraw. A boy of great promise, he had been destined to go to Dartmouth College, whose president presided at his funeral.[19]

Julia would be the matriarch of an extended family after Frederick's death, devoting untold hours to writing her many grandchildren, nephews and nieces, cousins, and surviving aunts and uncles, meting out advice and sound judgment, religious instruction, and reading lists. There would be six grandchildren, eleven nephews, and four nieces. Each year she would record in her diary the anniversary of Frederick's death, and when, in 1909, she carefully filled out a lifetime journal, she wrote at the close of the year, "Surely of my life I can say goodness & mercy have followed me & my kind & Frederick's have blest me in many ways & my children have given me devoted care."[20]

16

At a Crisis

IN FREDERICK'S ABSENCE Las Mariposas had become the money-eating monster he had feared. A quick visit to the Princeton Mine revealed that Park was paying out more than he was bringing in, and when Billings discovered that Park's monthly statements were grossly inaccurate, he concluded that his former law partner was guilty of serious mismanagement at the least and possibly of outright theft. The estate's creditors were owed $1,400,000 with no prospect of payment and they were compounding interest at 2 percent monthly. Parrott, one of the major partners, warned that Park's accounts were in error, and that his patience was near an end, especially since Park was continuing to take 5 percent of the Princeton Mine's proceeds in lieu of salary. Even Frémont urged Billings to take over from Park before he was "entirely robbed."[1]

Park defended himself vigorously. He could not be blamed for acts of God: the collapse of a mining shaft, the flooding of a river that broke through a crucial dam, the failure of machinery to arrive on time, a fire at the company store. Nor was he responsible for the wartime instability in gold or for the low productivity at the mine since the war depleted the supply of able miners. His records, he insisted, were as accurate as they could be. Frémont himself was an atrociously bad businessman, gambling recklessly with his share of the mine. In at least some of this defense Park was certainly correct, though one might argue—and in reply Billings did so—that late deliveries, a poor labor pool, and collapsing dams were due in some measure to management's decisions to postpone essential improvements or failure to work directly with the men. On October 31, 1862, Billings flatly charged Park with unethical behavior, and two weeks later demanded that Park come immediately to San Francisco with his record books and render a full accounting.[2]

What particularly angered Billings was his discovery that despite his instructions, Park was charging Frémont for bonds at par plus interest when they were buying other bonds at half the cost. Park also had factored in a large commission. He was, Billings apparently concluded, simply milking the cow until it was dry. "We are at a crisis," Billings told Park, and they must meet to decide how to get out of it. He referred not only to Las Mariposas but to their personal relationship.

Billings and Park met in San Francisco on November 19. No record from the meeting appears to have survived, but since Billings was soon negotiating the sale of Las Mariposas, their decision seems clear. John Parrott and his senior partner Nicholas Luning at first agreed to exchange claims on their substantial loan for $1,000,000 in stock and a portion of the estate, which was to be divided into eighteenths, that being the smallest fraction held by any one creditor, with Frémont to retain only three portions. Park was called down from Mariposa again, and upon his arrival, Parrott and Luning each demanded one-third of Frémont's interest. Hoping to avoid this erosion of Frémont's estate and prevent any one person from gaining more than four shares, Billings decided to look for potential purchasers in New York City, where he worked through his friend John Thomas Doyle, the scholarly lawyer who won Billings's trust by agreeing that Park had either cooked Las Mariposas' books or, at the least, had been guilty of grossly inadequate record-keeping.[3]

The sale of Las Mariposas was complicated by a number of encumbrances against the estate. Further, Park had hypothecated one of his shares in the estate to Parrott and Luning, so that they were, for all practical purposes, mortgagees. Park's share was by now quite large, for in addition to his original one-eighth, Frémont had transferred a second one-eighth for administrative services rendered, plus an additional mortgage for assistance with the sale of bonds. As manager and mortgagee Park held an additional claim for $400,000 against future earnings. Selover's interest was encumbered to four others, and a portion of Billings's interest had been assigned to Parrott and Luning to raise funds for operating the Princeton Mine. The entire estate was awash in debt.[4]

Communicating in code ("Alpha" was Frémont, "brandy" was stock, "English" meant "payable in bullion"), Doyle proved an effective intermediary. Billings told Frémont that he must accept any reasonable offer, gently reminding him that he was one of Frémont's creditors as well. In mid-November, when Park's remittances from the Princeton Mine fell even further behind his drafts, Billings warned that he would have to negotiate the best terms available regardless of Frémont's wishes. Then on November 25 Billings informed Frémont that a crisis was at hand, as all the debts were coming due under pressure from Parrott, who demanded full payment by the end of the year.[5]

At this point Delta stepped in. Delta was the code name for Morris Ketchum, a New York financier who had been involved in the Hall-Carbine affair. Park favored Ketchum as a purchaser, for he was close to Salmon Chase, and Park still had the notion that if Chase were to try for the presidency in 1864, he might choose Park as his running mate. Further, a Ketchum-Park alliance might well mean clear control of Las Mariposas. On December 29 Parrott and Luning, who had acquired assignment of another mortgage, thus giving them control over three shares, broke off negotiations, demanding full settlement in three days, and Billings warned Doyle that Frémont was about to lose his entire remaining interest to meet their million-dollar claim. Frémont pleaded with Billings to hold the San Francisco creditors at bay, though Doyle privately told Billings that another two months would be needed to raise credit in New York; on the same day Mark Brumagim, the hard-pressed and free-spending San Francisco financier, threatened to sell his interest to Ketchum. Selover learned of this on December 30 and immediately telegraphed Billings, declaring that Brumagim was seeking to benefit from the four-day "float"—the time across the New Year when he would, in fact, still have control over any money on which he authorized transfer. Doyle believed he could get a thirty-day delay from Frémont's creditors to end Brumagim's "wild plan," which was heedless of the needs of others. Fortunately for Billings, Brumagim's initiative created anger and confusion in both New York and San Francisco, leading Parrott to give Billings an additional week to clarify the situation.[6]

On January 3, 1863, Billings sent a telegram to Ketchum outlining terms, obtaining some coin for Luning—who had conceded only through that day, a Sunday—and, on January 6, for Parrott. A crucial coded message from Doyle was unintelligible (the dictionary he was using apparently did not match Billings's) and this may have bought the enterprise a bit more time, for when Billings instructed Doyle to abandon the code and repeat his message, Parrott and Luning agreed to wait "another day or two." Billings thereupon told Doyle to withdraw all offers unless Ketchum telegraphed at once that he had the means to meet the full Parrott and Luning obligation.[7]

Billings felt a great sense of stress, for he was standing on the brink without help from Frémont or Park. Julia was pregnant, expecting their first-born within the month. Personal friends, Parrott and Brumagim, were negotiating behind his back. Worst of all, Doyle now revealed that Ketchum had never intended to pay Parrott and Luning in gold, even after they agreed through Billings to wait yet another thirty days on precisely that condition. Billings had been prudent—the Montgomery Block had never been mortaged to this chimera, Las Mariposas—but everything else he owned could be taken from him. He was, for perhaps the only time in his life, on the verge of desperation; a ruthless cycle of financial risk and ill-health fed upon each new telegram from the East.[8]

On January 12, Frémont deeded to Morris Ketchum, James W. Pryor, and their business partners six-eighths of his portion of the Mariposa Estate (now subject to indebtedness of $1,500,000) for $6,562,500, payable in greenbacks and stock transfers. Parrott and Luning reluctantly accepted. Much of the money was used to clear encumbrances on the estate, or to pay Park and Billings what Frémont owed them. Frémont somehow managed to salvage for himself perhaps $1,500,000, most of which he subsequently lost. The record is not clear on what Park took out of Las Mariposas, though it probably was not much less than $2,000,000. Four days later Frémont revoked Billings's power of attorney, for Billings protested Frémont's decision allowing Ketchum to avoid payment in coin. Billings was determined that Parrott, Luning, and Brumagim, at least, should get their money ahead of him. For the new company, a one-eighth share in the Mariposa Estate became 12,500 new shares, and while Billings intended to sell most, he hoped to maintain a small interest in the company.[9]

Billings felt abandoned by Frémont. He was determined to carry out his pledge to pay his own debts in coin. Park, too, needed coin, and since he was in Sacramento looking after his senatorial interests, Billings got him down to San Francisco for a council of war. In the end Park and Ketchum proved cooperative. Park knew that any hope of becoming a senator would be hurt if he were thought to have dealt unfairly with Billings, who had powerful friends in the local Republican Party. Between them they were able to persuade Ketchum to pay $340,000 in coin to meet local obligations to Parrott and Luning.[10]

On February 28 Billings abruptly informed Doyle that his health had gone "very bad" and that he must prepare to leave California. Billings instructed Doyle to buy the Madison Square house owned by Charles Gould, one of Frémont's creditors, with its furniture "as it stands." He wanted an elegant home for himself and Julia, and wished to take his place in New York business and society.[11]

The nature of Billings's illness is unclear. To Doyle he wrote of kidney problems, inflammation of the bladder, and high fevers. Julia's diary records his bouts with the ague and headaches (that is, with malaria) and refers to cholera, by which she meant cholera morbus, the term for dysentery. Billings's California doctors prescribed visits to Napa Springs for his health (and he went for ten days in March and again in May). Nervous exhaustion, rheumatism, diabetes, pneumonia, and later epilepsy were adduced by various physicians. Despite Billings's marked tendency toward hypochondria, the problem was real enough, for others commented on it—no wonder, given both the business and social pace at which Billings drove himself. Large meals, relatively little exercise, worries over money, no awareness of his allergies, all contributed to a feeling of ill health. Each trip to Bear Valley involved a long steamer to Stockton, a dusty coach ride to the White House, and a horseback foray out to the mines. At the same time he felt obliged to show Julia as much of California as possible, and to introduce her to his many friends: there were days in San Francisco when someone came to breakfast, someone else to lunch, and he and Julia went to a dinner party in the evening. While any of the medical diagnoses may have made sense, one may also surmise simple exhaustion as the root cause of his troubled sleep.[12]

Relations between Billings and Park reached their public nadir in February 1863 when the rumor went round that Billings declared he intended to spend $30,000 supporting Park's opponents to his efforts at gaining a seat in the U.S. Senate. Park asked whether this were so, and in three crisp sentences Billings wrote that it was not. The two men would have intermittent business dealings concerning Las Mariposas, since several minor legal matters continued to require attention—indeed, the estate would not finally be cleared of all encumbrances of title until 1940, half a century after both were dead—but would remain socially distant from each other thereafter.[13]

Billings hoped that even yet something could be made of the Princeton Mine, thus to salvage more of his and Frémont's fortune. The new Mariposa Company had nominal stock of $10,000,000, and bonds for $1,500,000 were set aside to discharge the encumbrances. Billings, who had reexamined the records and knew he need not draw out all that was due him, concluded that $1,200,000 would be appropriate, leaving $300,000 as working capital for improvements at the mine. The mines had begun to yield far higher quantities of gold, and given its rising price success still seemed possible, especially since a new manager was to arrive soon. Chosen by Ketchum and the company's officers, the newcomer was said to be a professional estate manager, as Park had not been.

Frederick Law Olmsted was Trenor Park's successor as manager of Las Mariposas estate. He took the position in good measure to escape from the frustrations of being secretary to the Reverend Bellows's Sanitary Commission, which by 1863, though continuing to perform well, was caught up in a number of political squabbles and bedeviled by growing localism.

Olmsted was planning to leave the Sanitary Commission for a venture in journalism when a chance meeting with Charles A. Dana, former editor of Horace Greeley's New York Daily Tribune, led to his invitation to California. Dana had been offered the job and told the officers of the Mariposa Company that Olmsted might be available. Olmsted's politics were right and his experience relevant, for with Calvert Vaux he had planned New York's Central Park, a project that was now unfortunately languishing. Olmsted knew the Mariposa Company's New York officers, as he wrote to his wife, to be "respectable, steady, careful capitalists." Olmsted had no doubt that Las Mariposas would be dreary and did not expect to enjoy life among "tempestuous, gambling men"; he confessed that he hated "the wilderness & wild," a nicely self-aware remark from a man who would become the father of American landscape gardening. But the job paid too well and his unhappiness in New York was too great not to take it. In October Olmsted arrived in San Francisco to confer with Billings.[14]

Billings had been preparing to leave San Francisco and postponed his scheduled departure to meet with Olmsted. After reporting to him on the mine, and giving Olmsted a written report based on his revisions of Park's bookkeeping, Billings hurried up to Las Mariposas with the new estate manager, arriving on October 14. Olmsted soon concluded that the Mariposa Company was in grave trouble, since continued claims upon the estate threatened additional drains in legal fees and interest. He had been told that the mines would yield a great fortune, though in fact Olmsted found they were producing about $100,000 in gold a month only by exploiting the richest veins and postponing the development work that was so essential. The rate of overall production was 80 percent less than expected, and he soon had the task of telling those "steady, careful capitalists" that far larger sums would be needed for development. Billings had said that "radical and expensive" measures were needed to save the company, and Olmsted agreed, though the $300,000 Billings had set aside would not be nearly enough. Creditors were demanding payment in gold, which was more valuable than currency in real terms, and payment of debts absorbed all the working capital, including the reserve. Further, Brumagim had assigned his claim in an old Frémont debt to Cornelius K. Garrison, an associate of Cornelius Vanderbilt, and Garrison sued for possession of seven-eighths of the estate. By January 1865, the Mariposa Company was close to bankruptcy.[15]

Through the Las Mariposas matter Billings felt he had done the right thing by all and that few had done well by him. While he continued to do business with Parrott at least until 1873, and worked with Brumagim again, he felt both had betrayed his friendship. He had little further use for Frémont, who he felt had turned against him at a crucial moment. He could understand these people—this was one of his problems, Julia would tell him; he was too willing to understand the failings of others—but he no longer enjoyed their company. Above all, he was certain he wanted nothing more to do with Trenor Park, who he believed (correctly) was continuing to trade in Mariposa stock, having an inside source of information, most likely the president of the Mariposa Company himself.

Even so, Billings would cross paths with Park one more time, over the matter of a Nicaraguan canal, an interest he maintained since his arrival in San Francisco in 1849, convinced that a quicker route across the Central American isthmus would have saved Laura's life. This long-standing concern would be rekindled in the spring of 1879 by two events. The first was the prospective sale by Trenor Park of his controlling interest in the Panama Railroad, acquired in part with his profits from Las Mariposas and the infamous Emma Mine in Utah; the second was the excitement generated by the Congrès International d'Études du Canal Interocéanique, convened in Paris on May 15.

The Panamanian Canal route was passionately supported by the Vicomte Ferdinand de Lesseps, distinguished diplomat, scion of one of France's great (if somewhat impoverished) families, and father of the Suez Canal, which had opened in November 1869. De Lesseps was determined to repeat his triumph in Central America and had pressed forward on several fronts simultaneously, including the purchase of the Panama Railroad Company, of which Park had been president since 1875. Without the railroad the document by which the United States of Columbia had granted de Lessep's associates the exclusive privilege of constructing a canal across Panama would be useless. Park, who held fifteen thousand shares in the railroad, drove a hard bargain: he might, he said, sell at $200 a share, twice the market value.[16]

Soon after, the Congrès convened in Paris, with twenty-two countries sending delegations. Among the eleven American delegates were two men who had surveyed every inch of the proposed Nicaraguan route and knew the Panamanian route almost as well, and they were certain that the Nicaraguan way was the only logical choice. These were Admiral Daniel Ammen, chief of the Bureau of Navigation, and Aniceto Garcia Menocal, a Cuban-born civilian engineer who was co-author of the American Surveys of Nicaragua and Panama. At the Congrès the two nearly swayed the delegates in supporting the Nicaraguan route, for they had firm facts and concrete estimates and came prepared with clear maps and plans, which those favoring the alternate route had not done. In the end, however, de Lesseps was able to win a formal endorsement for the Panamanian—and the French—route; seventy-four of the delegates voted in favor of it, with sixty-two opposing, abstaining, or absenting themselves from the vote.

As work began at Panama the construction company quickly found itself utterly dependent on the Panama Railroad. The French soon came to believe that the Yankee operators were deliberately making life difficult, creating delays and misshipping supplies, in order to force de Lesseps to purchase the line. Convinced that Park was sending orders from New York to create all possible difficulties, de Lesseps decided to meet Park's price, only to find that it had now gone up to $250 a share, in cash. In June 1881 the canal company purchased 68,500 shares (of 70,000) for $17,000,000. Trenor Park made $7,000,000 from the deal and stepped down as president. However, the company was still incorporated under the laws of New York state, and Park's son-in-law, John G. McCullough, replaced him. This was, Billings felt, sharp trading and he did not like it, nor did he like Park or Panama. He made this known and Daniel Ammen was soon in touch.

Generally speaking, the builders of the transcontinental railroads opposed any and all canal schemes for crossing Central America unless they could control them, for they believed a successful canal would draw business from the rail lines. Three railroad men in particular, E. H. Harriman, James J. Hill, and J. P. Morgan, were thought to be promoting the Panama route both because they expected it to take a very long time and because it would forestall support for a Nicaraguan canal. By now Ammen and Menocal were a formidable team, however, and had won the support of the chairman of the Senate Committee on Interoceanic Canals, John Tyler Morgan of Alabama. Billings had read Menocal's report with care, studying its maps, and even though he was now president of the Northern Pacific Railroad, he continued to favor a trans-isthmian canal. In any event, his railway, being well to the north, was less likely to be deeply affected by a canal: it was the owners of the more southerly transcontinental lines who had the greater cause to worry.

Late in 1879 several Americans organized a Provisional Interoceanic Canal Society. Among the founders were Ammen, Menocal, General George B. McClellan, and Captain Seth Ledyard Phelps. In May 1880 Menocal obtained from the Nicaraguan government a ninety-nine year concession for a canal. As the contract required, the society reorganized as the Maritime Canal Company of Nicaragua and began at once to lobby Congress. Billings joined the group, of which Phelps was now the president, and was a member of the company's Executive Committee, which worked persistently to persuade Congress to help the enterprise, with Billings's old friend from Vermont, John A. Kasson, congressman from Iowa, as principal ally.[17]

In the end there was no Nicaraguan canal. The Maritime Canal Company of Nicaragua, with Billings's name still subscribed at its head, expired in October 1899, nine years after Billings's death. Thousands upon thousands of pages of reports had been churned out in support of the Nicaraguan canal route—if pens were spades, one newspaper remarked, the canal would have been dug long before—yet as both diplomatic realities and technological possibilities changed, the Panamanian route came to be considered—indeed, almost certainly was—the better choice. For once Billings had not seen the future plainly. In 1849 and 1850 he had been prescient of much, but in this instance his unpleasant memories of the Panama crossing and his antipathy to Trenor Park had led him astray.

There was no rapprochement between Billings and Park, though they exchanged brief notes on Vermont matters in the 1870s. In 1883 Frederick (or Julia) duly noted Park's death aboard a steamship on his way to the Pacific, pasting an obituary into the family scrapbook. The funeral, delayed because of unexpected problems in getting Park's body back to the States, was held in New York's Marble Collegiate Reformed Church on January 4, with an unexpectedly small attendance, in part due to snow. Billings attended and sat quietly in the back of the church, noticed by only one of the several newspapers that reported on the scene.[18]

17

"Home Again
and Home Again"

QUITE COMMONLY BIOGRAPHERS, in their godly way, move their subjects from place to place with a logic of their own. Certainly this is
true in many business histories, as though individuals invariably acted solely
or at least largely on the basis of rationally measured economic judgments.
Perhaps some men of business are so single-minded that one need not inquire
deeply into their motivations, but Frederick Billings was not. Yet if we are to
understand the individual as much as the business he was part of by viewing
that individual against the social nature of his time, then it is no small question as to why Billings left his beloved California to return permanently to
the East.

Family tradition holds that the reason he left was Julia's homesickness, but
it is clear that the state of Frederick's health, his attitude toward his work, and
his sense of home were far more important. Indeed, he always said he left
California because of his health. He had been ill on occasion in the years
before, and had been warned by friends and physicians that the California
climate was not good for him. He was intrigued by the development of ho-
meostatic medicine, which attributed many illnesses to the multiple functions
of the liver and to digestive activity, and found no one on the West Coast
who could do more than prescribe the ritual of the sitz bath. He had been
under exceptional stress, working day and night over the sale of Las Mariposas
and saving his own fortune. In the end, exhausted, filled with ill-defined and
recurrent anxieties, and angry at individuals he had counted as good friends,
he resolved to leave California.

No doubt Julia was pleased with this decision, both for her and for the
infant Parmly, and she probably did not hide this from Frederick. She missed
her close family and felt that her father needed her, especially after he was
injured in a railway accident in France. She also reminded her husband that
his parents were growing frail and he had obligations to his siblings that could
be better met were he living in the East.

What Frederick Billings knew best was land law. His whole legal reputation rested on it: property litigation defines his place in California history. As a central figure in the largest legal firm in the state during the most critical years of land adjudication, Billings was in part responsible for much of the continuity of argument concerning the legitimacy of hundreds of claims. HPB had tended to take on the largest clients, those most able to afford large fees and long battles. Being on retainer, the firm had a vested interest in the survival of such clients. But by 1863 matters had changed. State politics increasingly favored the view of the squattocracy, who were less likely to come to someone like Billings. HPB itself hardly existed, though Grannis maintained offices in the Montgomery Block on behalf of the firm. The unwavering hostility of the federal government to the kinds of claimants represented by the heirs of Andrés Castillero meant that the heyday of large land cases was over. Billings's association with Frémont continued to hurt him in the mining counties and in Washington. The question, then, was where Billings most effectively might apply what he knew best and most productively employ his small fortune.

Clearly the best opportunities lay in railroads, for railroad companies were, in fact, land companies. The Federal Railroad Act of 1850, the Right-of-Way Act of 1852, and the grant of twenty-one million acres of the public domain to some fifty railroad companies in ten public-land states between 1850 and 1860 brought an entirely new player onto the stage. This player used land differently, held it in quite different parcels, and for the most part kept that land by congressional act, so that proper use, not litigation, was the primary issue, and exploitation for private gain or development for the public good, or some combination of the two, was the challenge. To be able to play a major role on this stage, Billings concluded, he must be on the East Coast.

During Frederick's time in California four brothers and two sisters commanded his attention. Charles, born the year before Frederick, was the quiet one. He would live longer than any of the boys, dying in 1896. Though he was clearly proud of Frederick—he named his first son after him—there seems to have been little warmth between them and equally little correspondence. Aside from a loan or two, Charles seems not to have leaned on Frederick for money, advice, or jobs. Charles moved to Fitchburg, Massachusetts, while Frederick was working in Montpelier, and became a bank clerk. He remained with the bank for forty-nine years, always as "true as machinery," rising to vice president, living comfortably with his wife, Sarah Alice Towne, daughter of a prosperous woolen manufacturer.[1]

From distant Woodstock Sophia shared her concerns about his sisters with Frederick. She trembled, she wrote, for Sophie and Lizzie, for Laura had made such a mistake in marriage. Sophie (actually Sophia, but seldom addressed as such), who was the oldest surviving sister, seemed to have no prospects, but in the spring of 1851 a suitor appeared for Elizabeth, the youngest daughter, who was just eighteen—too young, perhaps, to marry. George Washington Allen, a twenty-four-year-old hardware merchant from Burlington, was "an excellent young man" who neither smoked nor drank.

Lizzie finally married him in May 1853, and, seizing the need to return east to raise funds for the Montgomery Block, Frederick attended the wedding. Unhappily, however, George Allen did not prove to be the successful husband that Mother Billings had predicted. Though honest and independent, he was plagued by bad luck and poor choices. In the winter of 1855–1856, when Burlington was in the midst of a severe depression, he discovered that his partner in his medicinal supply business had been cheating him, and he got out with only $500 to his name. Allen then opened a drug store, with a local doctor as a silent partner. When this enterprise did not go well either, he decided to move to Chicago, where he met Frank Simmons, back from California, and set up a hardware store with him, taking a loan he believed to be secured by his Burlington store. When he could obtain no money from the East because the doctor called in the debt, Allen had to turn to Frederick to repay the loan. This business was wiped out by the financial crisis of 1857, with Frank failing to carry his share of the burden, his honor having "proved very elastic," at least in George's view, who wrote to Frederick asking that he bail them out to the tune of $7,000.

All this fecklessness enraged Billings, who had little use for Frank Simmons and felt Allen was a naive and trusting soul with a dark cloud hovering over all that he did. Still, Frederick supplied capital for Allen to go into a partnership with his brother Franklin, and late in 1857 Allen & Billings, hardware merchants, opened in Oshkosh, Wisconsin. This went badly, the partners had a falling out, and Franklin took $650 and his hat and bid the firm goodbye. On June 10, 1859, George was wiped out once again by a great fire that swept through Oshkosh, taking 125 buildings with it. Once more George appealed for help, confessing that one of his insurance policies had proved worthless. With a sense of resignation, and determined that Lizzie was to have a happy home, Frederick provided the money for a fifth start, this time in Woodstock.

But George Allen continued to make his own bad fortune. Taking a loan from Sophia Billings's family, the Wetherbees, he failed to repay it and Frederick again came to his aid. He then obtained a loan from Frederick's old benefactor, Albert Catlin, which Frederick learned of only inadvertently. When he taxed George on the point, making it clear that he was offended, George replied—as he had with respect to his silent partner in Burlington—that Catlin did not expect to be repaid. This was pure charity, and it bothered Frederick even more. There must have come a point when George Allen's unrelieved cheerfulness was too much for Frederick—was it when a child was stillborn in

1860 or when the Allens's year-old son died in 1863 and George—both times—remained content with God's ways? Or when his store was burned out early in 1864 and he wrote that this was a benefit because it opened up space for expansion and hurt his competitors more? George was a victim, Frederick wrote, of a sense of "overabundant possibilities." And then, in May 1864, the patient George Allen died, and Lizzie, who had remained content in domestic work, always supportive of whatever George wished to attempt, moved back into the senior Billings household and apparently was happy thereafter to serve as a babysitter for the growing brood of Billings grandchildren, becoming in the 1870s an auxiliary nanny to Frederick and Julia's children. Indeed, she clearly took nourishment from being a surrogate mother, a role she performed with great dignity. Lizzie would outlive them all (except for Julia), and would die in Woodstock in 1905.[2]

Frederick's sister Sophie, three years his junior, appears to have married in some measure because Lizzie, though seven years younger, had taken a husband and she did not want to remain at home to become everyone's nursemaid, as was the practice for spinster daughters. On December 21, 1853, Sophie married Goldsmith Fox Bailey, a pleasant young man with a good sense of humor. The marriage was a happy one, for Goldsmith was a successful lawyer who, after service in the state senate, was elected to the U.S. Congress in 1860 as one of Massachusett's representatives. This solid and loving marriage was broken by Goldsmith's death in May 1862. Sophie returned to Woodstock, with Frederick acting as godfather to young Edward Goldsmith, now another claim on his presence in the East. After several years of watching over other people's children, Sophie moved to Fitchburg and in 1876 married a widower, Rodney Wallace, owner of the Fitchburg Paper Company. He, too, was elected to Congress, and Sophie lived in Washington with him, returning to Fitchburg to die there of pneumonia in 1895.[3]

Of his younger brothers, Frederick could not say that Richard was his greatest disappointment, for he had never expected much of him. Born seven and a half years after Frederick, Richard seems always to have been in a bit of trouble. Oel and Sophia often reported their problems with him—used to getting up at five o'clock in the morning, they found him indolent, for he remained abed until seven. He was a bad correspondent in a family that brewed up a writing storm every month, and when he did write, his letters were irritatingly devoid of real news. Those few letters reveal a lonely boy who at twenty-two still saw himself as sixteen, insecure, ready to believe, when he got his first job, that it was not on his own merits but because Frederick had intervened for him. He once confessed to Lizzie that he had been wrong not to take the chance of the college education offered him, for he had difficulty in putting his thoughts on paper. He was, Oel said, "a strange boy, not like any [of the] rest of the family."

In the spring of 1850, nineteen-year-old Richard set out on his own, becoming a levelman for William B. Gilbert, chief engineer of the Western Vermont Railroad Company, who was responsible for laying out a line from Swanton to Derby in northern Vermont. It was a hard and cold job—he had to live in his flannels well into June, and, as he wrote with an unaccustomed touch of wit, he needed an anchor to keep the mosquitos from carrying him off. But he clearly was happy in his independence. However, when Gilbert lost his job, Richard set out, with $100 of Frederick's money in his pocket, for Springfield, Ohio, where he hoped more railway work awaited him. After a brief stint as a clerk on a Mississippi River steamboat, Richard hooked up again with Gilbert, who had taken charge of a section of the Syracuse and Binghamton Railroad in New York.

Oel was a stern taskmaster, and Frederick seems not to have shown his customary forbearance when dealing with Richard. When Richard spoke of getting married, his father discouraged him, saying that he had hardly shown ability to take care of himself yet. On the last day of 1853 Gilbert wrote Oel that Richard had lost his job and had no money to come home; indeed, Gilbert said that Richard reported his family had cut him off. Complaining that Richard never answered his letters, Oel nonetheless sent money, and Richard returned to Woodstock, to work in the telegraph office and drink too much. Given another chance, Richard sobered up and, with $150 and a new suit of clothes from his father, set out for Illinois to work with Franklin. Then, to his surprise, Oel learned that Richard was in St. Louis, planning to go to California via the plains, driving cattle and sheep to Benicia, with a $60 loan in his pocket. Frederick had made it clear he did not want Richard in California, and Oel did not know what to do, for he suspected the young man already was well into Kansas.

As it happened, he was not. Taken ill eighty miles into Kansas, Richard set off for home once again, though when he reached Chicago he felt well enough to join Franklin in Freeport. Oel rather curtly told Richard to seek his Redeemer and not to come home since there was no work for him in Woodstock. By the time Richard had this cheering word he was sinking from typhoid. In the meantime Frederick had written to Catlin, asking whether he ought to assist Richard, and Catlin advised against it: Richard was a "hard boy," addicted to rum, certainly not a good risk, and he ought not to be killed with kindness.

Instead, Richard was killed by the railroad. After recovering sufficiently to work in Franklin's store, Richard again began to "spree it"; his conduct was costing Franklin customers, and so he sent him out with a wagon to sell surplus goods, only to have him return three weeks later with only one sale and a damaged wagon. Next Richard began to help himself from Franklin's cash drawer. Despairing, Franklin decided to sell out and go elsewhere without telling Richard. Franklin was still clearing up his local obligations when, on October 27, 1854, Richard, now a brakeman on the railroad, was run over by a construction gravel train, losing both his legs and an arm. He died before medical assistance could reach him, "firm as a tiger to the last," speaking final words of affection for his family.[4]

Frederick was, to some extent, victimized by his own sense of integrity, on which friends and kin alike were inclined to play. He was proud that he had paid all of his brother Edward's debts after his drowning in Baltimore, and that the people of Woodstock and Burlington still remarked on the fact years later. He now stubbornly paid off Richard's obligations. He was unwilling to see any of his sisters suffer, and he was determined that at least one of his brothers should be at hand in Woodstock, stable, dependable, and able to look after his parents while he was on the West Coast.

The choice was Franklin, who had a good business sense. Frank was likely to understand what Frederick was up to in the West, having worked there for Simmons, Hutchinson & Company. He came back to Woodstock with Frederick in the spring of 1853 to attend Lizzie's wedding and did not return to California. That fall, with loans from Frederick and the ever-helpful Albert Catlin, Frank had gone to Illinois and set up his shop in Freeport.

After bringing Richard's body home to Vermont, Frank settled in to become a solid merchant and, in time, banker. By 1859 he was looking after the elder Billingses and was Frederick's agent in Woodstock, negotiating the purchase of properties for which Frederick supplied the money. Frederick bought the handsome old Mower house, taking it off the hands of his old patron O. P. Chandler, and Oel and Sophia moved in that summer as Franklin renovated the house around them and Chandler, who stayed on in two rooms.

Marriage settled Frank down. On a visit to the Simmonses in New Bedford, Frank met Nancy Swift Hitch, a sensible widow a good bit older than he who had some money of her own, and he proposed marriage. Frederick added his seal of approval, writing Miss Hitch (or Nannie, as she quickly became) that he was delighted to welcome her into the family. He returned East in 1859 for the wedding. Franklin carried out Frederick's instructions carefully, managed the family money well, and was Frederick's confidant, being the first to know, in 1862, when Frederick himself made the decision to marry. There would be a single son from his marriage to Nancy, who carried on the Billings name and served as governor of Vermont. Franklin died in 1894.[5]

As dependable as he found Frank, however, Frederick reserved a special place in his heart for his brother Oliver, formally Oliver Phelps Chandler but always known as Ollie. Bright, cheerful, fascinated by all that went on around him (in particular, for a time, Mormon thought and activities, an interest Frederick shared), Oel and Sophia's youngest child, born when Sophia was forty, was the cliché of the large Victorian household: everyone's pet. Also a prolific correspondent, with a wry sense of humor about the often charmless obligations of a Victorian childhood, Oliver did on occasion go off on a spree of his own, but he was always forgiven, for he was making something of himself. He long remained the baby of the family. When he was twenty-seven, he was still being told that he ought to grow up. Yet here his parents seem to have misunderstood him, for he appears to have known what he wanted to do with his life from an early age, he never lacked the discipline to write his parents faithfully every Friday, and he easily crossed the generation gap between himself and his older siblings. He also admired Frederick extravagantly.

Sent to Phillips Academy in Andover by Frederick, then to the University of Vermont, with the notion of going into the ministry, Oliver did well, graduating fourth in his class. Frederick wanted Ollie to study theology; he replied that all men were different and that for as long as he could remember he wanted to be a lawyer. After graduation in 1857 he read law in William Collamer's Woodstock office for a year, and then Frederick sent him to Harvard Law School.

All large families have a go-between, someone who becomes a successful adult by remembering the child inside, and this was Oliver's function. It was he who told Frederick that Lizzie wanted, more than anything else, music lessons and a piano, and that Franklin must marry Nancy Hitch, who would make something of him; he who wrote the letters that revealed the true state of affairs with George and Lizzie in Oshkosh. It was to Ollie that Frederick wrote of his romantic interests, and he was the first to know that his sister Sophie had lost her heart to Goldsmith Fox Bailey, exchanging notes discreetly that referred to her growing interest in the "writings of Goldsmith." Unmarried, "Little Ollie" would continue as the family messenger until Frederick permanently returned to the East.

Oliver married Charlotte Lane in 1868, and thereafter appears to have been the one child who, while remaining warmly devoted to all the clan, cut himself loose to go off completely on his own. He became a partner in a thriving New York law firm, Billings and Cardozo, and a three-term alderman. His three children bore names from his wife's rather than from his family. Oliver died in New York, weighted with honors and club memberships, in 1894.[6]

Though Oliver may have been the most lovable member of the family, Frederick was expected to be the perfect one, and the burden of being the model son can never have been easy. In every weekly letter Frederick's parents reminded him of their high hopes for him. He reciprocated by sending them money, so that they might refurbish the Maples, and later the Mower house, and hold their own in Woodstock society. He plied them with trips to Boston, New York City, and the White Mountains, and on one memorable occasion to Niagara Falls. He made the long journey home four times between

1853 and 1861. Still, he preferred his independence in California, where he could deal with all these family stresses at a distance, with the bias of delayed communication softening bad news and intensifying good. While his depressions and headaches may well have been due, in some measure, to overwork or to a mild epilepsy, it is also apparent that both came on shortly after the receipt of another unhappy tale from the East. That Frederick Billings missed Woodstock and felt himself a Vermonter who was bound to go home one day to stay is quite clear, but it is no less clear that his feeling was close to that of St. Augustine when he prayed for release from the sins of the flesh: Oh Lord, save me, but not just yet.

In September 1863, Frederick and Julia began preparations for leaving California, making rounds of calls, having newly fashionable *cartes de visite* prepared for their friends, and revisiting treasured sights. Frederick slipped away to the Yosemite Valley one last time, his fourth trip since 1856. He gave what might be final gifts to several charities, including a substantial sum to the Colored Baptist Church, which Maria "Ridey" Barnes, Julia's maid, attended. In late October, Frederick, Julia, Parmly, and Ridey sailed out the Golden Gate for home.

Through 1864 Billings received many communications from California suggesting that he should return, both to protect his business interests and test the political waters. He was still remembered for his unionist oratory and the Sanitary Commission. None of these appeals moved him until early in 1865, when a proposal far more to his liking was put to him: that a group of well-placed Californians might get Billings named to President Lincoln's second Cabinet. This he wanted, for he believed he was ready for high office, and that he still could perform a major service for the nation without the many compromises he was convinced virtually all men who ran for elective office had to make. Further, he and Julia had many friends in Washington, and while he knew that she did not wish to return to the Pacific Coast, he felt she would happily move to the nation's capital.

To make Billings's appointment as a Californian creditable, he would need to reestablish himself there. So on March 30, 1865, he set out on the long voyage around Cape Horn to "revisit his California home," having found a doctor who recommended that his health required precisely the climate he had been instructed to leave less than two years earlier. A second Billings child, Laura, born in Woodstock on August 20, 1864, had fixed Julia even more firmly to the hearth, while Frederick was discovering that children could be an irritant to the highly strung. Julia moved back into her father's house in New York, and Frederick left with only Oliver to see him off.

Billings wanted to be secretary of the navy, but this position was held by Gideon Welles of Connecticut, one of the ablest members of Lincoln's Cabinet, and his displacement would depend upon intriguing with Salmon Chase, the former (and very powerful) senator from Ohio who was Lincoln's first secretary of the treasury. Chase had resigned in June 1864 and Lincoln had appointed him chief justice of the U.S. Supreme Court. Chase was important to Billings on other grounds, however, because matters relating to Mariposa were before the court and Billings wanted Chase's good will, so he hesitated to make any overtures.

Billings's California supporters had another position in mind: secretary of the interior. They approached Lincoln, reminding him of his earlier promise (or so they took it to be) to appoint a Californian to his Cabinet, and Lincoln appeared to be giving serious thought to the matter. Thus, as Billings sailed, he had reason to believe that his arrival in California might soon be followed by the coveted offer of a Cabinet post.[7]

Billings's voyage on *Colorado* was all that he hoped for: rejuvenating, demanding, and informative. It forcefully reminded him of how much he enjoyed his leisure time, free from the obligations of office. His companions aboard the ship included the noted Swiss-born scientist Louis Agassiz, who held a post at Harvard where he was assembling a remarkable museum. Agassiz was on his way to Brazil to investigate local fish as a guest of the Pacific Mail Steamship Company. Billings and Agassiz had corresponded the year before about the wonders of the Yosemite Valley, and now Billings, also a guest of the company, counted himself quite lucky, for he could attend Agassiz's shipboard lectures, given for the benefit of six Harvard students who accompanied his expedition, and take advantage of the *entrée* that *Colorado* provided en route, since Secretary Welles had ordered all naval officers to aid the expedition in any way they could.[8]

Throughout most of the voyage Billings enjoyed unaccustomed good health, in part because he had been put on a special diet of hominy and milk for breakfast and soup and rice for dinner. He was able to bring his own cow with him on board, assuring a dependable supply of fresh milk for his hominy. (Frederick always saw to it that he travelled well on these long voyages: as Olmsted reported to his wife, on a previous journey Billings had found the barber's chair the most comfortable place on the steamer, so he asked the barber how much he made during the trip, paid that sum, and had sole use of the chair thereafter, to the disgruntlement of unshaved fellow passengers.)[9]

There was also an interesting young man aboard named William James. Then only twenty-three, James was a long way from becoming the father of American pragmatism and author of *The Varieties of Religious Experience*, and if Billings's journal is sufficient evidence, he recognized no special brilliance in James at the time, perhaps because the young man was laid out with seasickness for twelve consecutive days. James did mention Billings to his parents as a "rich man" who paid for a handsome Brazilian dinner on shore.

James was somewhat scandalized on Billings's behalf when Agassiz, after holding forth for some time, asked whether he might not go into Billings's stateroom and take some of the supply of books he had brought for reading on the voyage. "Sir, all that I have is yours!" Billings replied in his most stately manner, to which Agassiz replied, shaking his finger at Billings, "Look out, sir, that I take not your skin!"[10]

Rio de Janiero provided a rewarding stopover. The group went ninety miles into the interior, partly on the Brazilian emperor's own rail line, and spent the day fishing; Billings caught the first fish, which proved to be one Agassiz especially wanted. The emperor, Dom Pedro II, joined them on board ship, much impressing Billings with the quickness of his mind. Billings paid for everyone to be taken by horseback to the top of Corcovada, where they were left alone, and reflected on how this remarkable view must be preserved, and how it was always Americans who went to the top of every mountain because the Brazilians thought it silly to expend energy in such a way. He was so excited by Brazil—its people, its coffee plantations, its picturesque landscapes—that he could scarcely sleep during the seventeen days he was there (*Colorado* was delayed for repairs). All the passengers were worried, however, for they could not forget the black, ominous cloud seen from *Colorado* as it passed down the Virginia shore; they had feared this was evidence of some climactic battle in the Civil War. As they would learn from another vessel before leaving Brazil, they had seen either the spume of the devastating battle of Petersburg, as the captain guessed, or (more likely) the burning of Richmond.

Callao—where sightseeing was curtailed, as Peru was in civil conflict—brought the astonishing news that, on April 14, Lincoln had been assassinated. For the president to be murdered by rebels to whom he was in truth a friend, Billings wrote, was a stunning disaster for the nation. He did not comment on his personal disaster, for Lincoln's death had wiped out any immediate prospect of taking a seat in the Cabinet.

The remainder of the voyage was made pleasant by the company of the Episcopal bishop of Philadelphia, Alonzo Potter, and his wife. Potter gave a series of lectures that, being low-church and liberal, Billings quite enjoyed. He buried himself further in books, in part to avoid the captain's wife, who was full of childish talk and whose yapping small dog Billings dearly wished to throw overboard. At Panama he received a bundle of letters from Julia, a massive sixty-four-page outpouring from Oliver, and a packet of newspapers, and from there to San Francisco he read avidly. This had been a wonderful holiday, as though he were young again and arriving in California for the first time, with no notion of the burdens that success could bring. Then, quite unexpectedly, this feeling of well-being and anticipation appears to have been shattered by Bishop Potter's death. Taken ill shortly out of San Francisco,

Potter died aboard ship. Billings sat up with him on his last night, and Mrs. Potter threw herself upon him for solace, so that he felt a burden of ironic responsibility; a second California arrival had been contaminated by the dreaded Panama fever, apparently contracted when the bishop helped officiate at the consecration of the first Episcopal church in Aspinwall.[11]

Checking in to his old suite at the Occidental, Billings was thrilled to discover how many people wanted to pay their compliments—it took him two hours to walk the six blocks from the hotel to HPB's old offices. Several seemed to think he was back to run for office. But he felt ill, having fallen into a depression over Potter's death, and was somewhat morbidly contemplating the relationship of the natural to the supernatural and pondering the inevitability of his own death. He knew Julia wanted him to return quickly. Even though the *Alta California* was reporting that he would be a candidate for the Senate, he promised her he would not be.[12]

Billings may have surprised himself by this promise, for when he sailed for California he was entertaining the notion of running for office in California and intended to talk with friends and potential supporters. As secretary of the state Republican Central Committee, Alfred Barstow had continued to urge Billings to return and seek the prize, filling him in on shifting party allegiances. Barstow predicted that the unionists would split into warring factions as soon as the war was over and that Billings ought to commit himself to one group or another quickly, though Billings continued to hope that the party might be kept intact. The only prospect of this, he reasoned, lay with John Conness, the former Union Democrat who, through control of the state's Union League clubs (in which Billings also enjoyed a following), had built a strong machine. When the party split in 1864, Billings stood behind Conness, who in turn worked, though to no avail, to get Billings some relief on the New Almaden matter.

Now, back in California, and with no cabinet prospects, Billings allowed his friends at the Republican state convention in Sacramento to put his name forward as a candidate for Congress. He clearly lacked the requisite fire in the belly, however, for instead of working the corridors of power, he made his promise to Julia and gave no effort to pursuing the prize. It went instead to a good friend, Donald C. McRuer, a former San Francisco harbor commissioner, who took the congressional seat that Barstow insisted Billings could have if he would bestir himself.

Without Julia, and without his youth, Billings did not find California the gay place he once thought it. A round of dinner parties given by old friends, with the inevitable talk of who had died, lost a son or wife, or been crippled, seems to have made him feel insecure about his children, for he hurried off a note to Julia urging her to have their picture taken every six months lest God claim them and their faces be forgotten. He confessed that he did not find the

people so affable as he remembered and that he had never seen the city looking "so ragged & smutty & forlorn." Starr King was dead, Jessie Frémont
back East, no one of the band of brothers left. Halleck would not return, and
Peachy was distant; Billings did not see him for some days, and then his
former partner unexpectedly walked up and held out his hand. "I was cor-
dial," Billings reported, "but did not melt."

In September, after a nostalgic trip to Yosemite, Billings began his travels
once again, this time to Oregon, ostensibly in search of his elusive good health,
actually with an eye to whether the region would benefit from a railroad. He
went north to Mount Shasta, thence via the Rogue River, on to Salem, where
he had a clear view of Mount Hood. He was pressing hard, not taking the
time to savor the sights as he usually did, giving himself only two days in
Portland and then hurrying up the Columbia gorge. He was in Umatilla for
his birthday, then turned back and took the steamer to Portland, which he left
after an overnight's rest for Seattle by road. Almost every day his diary contained the notation "not well."

Billings's heart was not in this trip. Still he soldiered on, not missing any
part of the Northwestern Grand Tour: up the Fraser River to Yale in an October fog, thence to Lillooet, and back to Victoria, where he lost interest in
his Northwest journal. While he had seen what he came to see and had some
ideas about how the Columbia gorge might serve as a railway route, his mind,
if one is to trust his jottings, was far more on Julia. This was almost automatic
travelling, movement for movement's sake, frantic, without his accustomed
background reading and leisurely stops: before he boarded the steamer to sail
back down the coast, he had travelled over fourteen hundred miles by river
boat, Concord coach, and horseback in less than a month. By November Billings was back in San Francisco, to find any office for which he might have
been nominated already taken—Barstow concluded Billings had intended such
a result—and he was clear in his mind that he would return, if at all, only
when a transcontinental railroad might make summer visits possible.[13]

Billings thought he was likely to die, and he wanted to get back to his
family, though he also intended some philanthropies for San Francisco first.
He funded young men who wished to go East to study just as Catlin had
helped him to go West long ago, he helped the fledgling University of California (it was at this time he contributed the name of Berkeley), and together
with William Ashburner, one of Yosemite's commissioners, he encouraged
Olmsted to plan a great boulevard and public park for San Francisco, the three
men meeting with members of the mayor's office. He spoke out again of the
need to protect not only Yosemite but other natural landscapes in California.
When Julia warned him that he must not try to relieve the First Presbyterian
Church of all its debts, he replied that he had nonetheless done just that and,
were he to die, he could now feel that his work had been done.[14]

In March 1866, Billings was back in New York. For all practical purposes he had, in fact, moved to New York and Vermont at the end of 1863, and his return to San Francisco had simply proved a last visit, to wind up affairs. But he was reluctant to give up California, despite his conviction that "the values of Vermont," as he understood them, were basic to his nature. (This was not a contradiction: after all, that figure taken to be the embodiment of twentieth-century Vermont, Robert Frost, was born in San Francisco.) Yet Billings still harbored a hope that a Cabinet appointment would come his way as a Californian.

On April 7, 1866, the California Senate transmitted a resolution urging President Andrew Johnson, who had succeeded the slain Lincoln, to move forward with the speedy development of the Pacific States and place in the Cabinet a person who understood the needs of the West. If the resolution were accepted, the legislature recommended that Billings be appointed to the position. Clearly they had the Department of the Interior in mind, and this time, given the strength of the resolution, and the fact that Billings still had the backing of the California Republican state committee, there seemed a real possibility of success.[15]

Once more, Billings failed to reach for the ring. He was seriously ill, and in the hands of a noted doctor, C. E. Brown-Séquard, who diagnosed his mild epilepsy, which manifested itself in the most intense and disabling migraines, complicated by the onset of what would prove to be a succession of debilitating depressions. These depressions were held at bay by Dr. Brown-Séquard, whose fashionable theories about mental magnetism appealed to him, but this meant living near the doctor, who was either in Cambridge, Massachusetts or New York City most of the time, and Billings's sense of dependency further crippled his energies.[16] In December 1866, the moment past, Billings wrote to Barstow to explain why he had not thrown himself into pursuit of office, confessing that he wished he could have taken the flood at high tide to see where it carried him, admitting that bad health had obscured his judgment and blunted his vigor.[17]

Even had Billings's health been fine, and he had pursued the secretaryship, it is not likely he would have attained it, for by the summer of 1866 President Johnson's policies had split the Republican Party. As a Tennesseean he was suspected of being soft on the defeated secessionists, for he did not want to rub salt in the nation's wounds as he felt the Radical Republicans, with their harsh treatment of the defeated states, wished to do. Holding to Lincoln's Reconstruction plan with some minor changes, Johnson had taken advantage of the congressional recess to recognize the loyal governments of Arkansas, Louisiana, Tennessee, and Virginia, as set up by Lincoln, and granted amnesty to Confederates who took an oath of allegiance, with some excepted classes. In April a Civil Rights Act, giving Negroes citizenship, was passed over Johnson's veto, bringing the union congressional majority into open war with the president. The California Republican party divided into Johnson loyalists and unionists—that is, those opposed to Johnson—and each selected its own delegates to national conventions. By now Billings was a full-fledged Radical Republican, and since he was back on the East Coast he was appointed one of three men outside California's congressional delegation to represent the party at the union convention to be held in Philadelphia. By September, however, when the convention met, Billings once again was too ill to attend.[18]

Why did this charming man, popular, untouched by public scandal, with the wealth and dignity that so helped gain public office, never act with vigor when put forward by friends and supporters? Was he, as some critics concluded at the time, simply a dilettante in politics? Did he feel he was too good to campaign, awaiting an appointment that never came? Why did so many people continue to think he would be a strong candidate in the face of his repeated failures to reach out for the prize? He had played a moderately important role in California politics, though only his record as a fund-raising orator for the Sanitary Commission would earn him a place in the most detailed political histories of the state. He was capable of very hard work, he commanded the intense devotion of his friends, he possessed the oratorical skills so essential to the politicians of the day. As one of the Argonauts, his California roots were deeper than most, and he could command the respect of those who had endured the hardships of the isthmian crossing and the first-year rain and mud of San Francisco. Yet Billings never capitalized politically on his many attractive qualities.

Perhaps Alfred Barstow, who had his offices in the Montgomery Block and often lunched with Billings, understood him best. Frederick Billings of Woodstock simply could not, Barstow hinted, divest himself of the virtues of his Vermont ways. While preparing a draft obituary notice for the California Sons of the Pioneers in 1891, Barstow concluded that Billings, who was a life member, was genuinely happy in his solitude. He wanted to do good works, increasingly devoting the energy that many men of his generation gave to politics and to business to philanthropy. When a legal case gripped his mind, he pursued it with a scholar's relentless tenacity. He cared more about the preservation of California's natural wonders than the exploitation of them. He wished very much that his life should be, to the very end, "clean and entirely free from scandal." He was concerned that wealth be well used and that those who were fortunate enough to have money spend at least some portion of it on the public good. His very sense of citizenship, his admirers concluded, kept him from the relentless pursuit of office or mammon, the marks of success in the post–Civil War years.[19]

Further, a new vision of the ideal home had begun to grip Billings. When he and Julia had visited the Marsh house in Woodstock, taken through it room by room, he had remembered how Marsh Hill had symbolized strength and security to him as a small boy. During her stay in Vermont while Frederick was in the West, Julia walked often to the hill, gathering acorns with the children, and she reminded him that he frequently spoke of the house as desirable for them. She also told him that she did not want to be tied to Woodstock, that buying the house would be a mistake, for in the summer they could go to her father's place in Shrewsbury for the sea. Julia was used to the ocean and liked it, but Frederick did not: when she went bathing off the beach at San Francisco he always stayed behind with the horses. Nor did he wish to be beholden to Eleazar Parmly. He would, Frederick thought, buy the Marsh house and enlarge it, so that it might provide the roomy home Julia wanted, though not in New Jersey, which he did not much like, but in Woodstock, his spiritual home, to which he would now return.[20]

The idea of home was exceedingly important to Billings. Perhaps the forced and, for a young boy, not wholly explicable uprooting from Royalton to Woodstock years before had given the word unusual evocative power. In a rootless society, certainly Billings used the term with great frequency and emotion. When in 1874 Samuel Hopkins Willey sent him an essay on the founding of American California, Billings responded warmly that Willey's references to him had touched his heart and that he was honored to be considered a builder of California. "Nothing could have attracted me away," he told Willey. "It was broken health that drove me away—and no one will ever know how my heart broke as I turned away from California never more to be my home."[21] Yet when in California Billings routinely invoked Vermont as home, and Julia's shrewd play upon his sentiments concerning the Marsh house and his Woodstock roots touched Billings's divided soul. Even as he enjoyed the challenge of life in California, he believed himself longing for the serenity and security of life in Vermont. During this period of indecision, of looking for the right political or, more emphatically, business opportunity, he could well understand his friend Henry van Dyke's cry, "So it's home again, and home again," for one could have two homes.[22] In having them, however, one could not be fully content in either. As a sage has remarked, and as Billings discovered, one does not experience happiness, one only remembers it.

 18

Buying into Railroads

Throughout the 1850s one of the foremost issues of both national and California state politics had been the question of how to improve communication with the Pacific Coast, whether by trans-isthmian canal or by railroad. Billings had maintained an interest in both possibilities, though he thought a transcontinental railroad the most likely choice and certainly the best. In 1844 a New York merchant, Asa Whitney, had proposed a line from Milwaukee to the Columbia River. If Congress would sell a strip of land sixty miles wide along the route for sixteen cents an acre, he offered to build the line in twenty-five years (a figure Whitney later lowered to fifteen) by selling the land to settlers and using the proceeds to meet construction costs. Though Whitney's line was not built, the formula he proposed was widely accepted.

Less than a year later Stephen A. Douglas, then a first-term congressman, suggested an alternative route, with Chicago the eastern terminus and San Francisco the western. This idea naturally won support from those who wished to annex California to the United States, especially when Douglas reversed one element of Whitney's scheme and proposed pushing the line out in advance of settlement so as to attract permanent occupants. When California became a state in 1850, Douglas had a ready-made following, especially in the bay area, and remained faithful to his dream. After the Mexican War there was no session of Congress without one or more bills aimed at creating a transcontinental railroad.

Not everyone thought a railroad a good idea. Many conservatives felt the economy could not sustain such a massive land giveaway, or that the temptations of so large a project would bring in spoilsmen who would corrupt the governmental process. Others believed that internal improvements were unconstitutional and likely to tilt the delicate federal balance toward the central government. Many southerners wanted a line only if it originated in New Orleans and ran across Texas, to the disadvantage of free soilers, who of course wished just the opposite. Merchants on the East Coast were not certain that a rail line would benefit them as much as a canal would, since a canal would keep the port cities flourishing. Rail lines across the interior of the continent were bound to benefit midwestern locations disproportionately. But one would not have found many in the Mid- or Far West who held any of these views. While Washington might debate, these regions would get on with it.

Schemes for short lines were thrown up like gopher mounds throughout the settled West. A San Francisco to San Jose line would fill one obvious need and so in September, 1851, the Atlantic & Pacific Railroad—surely the most overreaching title any of the pugnacious little western lines ever took—was incorporated with Henry W. Halleck as its president. Whatever the line's title, its purpose was to get people, goods, and in particular the produce of the New Almaden Mine moved to the harbor at San Francisco. However, the A&P went broke during the financial crisis of 1855, and track was not completed to San Jose until January 1864.

The Sacramento Valley Railroad was rather more successful and certainly built more quickly. Proposed by Sacramento, Placerville, and Marysville merchants to link the state capital with the foothill communities, the SVRR had the early support of Colonel Charles L. Wilson, William Tecumseh Sherman, and other San Francisco bankers and lawyers, including Frederick Billings. Reorganizing an abandoned company, Wilson filed articles of association in October 1853, and sailed east to find an engineer to build the line. There he hired Theodore D. Judah, a young engineer who had just built a remarkable railroad through the Niagara gorge.

The SVRR was to be built on land so level, local wags said, that it could be laid out with a half-empty whisky flask; financially, however, it was on very hilly terrain. In 1855 it too suffered, and Wilson particularly so, for his funds had been tied up in Page, Bacon & Company, and he was replaced as president. As vice-president Sherman put the stability of Lucas & Turner behind the project. In August money began to come back, with Billings again among the investors. Nonetheless, the money was not enough, and in October the courts placed the SVRR under a deed of trust. J. Mora Moss, who had won a reputation for bringing ice from Alaska to cool champagne buckets at the Montgomery Block, was named trustee.

Billings shortly found himself quite temporarily and in name only the president of the line, since he had been involved at all stages and Moss knew him well. On Washington's Birthday, 1856, with the railway completed to the new town of Folsom, the SVRR held a grand opening party. In 1861 Billings noted how impressed he was by the line—in five years of operation no passenger had been injured and the loss of freight was remarkably low—and by Judah's vision, as expressed in a proposal Judah had published at his own expense in 1857, a *Practical Plan for Building the Pacific Railroad.* He had heard Judah speak almost mystically of his intention to cross the high Sierras with iron rails, and had endorsed Judah's proposal as presented to a transcontinental railroad convention in San Francisco in September 1859.[1]

Still, many San Francisco businessmen were of a saltwater orientation. Most freight and the majority of settlers came by sea, and the city's largest business was the distribution of goods that arrived at the port. In the 1850s local railroads were, like interior steamboats, stagecoaches, and express and wagon freight companies, merely feeders to this commerce. With nearly all goods coming through the harbor, San Francisco merchants had monopolistic control over the interior, far into Nevada. A railroad might penetrate this defensive line from the east and destroy the monopoly. Consequently local businessmen had little truck with the distant enemy; they were interested in local lines, not transcontinental dreams. Sacramento merchants eager to break San Francisco's hold on the mines, such as Charles Crocker, Mark Hopkins, Collis P. Huntington, and Leland Stanford, would think more imaginatively.

Billings was not a local merchant: he was a lawyer and increasingly a financier, and he thought San Francisco's merchants were shortsighted. True, communications to the East had greatly improved since he had first arrived in San Francisco, though not enough. Stagecoach service had taken a great leap forward when a government subsidy to the Butterfield Overland Mail Company, made possible by the Overland California Mail Bill of 1857, brought service from Saint Louis to San Francisco via El Paso, Tucson, Yuma, and Los Angeles. Still, the trip took twenty-five days, each passenger had to pay $200 and was limited to forty pounds of baggage, and the route was subject to disruption by weather, Indians, and (as the Civil War proved) the Confederate Army. The Pony Express brought mail quickly from the East—a letter posted in New York was supposed to reach San Francisco in thirteen days—but the cost was very high ($5.00 a half ounce), the riders could carry only fifteen pounds of mail, and the route was subject to even more disruptions than the stage line. Billings and others saw a railroad as the answer.

By 1860 the San Francisco merchants had come to be of the same mind. A transcontinental railroad would, they now realized, open up the interior of the continent, stimulate agriculture as markets increased, and could make San Francisco the financial center of European exchange with China. The *Alta California*, always a booster, predicted that a transcontinental line would turn San Francisco into a city of the first rank, "as a centre of population, wealth, commerce and luxury, and also, it is to be hoped and expected, of art, science and literature." A little hysterically, the newspaper declared that a railroad would increase intelligence, liberalize intercourse, improve industry, and drive out mere adventurers "as in the olden time the temple of God was cleared of money-changers by the presence of a superior spirit"—for which read the conviction that a railroad would bring to the city the right kind of industrious immigrants, men and women not attracted by delusions of gold.[2]

Even so, San Francisco's merchants did not want to finance a railroad. After all, the government was subsidizing internal improvements, especially during and after the Civil War, when a Republican-dominated and industrially inclined Congress was setting the national agenda. San Francisco had benefited from subsidized steamer lines, mail service, navigational aids, harbor facilities, and now an overland telegraph; surely railroads would follow. Crocker, Hopkins, Huntington, and Stanford—soon to be known as the Big Four—thought federal subsidy likely, but they also wanted to buy into the process, and they suggested that the city ought to subsidize a rail line. When a bond issue was defeated by the city's voters, the Big Four proceeded to build the railroad anyway; the Union Pacific and Central Pacific were completed west to San Francisco and east toward Ogden, Utah, connecting with the Union Pacific and, thus, the eastern United States, in 1869.

By the late 1860s the locomotive had become the symbol of a liberating force, of the next stage in the great industrial revolution that was putting the machine into the garden.[3] That the lines penetrated, crossed, conquered the West was itself a grand national adventure, intensely practical yet endowed with the grandeur of an epic achievement. Further, the corporation as a form of business organization, of management, of rationalizing a chaotic capitalism, held its own mysteries, its own romance as well. Billings knew he wanted to be part of all this.

Whenever Billings went to Vermont he found it further transformed by the creative—and destructive—power of the railroad. He had visited often enough to see the changes in progress, though relatively brief stays had not made him fully aware of just how much the Vermont he had known was lost, and when he returned permanently he was shocked. He did not grieve much, however, for Vermont in 1849 had been a land of little opportunity for anyone with education, a land apparently locked into the eternal cycles of farm, woodlot, and stream. Those were good cycles, but not sufficient ones, and while their predictability had been comforting in the stability they brought, they were also limiting. Now, Billings found, the world had reached deep inside his native Vermont, and on the whole he liked what he saw.

Throughout the 1850s and most of the 1860s Vermont had suffered from a drain of brain and brawn. The census report of 1850, the first after he had left for California, showed a total population of native Vermonters of 232,000. In the last ten years, Vermont had lost 97,000 people to the rest of the country, not including those who had moved to Lower Canada. The next census report, in 1860, listed 239,000 native-born Vermonters in the state and a continued drain elsewhere, especially to the West. In short, Vermont lost most of its natural growth to emigration. Earlier migration had been to New York, Pennsylvania, and Ohio, places near enough to allow family ties to remain intact; the migration of the 1850s had been to Missouri, Kansas, Colorado,

Utah, and the Pacific Coast. In 1860, 42 percent of Vermonters were living outside Vermont—indeed, more, for the figures did not take into account those outside the country. The population of Vermont had increased across the decade only three-tenths of 1 percent.

There were many reasons for this relative diminution, but the obvious one was the coming of the railroad. The first tracks were still being laid when Billings left in 1849; by 1860, even before the stimulus of the Civil War, there were 556 miles of track, a substantial mileage in so small a state. The railroads had virtually doubled the size of the lumber industry; inaccessible forests were being recklessly cut and sliced at dozens of steam sawmills to support a myriad of small local industries that, by using the railroads, poured their bedsteads and chairs, their wooden bowls and pails, their washboards, clothespins, and firkins onto a widening market. The railroads themselves gobbled up lumber in prodigious amounts: a steam train required a cord of wood to run twenty-one miles, the lines used thousands of ties and hundreds of grade-crossing signs. Dozens of local depots were built, and the trains passed through an ever more barren landscape above which rose ever more denuded hills. Farmers, idle in winter, were happy to chop and haul, delivering wood to the tracks for $2.50 a cord. The Vermont Billings had left behind had been heavily forested, though already in decline; the Vermont to which he returned had only 25 percent forest cover, thanks to the hungry locomotives that were abetted by the arrival of the charcoal burner for household heating. The Green Mountain state was green no longer, at least not with the heavy forests that had isolated it so effectively from lower New England.

Less dramatically, railroads had brought other changes. Vermont slate and marble were in steady demand once the quality of roadbeds allowed for the transport of such heavy cargo. A thriving summer hotel business had taken hold and though slowed by the Civil War, visitors soon returned on the expanding rail lines. The sawmills and local manufacturers, the quarries and rail sheds and hotels provided thousands of new jobs that might keep people at home or increase the likelihood of their replacement by others. True, they also blighted the landscape, destroyed watersheds, poured soil into the rivers and streams, and disturbed the balance between man and nature. Few cared or even noticed, however, for little was known about that balance, and there was always the pressure of the immediate pay packet, of the children at home, and the healing thought of even greater wildernesses to the west which, presumably, were not facing the ravages of change.

Agriculture, too, was changing. Railroads took produce to market for half the cost that the old wagon-freight lines had charged. Farmers could ship bulky items—potatoes, for example—at profit. The cattle business experienced a great leap forward; the railroads ran butter and cheese cars to Boston, and Vermont beef cattle, transported in special cattle cars, were outselling all others in New England. Farmers were now thinking in terms of quality, importing Ayrshire, Devon, and Durham strains to improve their herds. Sheep, although their numbers continued to decline, also improved in quality, with wool-growers specializing in producing prize rams, the Vermont Merino colonizing the flocks of New York, the Middle West, and eventually the Pacific Coast. Horse-breeding also benefited from the ability to move animals in quantity safely and quickly, and Vermont became famous for its Morgan Messenger, and Hambletonian strains. Oats and hay were up, maple sugar was in demand, agricultural societies were flourishing. Near railway lines farm values had risen about 50 percent. Vermont's economy was now one based on cash, not barter.[4]

Except for the destruction of the forests, Billings thought all this was to the good. He realized that many towns had fallen into decline when they were bypassed by the railroads, but he did not regard such decline as invariably bad. He had known small-town life and did not fully recognize the warm, traditional village life that, some commentators were beginning to say, was being destroyed by industrialization, the transportation revolution, and a dubious progress. He thought urban life was, on the whole, good. He believed railroads had helped to slow northern New England's decline, not contribute to it, and he thought that life was better; certainly it seemed so to him in Woodstock. He knew farmers had abandoned their hill farms and higher meadows because of market forces and thought that it was just as well that they had, since the railroads would bring the improved produce of the larger farms to the market and the rewards of urban life back to the countryside. While he strongly believed that Vermonters were special people, honest, industrious, thrifty, he did not accept the notion that it was better for them to labor on an unproductive and rocky farm than to work in a railroad's roundhouse.

Above all, having lived in both, Billings did not accept the growing view that city and country had to be antagonists. Bringing the machine into the garden had created a surplus rural population, and much of it had moved into cities, where it represented a new market for increasingly prosperous farmers who stayed behind. If church attendance was down, then make the churches more attractive to attend and the preaching better, and people would return; if rural life was grim, render it less so through education, improved transport, and enhanced breeds. Worldly men, professionals of education and experience like himself, had an obligation to help rural communities like Woodstock, and he was determined to do so. He wanted to create in Woodstock a model farm, to breed horses, to demonstrate that there could be an alliance between commercial and conservation, between town and country, to make his mark now on Vermont as he had in California. He would do this in two ways, he thought: on his estate and through railroads.

Railroading was enormously challenging, intellectually as well as financially. By 1866 the railroad industry was changing rapidly, for it had produced its own changes, feeding upon itself, so quickly and with such complexity as to require constant adjustments. Many men were attracted to railroading not only because it promised them fortunes but because of the demand for innovation and intelligence, for the almost constant air of crisis that hovered over the enterprise, for the sense that there were few precedents by which decisions could be guided. Certainly this was one reason that lured Billings.

The problem with Vermont railroads was that they were too small to be of sufficient significance in the larger scheme of things. Too often a Vermont line was simply one unit of a larger plan, and an ambitious man would want to seize upon that larger plan, not some small segment. Many Vermont railroad projects had been related to one or more dreams of building a single line from Portland, Maine, to "the West"—a West that was continually expanding so that it had meant, at first, simply upper New York State, then the Great Lakes, later Minnesota and the Dakotas, and eventually the Pacific Coast. By the time Billings returned to Vermont, most railroading opportunities in the state had been taken up and the lines were in the hands of local entrepreneurs who had, he was inclined to think, limited vision. There were exceptions, of course—John Smith and his son John Gregory Smith of St. Albans, or Horace Fairbanks of St. Johnsbury—though not many.[5]

Circumstance, opportunity, surplus capital, and friends had already drawn Billings into railroads. The first friend to thoroughly engage his attention was John C. Frémont, at a time when the two men were still making common cause over Las Mariposas and arms purchases. Frémont's explorations had been, in part, to scout out the best route for the transcontinental wagon road or railway line, and even before the Civil War ended he was casting about for ways to put the money salvaged out of the Mariposa debacle into railroads. In 1863 Frémont went into shaky partnership with an adventurous young freebooter, Samuel Hallett, in an effort to buy up the franchise for a projected rail line, the Leavenworth, Pawnee & Western, which they intended to rename the Union Pacific Railway, Eastern Division, in order to bask in the publicity being generated for the projected Union Pacific Railroad. Ever skirting the edge of propriety, Frémont hoped to be able to attract a loan of $5,000,000 to construct his copy-cat line, which was to become the Kansas Pacific.

This scheme died a natural death with Hallett, who in July 1864 was killed by an engineer he had fired. Worried about the resulting scandal, Frémont sold out, shortly to announce that he was president of a new line called the Atlantic & Pacific (the name he gave to the Missouri Pacific, which he bought with the last of his money). The A&P was to run from Springfield, Missouri, across the Indian Territory to the California border. Another line, to be called the Southwest Pacific Railroad, would be constructed from San Francisco along the coast to San Diego and then east to meet with the A&P.[6]

While Frémont appears not to have tried to draw Billings into the Leavenworth, Pawnee & Western scheme, he lobbied him hard about the Atlantic & Pacific, sending him a bundle of hand-colored maps, a prospectus, and an ambitious statement about the Southwest Pacific. Billings was interested: he had unemployed capital, he believed that a railroad might bring the defeated South back into the United States, and, oddly, he was still attracted to Frémont's dash, his ever-present high spirits and optimism, his willingness to take large risks.

Billings wanted to see for himself, and so in the fall of 1866 he travelled to Missouri along the projected route and into the Indian Territory. Satisfied, he committed funds to the scheme; more important, he hoped to play an active role in development, and in the spring of 1867 he often was in Washington trying to win friends for the railroad.

In Washington Billings heard much against Frémont, and concluded that the project was being badly mismanaged.[7] Unless he could get rid of both Frémont and another director, Levi Parsons, Billings wanted out. Early in June 1867, he resigned from the board of the Atlantic & Pacific. Later that month the state of Missouri seized the railroad for nonpayment of the first installment on the purchase price. About all that Billings would have to show for his journey to the southwest was a tiny town some fifteen miles beyond Springfield, which belatedly took his name when he offered the community, then called Elba, a thousand dollar gift so that it might build a union church.

Billings was still a director of the Southwest Pacific Railroad, however, and he felt he saw a way to salvage the situation. His friend Carlile Patterson had suggested that Billings discuss the company's problems with Gustavas Vasa Fox, who was known for efficiency and decisiveness as Lincoln's assistant secretary of the navy. Recently Fox had returned from an important mission to Russia, and though he had little railroad experience, he was regarded as a person who knew how to manage complex organizations. Billings wooed him intensively, declaring that the Southwest Pacific Railroad could succeed only if he became its president. Fox spent a good bit of time with Billings in Woodstock and at the Brevoort House in New York, with Billings meeting his bills while he looked for new investors, worked with Billings and Patterson to draw the early railway magnate John Murray Forbes into the net, and searched for ways to make the Southwest Pacific and the Missouri trackage work as one.

In the end Billings became disillusioned with Fox, and in September declared that he would have nothing further to do with the railroad. Fox, he had learned, was trying to create a new slate of directors, while Billings had wished to retain the core of the original group, minus—in particular—Frémont. Billings told Fox that dumping the old directors in favor of an entirely new group would be dishonorable, and apparently thought this would end the matter, but in October Fox did precisely that, and Billings at once wrote a stinging letter accusing Fox of treachery. Had Fox not gone over to the Missouri crowd, Billings thought, the Southwest Pacific might yet have attached California to Texas and Missouri well before the Union Pacific could learn how to conquer the mountains.

Fox was successful, at least up to a point, for in 1868 the Missouri legislature rechartered the railroad, with Francis B. Hayes as its president. In 1870 Fox sued the line for compensation and the recovery of expenses, and asked Billings for permission to submit seven of the letters the Vermonter had written him during their short friendship, all of them full of praise for his talents, in support of his case. Unhappily, in one of the letters Billings had also said that Hayes was not a man of the "right stripe" for the presidency and now this judgment would be made public. Billings curtly refused permission and told his lawyers to have nothing to do with Fox, whereupon Fox went ahead and added to his documentation recopied versions of nine, not seven, of Billings's communications, including one that made it clear Billings wanted Frémont off the board. When Fox's claim went to arbitration, he was awarded $25,000 and reimbursement of his business expenses. Billings concluded that he had behaved impulsively—that he had been too eager to "do something," had been too gullible about people he hardly knew, perhaps was too concerned with making an impression in California, which he had hoped would chose him as a delegate to the 1868 Republican national convention.[8]

Billings persisted in believing that railroads would be the route to riches, however. He became a director of the Southern Pacific Railroad, made a token purchase of stock in the Memphis, El Paso & Pacific, and was closely involved in a lilliputian venture, the Woodstock Railroad, though in this instance as a local benefactor and in support of his brother Franklin, since he expected the projected line would make few men wealthy. Agitation for a railroad from White River to Woodstock had led to the granting of a charter in 1847 to several Woodstock worthies, including O. P. Chandler, though nothing had come of the idea until Franklin, Charles Dana, and others extracted a fresh bill from Vermont's General Assembly in 1863. In 1867 Frederick was drawn in by Franklin, Dana, and his friends Peter T. Washburn and Albert G. Dewey— and possibly by the fact that Trenor Park was supporting a competing venture in Rutland—and began raising funds. Construction stalled at Quechee, but after much renegotiation, the track to Woodstock was completed in 1875. By this time Billings had found his new calling at the Northern Pacific.[9]

Looking back, Billings regarded 1869 as a year second in importance only to 1849, when he left Vermont for California, and 1862, when he met and married Julia. It was the year he purchased the Charles Marsh home from George Perkins Marsh, with farm and 270 acres of land; more important, in

June 1869 he instructed Colonel Granniss to begin the sale of his California properties. Though he retained some, he gave up his share in the Montgomery Block, his flagship, in order to focus seriously on railroads: 1869 also was the year Billings bought his first share in the Northern Pacific. Ten years later he would be its president.[10]

19

"A Splendid Adventure": The Northern Pacific

THE 1853 CONGRESS AUTHORIZED a series of expeditions to explore a feasible railroad route from the Mississippi River to the Pacific Ocean. Five surveys produced three acceptable possibilities: one became the Union Pacific/Central Pacific line, the second the Missouri Pacific/Texas & Pacific route, and the third was proposed for the northern tier of states westward to Puget Sound. In 1860 Josiah Perham, a Maine-born promoter from Boston, obtained from the Maine state legislature a charter for a People's Pacific Railroad Company. The company was to be financed by a million individuals, each to purchase one or more shares for $100 par value, with no one allowed to subscribe to more than one hundred shares. Perham wanted no money from Congress and sought to keep any small group of investors from dominating the railroad, which he hoped to run to San Francisco from some point between present-day Omaha and Kansas City.

When this central route was preempted by the Union Pacific–Central Pacific Railroad bills of 1862, Perham shifted his attention further north, hoping to build from St. Paul to the Pacific Northwest. With the aid of Thaddeus Stevens, a member of the House Pacific Railroad Committee and one of the most powerful voices in Congress, Perham got his bill placed before the House, only to see it defeated because many doubted the ability of a Maine company to do the job. On May 23, 1864, Stevens reported from the committee a new bill, to grant lands in aid of the construction of a railroad and telegraph line from Lake Superior to Puget Sound, and to create a federally chartered Northern Pacific Railroad Company. The bill provided for the largest land grant ever given to a railroad, an area larger by ten thousand square miles than all the New England states put together, initially 47,360,000 acres (and in the end, 60,000,000), but with no subsidy. In its final form, the bill bound the government to extinguish Indian titles to lands within the area of the grant and placed no limit on the number of shares an individual might buy. President Lincoln signed the bill on July 2, with the condition that the company must begin work within two years, complete not less than fifty miles a year after the second year, and finish the road by July 4, 1876, for the nation's bicentennial—a statement of faith in the future of the Union.[1]

The Northern Pacific Railroad was the single greatest American corporate undertaking of the nineteenth century. The bill gave the Northern Pacific alternate sections for twenty miles on both sides of the track within a state and for forty miles in a territory. At this time, of those areas through which portions of the Northern Pacific would pass, Wisconsin, Minnesota, and Oregon were states. The Dakotas, Montana, and Washington would be territories until November 1889, and Idaho until 1890. The railroad was to receive the odd-numbered sections of the townships and an indemnity area, often called lieu land, ten miles beyond the limits of the grant as compensation for lands already held by private corporations or individuals. The charter prohibited the issue of mortgage or construction bonds. Stock subscription books were opened in only two cities, Boston and Portland, Maine, to help keep control in the hands of the projectors.

Though unprecedented in extent, the Northern Pacific land grant was not otherwise unusual. With Southern Congressmen absent, opposition to lines that would not link the South with California had ended, and the conventional wisdom held that capitalists could not be expected to finance new lines in advance of traffic, though they would do so if the line gained land as a reward for completed construction. This would stimulate Western settlement and agricultural production, provide the army with rapid mobility during the period of the Indian wars, and tap sea lanes to the Orient, which was believed to be a nearly bottomless market. Just about 10 percent of the total land area of the continental United States would be granted to railroads.

The railroad acts did not give away land; rather, the railroad earned sections as construction was completed. For the NP, this was whenever twenty-five miles of line had been laid down and accepted by government inspectors. To allow a railroad time to construct lines, the Department of the Interior would withdraw both the base and the lieu lands from public entry once the railroad company had filed a map indicating the definite location on which it would build. Thus there was a premium on completing surveys as quickly as possible. Even so, while the railroad company did not, in fact, own a vast number of sections (it is more appropriate to refer to being given an opportunity to earn land instead of being given land outright), the company was able to tie up vast acreage, block settlement (or dictate its form), and negotiate loans predicated on an assumption of control over the land.[2]

Speed, persistence, a willingness to take risks, and very good connections in Congress were essential to success; so was good management, ample financing, and effective public promotions. Railroad companies could cut timber for ties and fuel and remove stone for constructing the line, but generally they did not have title to oil and mineral rights in the land. In exchange for this very substantial incentive system, railroads were expected to perform as quickly as possible and charge the federal government only 8 percent of regular rates for carrying the mail and 50 percent on other traffic, including military. These matters—winning friends in Congress and journalism, using (or protecting) the timber, stone, and mineral resources within the areas of the grants, and determining and applying rates—invited manipulation and collusion, and corruption.

While the Northern Pacific received an enormous land grant, it also faced severe problems, and in the end the grants did not, in the judgment of one close student of the situation, "come close to paying for investment."[3] A good bit of the terrain across which the line must be built was virtually treeless, so ties and fuel would have to be brought in. The line would also run through an area in which tensions with the Sioux were mounting, promising the need, at the least, to move more than the usual number of troops and supplies. Further, the line was not the first, and the initial excitement and romance of laying down iron from coast to coast would have to be regenerated. Speculation in land must be controlled, and a Land Department would need strong leadership to limit that speculation, get land brought quickly and abundantly into production, and promote immigration.

In the early running not all of these concerns were properly assessed, and matters did not go well for the company. In December 1865 J. Gregory Smith of the Vermont Central was elected president. Five years older than Billings, and also a graduate of the University of Vermont, Smith had attended the Yale Law School, served in the Vermont Senate and later as speaker of the Vermont House, and completed two terms as governor. The Northern Pacific was granted an extension of two more years before construction must begin, and Smith, preoccupied with the Vermont Central line, turned effective authority over to Thomas Hawley Canfield, a shrewd Vermonter with long experience in railroads. Canfield earned a fine reputation for the way he kept the Baltimore & Ohio Railroad line open to Washington even though the territory through which it passed was sympathetic to the Confederacy, and like Smith was well known in Canada, a decided asset since an early thought was to combine the Vermont Central with the Grand Trunk, which ran through Canada West (present-day Ontario), to connect with St. Paul. Canfield persuaded William B. Ogden, president of the Chicago & North Western Railroad, to back the Northern Pacific, and he then formed a syndicate of twelve men, each representing a major company and each to buy one-twelfth interest in the enterprise.

This group, known as the Original Interests, took over direction of the Northern Pacific in May 1867. A thirteenth interest was reserved for someone from the Pacific Coast. Smith was appointed president and Canfield general agent. Each man who held an interest could take a seat on the board or appoint someone to represent him.[4]

Billings knew Smith and Canfield well, and in the spring of 1867 went to Washington to join Canfield in an effort to wring a belated bond subsidy from Congress. They failed, though the railroad was granted a further reprieve; the deadline for beginning construction was extended to July 2, 1870, and for completion to July 4, 1877. In 1869, so that construction might start, Congress permitted the company to issue bonds secured by a mortgage on the proposed railroad and telegraph line, and the directors turned to Jay Cooke & Company to raise the money. The number of shares was increased to eighteen, with six assigned to Cooke & Company, while the board of directors was increased to thirteen, with Jay Cooke having the right to name two members. Cooke & Company confirmed with Congress permission to begin issuing bonds, and on February 15, 1870, ground was broken twenty-three miles southwest of Duluth, Minnesota.

In the meantime the Original Interests Agreement had been modified with some redistribution of interests, which were not held equally since Smith and a group associated with him held 4⅔ shares. Portions of a share could be disposed of, and in 1869, at Smith's urging, Frederick Billings purchased one from Hiram Walbridge, a Vermont-born Toledo and New York merchant and former Democratic Congressman with close ties to Salmon P. Chase. On March 9, 1870, at the first meeting of the NP board in two years, Billings was elected a director.[5]

There are two competing views of the great railroads of this time. One was that the men associated with them were heroes of capitalism, the epitome of American get-up-and-go. The opposing view, already developing in the 1870s, was that the railroads represented, or revealed, the worst abuses of capitalism, that railroad men ruthlessly pillaged the public purse purely for personal gain, that the lines held helpless farmers in thrall, destroyed the landscapes through which they passed, and made possible the Gilded Age, the robber barons, and a time of untold political corruption.

The truth, surely, is somewhere between and when a specific railroad enterprise is under the microscope, or an individual railroad leader the subject of scrutiny, it is likely to be some degree to the left or right of any hypothetical midpoint. The Northern Pacific was the only railroad planned from the outset to be managed as one enterprise, stretching from the Mississippi valley to the Pacific Coast, and as such the potential for public good or ill was enormous. Jay Cooke and the Northern Pacific precipitated the Panic of 1873, and President Henry Villard's subsequent risk-taking threw the line and many dependent upon it into bankruptcy in 1883. With the Great Northern and Burlington lines, it was the focus of the Northern Securities case which, in 1904, breathed new life into the Sherman Anti-trust Act. These three matters alone assured the Northern Pacific of a relatively bad press among historians. On the other hand, without the railroad, North Dakota and Montana would scarcely have been settled and hard spring wheat would not so quickly have become a staple of American agriculture.

This was not a time of extensive government regulations, of the bureaucratic or corporate state, though there were voices already speaking out in favor of preventing abuses through governmental intervention. While Billings might well have been one of these—he had argued for greater government regulation of banks when he was in California—for the most part he appears to have believed that business regulations best came at the state level and that the courts were the most effective weapons against abuse. As a lawyer, he leaned toward law made by judges rather than by legislatures, for the latter would, he thought, almost inevitably embrace one economic interest at the cost of another, making selections based more on emotional than on rational grounds. Certainly he felt there should be laws under which the railroads conducted their business, but he thought that legislated law must, like judge-made law, grow from experience.

Billings dealt with business matters directly, and for the most part openly, his style being one of straightforward discussion of all the issues in a dispute: when major lieutenants spoke behind each other's backs against one another, he simply went to them and told them to reconcile their differences. Given this approach to dissent and gossip, he was unlikely to feel that an abstraction like the government would be able to discern the truth of a matter that judges, with their ability to pose difficult questions, could learn, and by learning, most likely correct. That this was not a sufficient theory of management is abundantly clear with hindsight, and to some apparent at the time, but Billings's simple belief that truth was to be preferred to concealment and that what management usually wanted was best for a company rather than solely for themselves (and that if the company did not want what was best for the people, there would be remedies in the court to set the situation right) meant that he almost never commented on the kinds of issues that now preoccupy many business historians.

Cooke's plan—for it was essentially his—was audaciously simple. The Northern Pacific would be built from both ends, beginning in the east at Duluth, on Lake Superior, to and from which cargo would be transported across the Great Lakes, and in the west at some point at the head of navigation on the Columbia River north to Puget Sound. Not until 1927, when a great rock was blasted out of the Columbia east of present-day St. Helen's, was it possible to bring heavy shipping very far up the Columbia, so the eastern line would be constructed across northern Dakota, through Montana, over as yet unexplored mountains, and down to the point on the upper Columbia where its confluence with the Snake River defined a navigable stretch. Cargo would then be brought down the Columbia to link up with the shorter line north to Puget Sound, at least until the railroad could be built from the Snake across the Cascade Range directly to the sound. A Duluth terminus would give the Northern Pacific control not only over northern Minnesota but quite possibly over western Canada, since the Red River flowed north between Minnesota and northern Dakota into Lake Winnipeg. This would also promote Duluth, in which Cooke held much property, at the expense of cities further south.

The initial success of the Northern Pacific would rest upon quick and steady returns from settled lands, and an efficient Land Department would be essential. Accordingly, Cooke advised the NP to open land offices in the United States, Germany, Holland, and Scandinavia, to assure a steady stream of hardworking immigrants, and proposed that the company extend liberal credit to settlers and put up small houses in advance of the immigrants' arrival, so that their energy might go immediately into agricultural production. He urged the NP to assign agents to accompany immigrants, who ought to be recruited in groups so that they would at once have a sense of community, to protect them from "sharpers," and he recommended the construction of large hostelries at key points across Minnesota, where prospective settlers might be housed temporarily.[6]

Cooke also understood, having learned the lesson well during the war, that the press must be enlisted in support of the enterprise. Editors of key newspapers would be offered free trips to see for themselves the rich soils of Minnesota and the Dakota Territory. Humidors filled with the best cigars would be sent around to the editors' offices. Where a newspaper remained adamantly opposed to the line, usually because it had already been committed to a competitor, he would use persuasion first; if that failed, he would hope to see the owners, perhaps moved by a promise of an advertising contract, unseat the negative editor in favor of someone more amenable to Cooke's ambitions.

Most important, of course, was the money. Cooke organized a pool: subscribers would purchase at par $5,000,000 of Northern Pacific bonds, at 7.30 percent interest in gold. The interests allotted to Jay Cooke & Company ($600,000) were also for sale. The pool was broken into twelve shares, each costing $466,667. Shares could be subdivided, and payments spread over fifteen months. A private land company would buy and sell town sites, with interests in the company to be divided in the same manner. Stock would be allotted as sections of construction were finished and approved, with the pool finally to receive $41,000,000.

Many prominent men found these arrangements attractive. Rutherford B. Hayes, governor of Ohio and later president of the United States, General Robert C. Schenck, President Grant's ambassador to Great Britain and not yet disgraced by the Emma Mine, Vice-President Schuyler Colfax, Horace Greeley, Henry Ward Beecher, and many others joined in. By January 24 the entire pool of $5,600,000 was taken up. Congress passed the essential enabling legislation on May 26.

In Europe less interest was generated than expected since American railroads were then in poor repute with investors. Despite efforts by the London branch banking house, Jay Cooke, and McCulloch & Company (the resident partner was Hugh McCulloch, who recently resigned as U.S. secretary of the treasury), little was accomplished. In July the outbreak of war between France and Prussia (in which the Northern Pacific party unabashedly took the Prussian side) frustrated all efforts.

By the end of 1870, the Northern Pacific's funds were nearing exhaustion. Much money had gone to advertising and commissions. Completion of part of the line in Minnesota had made necessary the purchase of rolling stock. Cooke began the public sale of NP bonds, on which sales proved sluggish, as of July 1, 1871, bringing in only an additional $2,500,000. In June, a grand meeting, attended by four thousand people and presided over by John B. Geary, new governor of Pennsylvania, was held at the Academy of Music in Philadelphia. From July to September sales picked up, then fell off again. For the first time Cooke admitted to serious apprehensions about the possibility of failure.

Billings shared this worry, perhaps more than most directors, for he was in closer communication with Cooke than either Smith or Canfield. Billings was, by now, chairman of the NP Land Committee, and was aware that settlement was not proceeding as rapidly as hoped. One problem was that the Land Department had not been organized until the spring of 1871, and the first land commissioner, John S. Loomis, was getting bogged down in paperwork. Billings wanted to review the area of potential settlement himself, and in August 1871, Canfield led a group of directors, with Vice-President Colfax, in his role as investor, to the Dakota Territory.

Traveling to Brainerd, Minnesota, then the head of rail, with the old métis fur trader Pierre Bottineau as their guide, the party viewed the land around the Crow Wing River and the Detroit Lakes, and along the Red River. Camping occasionally and traveling by stage for most of the journey, the directors visited a Hudson's Bay Company post at Georgetown, where they reflected on how a successful line, quickly completed, might annex the western part of the confederation of Canada to the United States, and ventured out onto the Dakota prairie before returning via Fort Abercrombie and Sauk Centre. Frederick told Julia that he had visited the most beautiful country he had ever seen.[7]

Cooke and other were intrigued by the geopolitical possibilities of annexing western Canada economically to Minnesota, and on December 23, 1871, several NP directors, among them Frederick Billings, signed a secret agreement with Sir Hugh Allan, a Scottish-born railway and steamship entrepreneur. The signatories undertook to form a company to build a Canadian Pacific railroad, with $10,000,000 in stock to be issued to subscribers, most of whom were American. A Canada Land and Improvement Company would carry out the construction work and handle land, which was to consist of a grant of twenty-five thousand acres per mile east of Fort Garry (Winnipeg) and twenty thousand acres west of that point. The Canadian government was also expected to supply a cash subsidy of $15,000 per mile of construction.[8]

While the secret agreement was not illegal, since it simply stated what the group would intend to do if the Canadian government met their terms, it was both unrealistic and, in the eyes of some subsequent historians, hypocritical, since Cooke insisted that he had Canada's best interests in mind. When it became apparent that the Canadian Parliament was unlikely to give away control over the narrow corridor by which a new Canadian nation might be strung together, Cooke and Allan sought to set up a new company that would, on the surface, be entirely Canadian in origins though in fact controlled by the Northern Pacific. In 1873 the collapse of Jay Cooke & Company, the defeat of the Canadian prime minister, who had been leaning to Allan's plan, and a heightened sense of nationalism in Canada killed all prospects of such an arrangement.

During this time Cooke had become impatient with most of the Vermonters on the Northern Pacific's board, for he felt they had a limited vision. Though in fact Vermont had supplied a disproportionate number of leaders in the railroad industry, he was angry with Smith and Canfield. They were, Cooke thought, men who were too small, too insular, to grasp the enormity of the enterprise and the accompanying magnitude of the dangers.

Cooke found Smith dilatory on matters that counted and impulsive with respect to finances, and he thought both men were inattentive to detail. Time and again Cooke told Smith to be cautious, and time and again, in Cooke's eyes, Smith forged ahead with some new commitment when he should have waited, buying iron for use a year before it was needed or purchasing a steamboat on the Missouri far in advance of the line's reaching the river. Further, "the Vermont clique" now appeared to be trying to promote the town of Superior as a terminus to the harm of Duluth, and Cooke was committed to the latter.[9]

Billings believed that Cooke must not be hurt, for anything that diminished his reputation also would diminish the Northern Pacific, so closely linked were Cooke and the NP in the public's mind. By late in 1871 Cooke concluded that there were only three honest men that he could count on at the railroad: George W. Cass, president of the Pittsburgh, Fort Wayne & Chicago Railroad, Billings, and A. H. Barney, the treasurer. Cooke decided that he must get Smith removed from the presidency.[10]

Billings had proved fair and effective in his work with the Land Committee and, unlike the dilatory Smith, had moved with dispatch to complete the necessary paperwork by which the NP might come into its property. Billings wanted the company to get endorsements from the government on every twenty-five miles of track that was accepted, rather than waiting for larger lengths of rail, as Smith was doing. To be sure, Smith had a reason, and possibly a good one, for he protested that no portion of railroad should be accepted from the contractors until some months had passed so that deficient work might show up, thus forcing maintenance back onto the contractor. Smith also thought that the eastern portion of the Minnesota Division ought not to be opened to settlers until the western portion was available, since the eastern lands were swampy and heavily timbered, and premature settlement there would lead to bad reports, thus discouraging settlement on the more productive land to the west. Billings (and Cooke) argued that this reasoning was quite wrong, for the tradeoff in being able to place a temporary burden of maintenance on a contractor was not attractive if land was not conveyed to the company in a timely fashion, and more rapid construction would bring settlers onto the better lands before those who took up the poorer land could, if they did complain, be heard. Further, Billings thought that the heavy timber was not a disadvantage, for the settlers could cut it for ties, which would be transported over the railroad's own lines, or could ship timber east from Duluth in advance of the promised wheat crops.[11]

In April 1872 Cooke kept Billings from resigning as chairman of the Land Committee. Billings had been grumbling for some time about the workload, and misunderstanding Billings's concern by thinking he merely wanted a secretary, Cooke told him to hire someone to do the correspondence. In fact Billings generally preferred to do his own letters (though, fortunately for those receiving his missives, since his handwriting was deteriorating, he did hire someone); he was complaining of overwork brought on not by the volume of correspondence but by the slow responses he was receiving from Smith, with whom he too was becoming daily more disgruntled.

Billings was also in a trough of depression, for disease and death had been frequent visitors over these years. He often used his work to mask grief, while the intensity of the work itself seems to have given him bouts of querulousness. Late in March 1870, his mother Sophia took ill while on a trip to Boston, and throughout April, while she stayed at the Billings home on Madison Avenue, she suffered from a cold and earache that suddenly took a turn for the worse. On May 1 she died. Oel fell ill soon afterwards, and though he rallied, in November 1871 he began to sink rapidly. A hurried note from Julia summoned Frederick to Woodstock from his New York office, and he arrived at his father's bedside only an hour before he died. Then there were deaths of the friends—in January 1872 Henry Halleck, only fifty-six, died within a week of becoming ill, and was interred in Brooklyn—and the fears for his family:

Julia was sick a good bit of the time and pregnant much of the rest, with Elizabeth, born in 1871, and Ehrick in 1872, each a month early and sickly. Aunts and uncles, in-laws and cousins were dropping away, with funerals to attend and condolences to express. The point had been reached when Frederick, just forty-nine in 1872, turned to the obituaries in the newspaper first. He felt he ought to be home in Woodstock, enjoying what might be the last months with his family, rather than at his New York office.[12]

When Billings sought to escape from the Land Committee, he was suffering something of an anxiety attack and gave ill health as his reason for resigning. Cooke had heard, correctly, that Billings had been thinking of running for the governorship of Vermont, and suspected this was the real cause of the offer of resignation, apparently not understanding that Billings was in search of something to legitimize his withdrawal into Vermont. Cooke persuaded Billings to stay on, somewhat bluntly declaring that Billings could be both governor and chairman: "That little two penny state ain't one-fortieth as big or as important as the other office & I would rather have yr. little toe than any body else." Indeed, Cooke wanted to see Billings come in as vice-president of the line, with George Cass taking over as president.[13]

By September 1872 it was evident to all that the Northern Pacific was in serious trouble through overcommitments, rising debt, and poor management. Cooke told Barney to sell all unused iron and suspend payments to creditors. Salaries were cut and engineers and laborers laid off. He had tinkered with the railroad's administration earlier, forcing a reduction in the executive committee to five members, so that quick action might be taken, and had seen to it that, together with President Smith, the executive committee consisted of men he trusted: Cass, Billings, William G. Moorhead, his principal partner, and Charles B. Wright, a Philadelphia banker. Cooke secured Smith's resignation, but though offered in June, it did not take effect until October, which meant that Smith, even less attentive than usual, still blocked the position Cooke was eager to fill with Cass.

Then, on September 4, the New York *Sun* accused Vice-President Colfax and other prominent politicians of having accepted stock in the Crédit Mobilier construction company. Congress was in an uproar, and there was no prospect of government help for any Pacific railroad, and certainly not one Colfax was closely identified with. The bond market was stagnant, the Europeans showed no interest in shoring up what was widely regarded as a corrupt industry, and the public was unlikely to be wooed further by applications of newsprint and snake oil.

Cass became president on October 1, when he was at Puget Sound with five other directors, Billings among them, to decide on the western location for the railroad's terminus. Accompanied by the NP's chief engineer, W. Milnor Roberts, the group steamed up and down the sound looking for a spot that could accommodate the projected new city. They began with Olympia and found that the tide receded too far, making the port useless for half of every day. Steilacoom was ruled out as lacking a good roadstead. Seattle was dropped from the list because of its steep hills, while anything much further was too distant from the Columbia River. The committee settled on Commencement Bay, where a few houses had taken the name of Tacoma, and behind which rose the majesty of Mount Rainier, a location Billings had favored since his own trip to the sound. The Northern Pacific formed the Tacoma Land Company to establish the city, sell lots, and grant wharf privileges.

After he returned to New York, Cass moved the corporate headquarters from Broadway to Fifth Avenue, combining them with Billings's land offices. Free passes were called in, high-priced talent was turned to other work at lower pay, and with great difficulty, Canfield was dropped from the board, though he was named as head of the western land company, which Cooke hoped to see brought under Billings's wing at the Land Department. Charles Wright became vice-president. For the coast, Cass appointed a committee of three, headed by Judge Richard D. Rice, the president of the Maine Central, and joined by Moorhead and Billings, and asked them to go to Puget Sound once more. Thus twice in less than a year, Billings undertook the difficult transcontinental journey.

The committee arrived in San Francisco in March 1873 and quickly went to the northwest to examine the line. Rails had been laid from Kalama, on the Columbia River, almost to Olympia. Next the NP planned to build along the sound to Tacoma and Seattle, and also to link up Wallula, at the head of navigation on the Columbia, with Lake Pend d'Oreille, 208 miles toward Montana. Between Wallula and Kalama, vessels of the Oregon Steam Navigation Company, which the NP had acquired that spring, would be used. The committee approved construction of the line to Tacoma and returned to the East Coast.[14]

On June 3, 1873, the Northern Pacific Railroad reached the Missouri River. Now the resources of Montana and Dakota could be tapped, troops and supplies could be transported, and the entire project might at last be rejuvenated. The point on the river was named Bismarck, with the intent of encouraging German immigrants to the area, and the proposed route on to the Yellowstone River was refined in readiness for the next major push. The company was now due 10 million acres, and if it could move quickly enough, it might begin to meet overhead, resume payment of construction costs, and pay interest on its bonds. In April, knowing that he must gamble, Cooke had tried to conclude the bond issue at $30,000,000 by organizing a syndicate to take those still unsold. By closing out bonds issued at 7.3 percent, the railroad could issue future bonds at 6 percent. But Cooke failed, raising only $2,000,000 of a needed

$9,000,000, largely in small subscriptions, and by mid-August Cooke & Company's indebtedness nearly reached $7,000,000.

Still, Cooke seems to have been unaware of just how close and how deep the impending disaster was. Though Billings was greatly worried, falling back into his old pattern of sleepless nights and anxious days, he too does not appear to have foreseen the extent of the approaching crisis. He repeatedly urged Cooke to cut costs by paying lower salaries and curbing his tendency to hand out free passes and expensive gifts, but Billings never reached the point of near (and as it turned out, justifiable) hysteria voiced by Cooke's Washington partner, Harris C. Fahnestock.

To be successful, and be seen to be, and quickly, Cooke had employed a yeasty mix of propaganda, hardsell politics, and economic risk-taking. He had expected to use the Northern Pacific lands as collateral for loans so that he might begin building. This was not illegal, and indeed it was nothing more than the usual chicken-and-egg problem that faced a business enterprise at the time: how to take the first steps. If the investment climate remained good, if nothing disrupted European or Eastern capitalists, if the government remained intent upon getting a railroad out onto the northern prairies, if the fear of Indian attacks remained just right (neither so little as to dampen enthusiasm for a militarily significant line or so great as to frighten away potential settlers from the eastern fringes of the undertaking), if money would flow rapidly enough to allow steady construction of the line so that more money would flow as more land was sold, if the tide of immigrants would rise ever higher so that there would be plenty of cheap labor, if the public would hold to its fascination with railroads despite scandals and daily exposures of greed—if all these ifs, or most of them at least, came out right, there would be a grand new railroad and many rich men, an enlivened economy, and extensive new settlements, and Cooke would be cock of the walk.

But the "ifs" did not work out. Yes, the Indian threat remained at just the right level, with President Ulysses S. Grant continuing to support military expeditions into the Dakotas and reminding the public of the need for faster transport. Yes, labor remained cheap, and even though the railroad was being constructed in advance of settlement, in the first two hundred miles or so beyond Duluth immigrants flowed quickly onto the land. But success required more than this and more did not come. Labor was less and less docile. European investors remained cautious. The Crédit Mobilier scandal discredited the vice president and other prominent Republicans. Cooke's propaganda was not always successful, and critics were able to make the derisive label, "Jay Cooke's Banana Belt," stick on a good bit of the northland as Cooke's propagandists overreached themselves in their descriptions of the salubrious climate. Most important, the figures did not add up: that balance between revenue from land that belonged to the railroad by virtue of completed construction, of loans outstanding against other land that construction had not yet reached, and of rising interest rates in the agricultural sector as European demand for American farm products declined, produced a heart-stopping squeeze, most particularly on railroads, and most especially on Jay Cooke & Company.[15]

The first tremor began in New York. The Northern Pacific Company owed Jay Cooke & Company $5,000,000 and could not pay. One could not rob Peter to pay Paul; deposits in New York had been declining steadily while the Philadelphia branch had continued to draw on New York. When the tremor came, the New York house had nothing to fall back upon. On September 18, Fahnestock, now in New York, asked the heads of the major banks in the city to tell him what he must do, and shortly before 11:00 A.M. that day he closed the house. The Philadelphia house followed in a few minutes, and then, like falling dominoes, the Washington house and Henry Cooke's First National Bank of Washington. Wall Street reacted in a frenzy, and on September 20, for the first time ever, the New York Stock Exchange closed. Cooke had not been warned in advance that the New York house was closing, reading of it by telegram after the fact. Only the night before he had entertained the president of the United States at his estate outside Philadelphia. Now he sat in his office and wept.

The Panic of 1873 was without parallel, a crash of the capital market that brought on a nationwide depression of previously unknown severity, coming at the end of a period of aggressive and unrealistic optimism. The sense of panic was far more than financial. The idea that someone—the politicians, the bankers, the newspaper editors suborned with gifts of stock by the financiers, the railroads—had betrayed a national dream, had put to an end the rising expectations of a vast and mobile number of men and women, changed the climate for railroads thereafter. All railroad building was suspended, and the iron industry was stopped in its tracks. Not only had Cooke failed: his failure would haunt the Northern Pacific for years to come and make it extremely difficult to take up construction once again.

From the outset Cooke and Billings saw the potential land grant as the key to the NP's success. They wanted to get settlers onto the land quickly, both to grow crops to be sent back over the line and to serve as a market and inducement at the growing end of track, so that, as Billings often wrote, the company could be a "growing thing." Settlers who were already adjusted to the New World were preferred, though major efforts were to be made in Britain, Scandinavia, Germany, and Switzerland, where agricultural labor was depressed and the climate was at least somewhat similar to that found along the NP's route. The NP offered special incentives, including cooperative programs with steamship companies and reduced transportation costs, and comfortable reception houses were to be built in Minnesota, where detailed local information and tools at discount were available. Intensive efforts were made through Civil War veterans organizations, and the NP persuaded Congress to increase from eighty to a hundred and sixty acres the amount of government land within the limits of railroad grants that veterans might take up and to count the period of service toward the five-year residence period for perfecting title.[16]

The Northern Pacific's board named a Land Committee on February 1, 1871, with Frederick Billings as its chairman, to set policy and provide guidance to the Land Department. Though the committee worked closely with Jay Cooke & Company, it was independent on all matters relating to land. Under Billings, and assisted by a permanent clerk, there was to be a land commissioner who would administer the department and work in trying to influence legislation, obtain favorable rulings from the courts and the Department of the Interior, and cooperate with Cooke's promotional department in attracting settlers and shaping news reports.

There was also a full-time lobbyist, Benjamin F. Wade. A former Senator from Ohio who had served from 1851 to 1869, Wade was an early anti-slavery Whig, a Radical Republican, and an advocate of harsh punishment for the South. In choosing Wade, the Northern Pacific knew that it was cutting itself off from any potential Southern support, but since it seemed unlikely that many congressmen from the South would vote in favor of the most northernly of the transcontinental lines, and since Wade knew well many of the most powerful men in Washington, he was a logical choice. Wade proved especially valuable during Zachariah Chandler's term as President Grant's third secretary of the interior, for the two had worked closely together against the South before the war. Wade felt warmly toward Billings, whom he met in conjunction with Frémont's unsuccessful presidential campaign in 1856, for Billings supported Wade in 1867 when it appeared that he would succeed an impeached Andrew Johnson as president. Defeated in 1869, Wade had returned to Ohio to lick his wounds, and was delighted to be linked with Cooke, Billings, and the "Pennsylvania group" at the NP in this way.[17]

The work of the Land Department was threefold, with each activity meant to take place simultaneously. There was the necessity of getting the lands in hand: mapping, platting, appraisal, clearing out fraudulent claims, processing patents, and opening local offices for title searches. There was "the educational or missionary work," which embraced advertising, publications, and the hiring and deployment of lecturers and agents. And there was the emigration work, setting up the contemplated special facilities so that settlers might reach the lands. The key, Billings knew, was in coordination, for while the first could begin at once, before the land was confirmed, the second and third phases must proceed together and only very shortly behind the first.[18]

The Land Department got off to a slow start, as the Northern Pacific's executive committee did not meet promptly. With Smith away and Canfield failing to provide the required statement of organization, Billings found he lacked authority to hire or to act. Throughout February and March 1871 he complained to Cooke about the time being lost while his requests for a meeting of the board were ignored. Cooke wanted Billings to go to Washington to lobby for the company; Billings was adamant that he must remain in New York to organize the department—which, he added, probably ought not to be in New York at all, but in the field, close to the land actually to be granted. He wanted to get out to see it for himself and disliked being tied down to paperwork that others could be doing. Finally, at the end of March, Billings received the powers he needed, and the work of the department began.

After casting about, Billings settled on John S. Loomis as land commissioner. Loomis had been president of the National Land Company, which had been effective in colonizing the lands along the Kansas Pacific Railroad, and he had been secretary to a national immigration convention that met in Indianapolis in November 1870. At Billings's request, Loomis promptly drew up a plan for operating the Land Department, substantially refining the broad outline as worked out by Cooke, Billings, and others. He then set to work on the multitude of tedious small tasks he knew were essential in preparing for settlement, ultimately becoming so engrossed in these that he lost Billings's confidence.[19]

The NP soon added a variety of inducements to attract settlers. Emigrants were allowed to deduct from the price of land they purchased the cost of their ticket to reach that land, provided they travelled on a route approved by the NP, bought at least forty acres, and did so within sixty days of arrival. The company allowed a one-third deduction of costs on all freight. It agreed to purchase all wood cut by settlers who took timber land; the wood was then used to warm the tiny stations that sprang up across Minnesota, and stacks of logs were sold at cost at the stations to settlers who arrived too late in the season to lay in a sufficient supply of their own. Land prices were held below comparable Eastern costs, and contracts were attractive: 10 percent down, with 10 percent a year for three years and 15 percent a year for a final four years at 7 percent annually.[20]

Trusting Loomis to handle the technical work, Billings saw his first task as chairman of the Land Committee to be that of a skeptical inquisitor who must keep various nephews of directors and well-placed politicians from being given sinecures and prevent Cooke from paying salaries that were too high. His old friend from California, J. Ross Browne, asked Billings to take him on as a publicist, and he declined. When John Russell Young, a well-regarded journalist who was close to President Grant, was under serious consideration, Billings argued that Young lacked business skills, however well he wrote. When the idea came up that Mrs. Frances Fuller Victor, who was writing a book about Oregon and Washington, might be paid, Billings thought not. Anything written about the Northern Pacific would have more credibility if not done by an employee; the better way was to commit the company to the purchase and distribution of a bulk order of copies, for this would go some way toward assuring favorable comments. Special arrangements would always become known, Billings said, and would detract from the credibility of the Land Department's claims.[21]

Then there was oil—the term for making people who appeared to feel ill-disposed toward the railroad take a happier view of it. Much oil continued to come from Cooke, and Billings left the gifts of cigars and wine (and, though never proven, probably money) to others. He did, however, readily intervene when he thought that friends had gone astray; his usual manner was one of straightforward if friendly confrontation, going directly to someone's office and insisting that, as a gentleman, he be given a hearing. His old friend Samuel Bowles took to attacking Cooke in the influential Springfield *Republican*, and Billings asked him to stop it, which (for whatever reason) Bowles did, though the testy editor remained a friendly critic, cautioning Billings when he felt that the Northern Pacific people were demanding too much from Congress. Cooke asked Billings to intervene in a furious feud that broke out between General A. B. Nettleton, head of the propaganda department, and Samuel Wilkeson, a writer for Horace Greeley's *Tribune* hired to publicize the key bond issue by which the railroad would gain its essential loan. Billings was unhappy with both men—he thought Wilkeson was dangerously exaggerating the quality of Minnesota wheat lands and he was angry with Nettleton, who had invested a small sum in the Emma Mine and helped draw Ambassador Schenck into the 1872 scandal. Nothing was more important, Billings wrote, than both the reality and the appearance of financial probity at a time when the success of the Northern Pacific depended upon public acceptance of its bonds, its publicity, and its promises.[22]

There were those who felt that Billings was a bit too pious in such matters, for he, too, had bought property along the railroad line and was not above drawing politicians into the Northern Pacific's net. Certainly both charges, if charges they were, are true, though no one appears ever to have suggested that Billings used either money or gifts to create political alliances. Rather, he seems to have argued on the grounds of self-interest and, in some instances, from the threat that not being on the winning side—by backing another railroad, which would not succeed—could be costly in political terms.

This was especially so in Montana and Washington territories. In the latter, Cooke hired a Vermonter, Selucius Garfielde, to lecture on the benefits the railroad would bring to the territory. Cooke thought Billings would be pleased, since he had known Garfielde in California, where he had been a member of the state legislature with Peachy and, in 1852, had helped to codify the state's laws. However, Billings was displeased, since Garfielde was also Washington Territory's delegate to Congress and was "too much of a politician," with too many enemies who derided him as "Selucius the Babbler." In 1872, when Garfielde lost his bid for re-election despite efforts by the local NP faction to keep him in office, Billings advised the general agent for the Pacific Coast, John W. Sprague, that the defeat did not matter much, since the new delegate, Obadiah McFadden of Olympia, a former chief justice of the territory, would realize that it was in his own interest to come over to the Northern Pacific. Billings would apply the same hard-headed assessment to the territorial governor of Montana, Benjamin Franklin Potts, who in 1877 looked for a time as though he might defect to the Union Pacific.

Billings also wooed Vice-President Colfax, at least until he was implicated in the Crédit Mobilier scandal, promising to support him in any bid he might make for the presidency. Of course, Billings may have honestly felt that Colfax was the best man, for there were many who did not want to see Grant run for a second term. But when Colfax pressed Billings to begin to advertise the Northern Pacific lands in Dakota Territory, Billings told him that the company ought not to do so until the government had passed title, no doubt hoping that Colfax would use his influence to expedite matters.[23]

Promotion by its nature can easily lead to corrupt practices, and the Northern Pacific undoubtedly trod upon the edges of propriety even in so loose a time. If newspaper editors were open to bribery, they were bribed, with a free pass here, a pleasant junket there, a share, a lavish dinner, the offer of a guaranteed purchase of any favorable book they might write. Certainly Wilkeson allowed a mystical lyricism to creep into his prose, and when he described the wheat lands through which the NP was to pass as the best in the world, even Cooke objected. However, if reports lacked "sparkle," Cooke asked for more exciting prose, and if two faces could be put on a development—Indian activities to the west of the line, an ambiguous conclusion about the quality of spring wheat— he argued for the happier conclusion. While Cooke wanted the engineers' reports to "stick to the truth," he observed that "there are several ways of stating the truth" and invariably preferred the enthusiastic way.[24]

The NP produced so many informational brochures, so much material meant to inform the prospective settler, such a flood of statistical data, as to make parody quite easy. One unhappy New York broker capitalized on this by promising investors a new map that would provide "a correct census of the inhabitants, including Indians, foxes, muskrats, white bears, black bears, grizzly bears, green bears, polar bears, Wall-street bears, bisons, stationary herds of roving buffaloes . . . grasshoppers and wheat fields." In response to the somewhat ill-advised promotional statements that the prospective land grants were larger than all the New England states combined (true enough, but easy fodder for those who wanted to impress the public with how profligate the government had been in supporting the Northern Pacific), the parodist declared that the grant was larger than all nine (sic) New England states, combined with all the northern part of European Russia, all of France (except for Alsace and Lorraine), all of Turkey, the "two States of Delaware," and a portion of New Jersey, with Coney Island added—a pointed dig at the presumed ignorance of the European settlers at whom such literature was aimed.[25]

What Billings feared most from overly enthusiastic promotional brochures was that they would open the door to land speculators who would move in and forestall the kind of extensive settlement, based on yeoman farmers with families, he felt the line needed. His fear was realistic and his expectation that the Land Department could prevent speculation unrealistic. There were, after all, three land companies, all selling town sites, and as these companies frequently acted independently of the railroad, they often failed to promote Billings's three-phase plan.

Two of the companies, the Western Land Association and the Lake Superior and Puget Sound Land Company, were founded in 1870. The first was organized by officers of the Lake Superior & Mississippi Railroad, Cooke's line, which ran from St. Paul to Duluth, and the railroad's financial agents, E. W. Clark and Company. The association eventually held seven thousand Duluth town lots and four thousand acres of pineland, and reaction to it in the port city was quite negative. When the company did not limit itself to Duluth, however, and spread into Superior, across the bay in Wisconsin, it triggered even more hostile reactions, and when the Northern Pacific took control of the Lake Superior & Mississippi Railroad, the association's practices were changed.[26]

The Lake Superior and Puget Sound Company was far more important, since it claimed to handle sales from Duluth to the Pacific coast. The Northern Pacific controlled two-fifths of its stock; other individuals, including some who held NP interests, owned the other three-fifths. Thomas Canfield was president and three other railroad company directors were also directors of the land company. The Puget Sound company proved a serious drag on the Northern Pacific, and both Billings and Harris Fahnestock warned against its repeated secret selection of choice town sites. The two companies were so intertwined, it was impossible to separate their finances, and a few directors speculated with considerable success, to the detriment of larger goals.[27]

On the West Coast there was the Tacoma Land Company, set up in August 1873 to sell town lots near the Northern Pacific's western terminus, with Billings as vice-president and fellow director Charles B. Wright as president. John C. Ainsworth, owner of the Oregon Steam Navigation Company, was the third director. Billings had a particular interest in this enterprise, both because he immediately purchased lots on his own, as well as land in the Cowlitz Valley, and because he hoped to see an orderly settlement that would avoid the excesses he witnessed in San Francisco.[28]

The Tacoma Land Company was soon charged with speculation. In January 1872, general agent Sprague, together with John N. Goodwin, former territorial governor of Arizona and currently a local lobbyist for the line, announced that Olympia would be the road's terminus though in fact the selection committee was inclined toward Tacoma. In July when the New York office declared that to be the case, some felt Sprague had deliberately misled the public to forestall speculative price increases and, not incidentally, to allow Northern Pacific directors to get their own land selections in early.[29]

In December 1872, Billings did ask Sprague to choose sections for him along the line as it came up from Kelso, and subsequently when lignite coal was found to the east of Tacoma, he purchased land there as well. Three years after he bought 7,800 acres in Lewis County at $2.50 an acre, he sold most of it for $6.25 an acre, while retaining his Tacoma parcels, on which he anticipated a profit of $25,000. However, these were not large profits, and by December the public was well aware that Tacoma, not Olympia, was to be the terminus, though the line did not reach this point until a year later.[30]

Billings was quite intrigued with Tacoma, which he thought occupied a particularly beautiful position, with two large islands dead ahead from Commencement Bay (which he wanted the railroad to buy) and Mount Rainier rising less than forty miles behind. In the summer of 1873 he sent his sisters to San Francisco and then up the coast to see Tacoma, where they were looked after by Captain Ainsworth, and he instructed Sprague to clarify titles, establish clear policies, and act fairly toward all prospective settlers in order to protect the railroad's long-range interests. Billings commissioned Frederick Law Olmsted to prepare a master plan for Tacoma, and despite the collapse of Jay Cooke & Company, he and Wright received an attractive design from Olmsted, though early in 1874 they abandoned it in favor of the customary Western grid pattern.[31] (Subsequently Billings did hire Olmsted.)

Billings and Sprague also worked to protect the timber on Northern Pacific lands in Washington Territory. It was, Billings knew, prey to constant theft, especially if the railroad did not confirm its ownership as each segment came due. Local timber-stealers could observe that a portion of the line had been finished and cut the timber before either the railroad or the government could stop them. The land technically remained in government hands until the line was inspected and conveyed to the company, and protection of the timber legally fell to the government. But as a practical matter this clearly was not the case, since the government was in no position to enforce the law along most of the trackage. Garfielde sided with the timber thieves, Billings said, since they had the votes to reelect him. Of course, the government might be wise to wink at the theft, for the railroad could sell to the loggers, as it would do once title had been confirmed, and thus save the expense of trying to police the land. This was acceptable, however, only if the railroad promptly applied for title to the land under conditions that would definitely lead to confirmation of ownership, and this it was not doing, Billings warned. Indeed, the bulk of the work of the Kalama office from June 1871 until mid-1874 related to protecting the Northern Pacific's timber resources, working out stumpage agreements with loggers, and sending railway police out to seize logs cut illegally.[32]

Still, these matters were sideshows, as Billings frequently reported. What really counted was the "splendid adventure" of building the railroad and getting settlers, preferably in substantial colonies, onto the land west of Duluth. There was a complication: the St. Paul & Pacific Railroad, which had built from St. Paul as far as St. Cloud, only fifty miles from Brainerd, was siphoning off potential settlers and using most of the same inducements. Chartered to run northwest from St. Paul toward the Red River and British North America, the St. Paul & Pacific had played an important role in Cooke's scheme to control the lower Red River traffic in what, in 1870, became the province of Manitoba, and that year the Northern Pacific had taken over those portions of the SP&P known as the First Division and the St. Vincent Extension. This entanglement was dragging the Northern Pacific down—by April 1872 Jay Cooke & Company had advanced the St. Paul & Pacific over $1,000,000—and would be one of the causes of Cooke's failure in 1873. Because Cooke was committed to his dream of making Duluth the gateway to the Northwest, he did not see that St. Paul or Minneapolis were far more logical anchors for the Northern Pacific. Thus two lines, each with a different entry city, competed for settlers along two parallel routes which at one point were only forty-five miles apart. All the more must the Land Department emphasize large-scale colonies to speed the plow.

But colonies did not rescue the Northern Pacific, and the anticipated "colony fever" never occurred. Cooke's insistence on Duluth as the port of entry, Billings's loyalty to him on this point despite growing evidence that they were wrong to resist the attractions of St. Paul, a severe blizzard in the winter of 1872–1873, the worst grasshopper scourge even experienced settlers had ever seen in the summer of 1873, and the collapse of the entire NP enterprise that September, put settlement on hold well before the goal of fifty thousand settlers that Loomis and Billings had set could be achieved.[33]

Billings had now turned against John Loomis. Impatient, pressing, a workaholic himself, knowing that Smith and the Vermont clique were up to their eyebrows in debt and distraction over the Vermont Central, Billings had repeatedly told Cooke, Smith, Loomis, and anyone who was within reach of his pen that there was not a moment to be lost in inducing settlers. Still, they were not coming. And in February 1872, Billings discovered that his land commissioner had not done the paperwork nearly so meticulously as had been assumed, and that land under offer was not, in fact, ready to be occupied.

Billings had been unhappy with Loomis on grounds other than slow progress. Loomis had credited as settlers some of the agents of Canfield's Lake Superior and Puget Sound Land Company, and this had caused public complaint in Duluth and along the line. He allowed the company to price land in such a way as to bring on further complaints and encouraged squatters in order to get people onto lots quickly rather than proceeding in the orderly fashion his paperwork led Billings to expect. He had also allowed himself to be taken in by Lord Gordon Gordon, a bogus peer.

In retrospect, the story of Lord Gordon Gordon is one of the more amusing tales in the annals of railway chicanery, but it can hardly have seemed so to Billings and Cooke.[34] In the summer of 1871 a stranger, styling himself as G. Gordon, showed up in Minneapolis, opened a generous bank account, and began to frequent all the right places in the anxious city. Mr. Gordon did not declare himself to be of the nobility, but one day soon after his arrival his titled background was revealed when a letter came to his hotel addressed to Lord Gordon Gordon. Minneapolis society palpitated in the presence of a lord, and someone suggested to Loomis that it would be a good idea if the titled Scotsman were given the royal treatment. Soon after, Lord Gordon Gordon let it be known that he intended to buy several hundred thousand acres of land in order to move part of his tenantry off his overcrowded Scottish estates. Loomis promptly organized a buffalo hunt, complete with white-gloved waiters, a French chef, and two wall-tents for the noble lord's exclusive use. A government surveyor raced ahead of the lord's caravan marking choice sections of land on a map. The affair took two weeks and cost the railroad $45,000, cheap at the price since Loomis predicted that the Scot would spend $5,000,000.

When Lord Gordon Gordon proved to be a charlatan, Loomis lost all credibility in Billings's eyes. Still, the bogus lord must have been very convincing, for he took in not only Minneapolis society, which gave him a farewell banquet, but also for a time both Horace Greeley and Jay Gould, neither noted for their gullibility. In a time of colorful scoundrels, Lord Gordon Gordon was, until his suicide in Winnipeg in August 1874, one of the most successful.

Deeply worried, Billings was not persuaded by those who felt that Loomis had done about as well as anyone could under the circumstances. In a sense, Loomis paid the price of overly optimistic expectations, many of which he encouraged himself. He was, in the end, incurably naive—he told one agent sent to Europe that the agents of the steamship lines would no doubt provide information on prospective emigrants without payment—and inclined toward unrealistic inspirations, as when he decided after the great Chicago fire of 1871 that he would be able to get thousands of settlers out of the fire-ravaged city, an idea that helped the St. Paul & Pacific but the Northern Pacific hardly at all.

Billings dismissed Loomis in February 1872 and in March the board abolished the Land Committee. Billings instructed the clerks and surveyors to abandon Loomis's detailed platting program and to prepare rough township plats sufficient for getting settlers on to the ground quickly. For eight months Billings was managing director of the Land Department. However, he could not do the work of two men indefinitely, and as he turned his attention to the details of immigration, he looked for someone to take over the duties if not the title of land commissioner. In November 1872, he appointed William A. Howard, though Howard was unable to take up his office until the following March. Billings then reestablished the Land Committee and assumed the chairmanship again. The immigration department was, by then, offering the promise of results, and it was expected that 1873 would be a big year.

But 1873 followed 1872 as a failure. The Land Department's European agents had not begun to function well until the peak of the emigration season had passed. News of the severe winter was circulated throughout Europe and exaggerated by the Northern Pacific's competitors, so that much effort was needed to prime the pump in the spring of 1873. By September, the Land Department had sold only 41,000 acres of Minnesota land. The West Coast operation also was going badly, with Sprague committing much time to politics and preoccupied with timber contracts. And then, of course, came the collapse of September, with everything put on hold.

20

"A Great Triumph": Reorganization

I N 1873 THE NORTHERN PACIFIC APPEARED to have proved its critics correct. Neither Smith nor Cass had been able to lead it to success, and it
had, or so the public concluded, carried Jay Cooke & Company down with
it. Many who had praised the line when it held high promise now ridiculed
it. The Philadelphia *Ledger*, always an opponent, dismissed the original dream
as only "a wild scheme to build a railroad from Nowhere, through No-Man's-
Land to No Place." There was still a Northern Pacific Railroad Company,
there were trains to run over its route from Duluth to Bismarck, and there
was the prospect of at least some continued settlement on its lands. But far
more was called for than the belt-tightening of 1872. In the face of Cooke's
collapse employees were let go, all salaries were cut, most leases were abandoned along with stock in other companies, and bonds were pledged to obtain
operating funds, often at the rate of three dollars in bonds for one in cash.
On the West Coast, traffic was reduced to one mixed train of freight and
passenger cars operating in each direction on alternate days. Stock in the Oregon Steam Navigation Company was sold. The Northern Pacific's debt grew
$2,000,000 each year.

In this extremity, Frederick Billings rendered the Northern Pacific what
may have been his greatest service. He revitalized the Land Development, the
only hope for income, and in 1875 carried through a bold plan of reorganization that made the company's survival possible. Nothing he did later as president was as important.[1]

First there was the problem of the land. Only increased sales would give
the Northern Pacific the necessary credibility to make reorganization possible.
Only a vigorous new land commissioner could sell the land.

Billings took stock of the situation. Between May 1872 and his assessment,
the Land Department had sold not more than 46,000 acres. A third of this
was at one location, Glyndon. The average price had been $5.40 an acre,[2] but
since most settlers had made only their down payment and few had been on

the land long enough to owe their first installment, the railroad received very little cash. The summer had turned the soil so hard that from a distance it looked like granite. The grasshopper plague provoked many editirials in which Job figured prominently, and the press concluded that Dakota Territory, in particular, was one vast wasteland of dust, insects, heat in summer, and blizzard in winter. What was needed was a demonstration that the land was as productive as all those Jay Cooke brochures had said.

Even before the crash, Billings had begun to make land available in exchange for the Northern Pacific's $100 bonds, which in 1873 fell to below $20 in real value. The company now honored the bonds at $110 if exchanged for land, an offer which if accepted would help clear out the debt while getting more settlers onto the land. But colonists still were not coming and bills were due: why not, then, pursue the largest bondholders aggressively, for they could take up the most land, clearing the debt faster though not adding independent farmers along the line. If the large property holders would put their land into production, they would need hundreds of tenants, who would in time become permanent settlers and, in the meanwhile, would constitute a vital market.

With James B. Power, Billings found the man who could put all of this together.[3] Born in New York and schooled in Massachusetts, Power had gone to work for the New York & Erie Railroad as a ticket agent when he was seventeen. He followed railroads west to Minnesota where in 1867 he became chief draftsman in the St. Paul office of the U.S. Surveyor. In 1871 Power joined the Northern Pacific Railroad as chief clerk for the Minnesota-Dakota Division of the Land Department. Early in 1874 Billings appointed Power land commissioner, a position he would hold until the end of 1880.

Power hurried to New York and advised Cass that the company must reduce the price of its lands in proportion to the acreage a purchaser put into cultivation within a specified time. Cass immediately agreed to reductions ranging from 20 to 60 percent. Next Power told the directors that they must demonstrate their faith in Dakota Territory by exchanging their bonds for land in large amounts, and Cass and director Benjamin F. Cheney instructed Power to select suitable sections for them, as did Billings shortly after, taking four sections near Valley City.[4] Power chose eighteen sections west of Fargo, near present-day Casselton, and Cass and Cheney agreed to have not less than one-quarter of the land broken each year for three successive years.

In August 1874 the scheme received a grand boost: a settler who had homesteaded in the summer of 1871 rode into Fargo with a load of harvested No. 1 hard Red Scotch Fife wheat, the yield of forty acres. He sold his wheat at $1.25 a bushel, which put $1,900 in his pocket, and the word went out among farmers as though he had struck gold. Billings had been waiting for word of the summer harvest, and he immediately spread the news. Thereupon began the only sustained land boom in Dakota history. The company sold 190,000 acres in 1874, and in 1875 sales rose to over 475,000 acres.

The Cass-Cheney Farm was immediately seen as a demonstration project. In the first season three thousand acres were broken and backset. Power then invited an old friend and experienced wheat grower, Oliver Dalrymple, to become farm manager. Cass gave Dalrymple a contract for a conditional partnership, and by adding some Indian scrip and soldiers' "additionals," he acquired three sections of adjoining government lands, thus producing a farm of 13,440 acres. Dalrymple promptly seeded the three thousand acres already broken and plowed another three thousand. The Cass-Cheney Bonanza Wheat Farm, soon to be known as the Dalrymple Farm, was born.

Dalrymple understood farming, he had flair, he took risks, and he saw much of life as an act of theatre; he was ideal for the enterprise. While other and larger bonanza farms would follow, none would become an equivalent tourist attraction. Charlemagne Tower, pioneer of the iron industry and a director of the NP; Lord Francis Sykes, a genuine peer from England; Marshall Field, the prince of mercantile Chicago; and many others joined in the purchase of display farms. By 1878 the Northern Pacific realized nearly $8,000,000 from sales to some three thousand purchasers.

The Northern Pacific and Billings were also lucky. Ample rainfall broke the drought that hovered over the land until 1874, while European farmers faced crop failures in 1879–1881, precisely the crucial years for the NP and Billings's presidency. Desirable land to the east rose in price, and by 1879 taxes forced more farmers west. After 1875 the grasshopper stayed away for several years. In 1880–1881 there was another particularly hard winter, but this one proved a blessing, for the first snow fell on unfrozen ground, and because of the depth of the snow, the earth remained relatively warm. At snow melt, the runoff saturated the soil and filled the undrained depressions across the prairie, raising the water table and creating literally thousands of small lakes and ponds, which added to the humidity of the region so that in summer it appeared attractively green and remained so until 1886. Harvest yields continued high between 1880 and 1883, with that of 1882 perhaps the greatest in bushels per acre ever achieved in the territory. The territorial population shot up from 135,000 to 330,000. The long boom did not break until 1889, when three years without sufficient rain again took their toll.[5]

Critics of the Northern Pacific could make a case against the bonanza farms. Because large bondholders were favored, the small farmer was squeezed out, and northeastern Dakota became a land of huge spreads, quite commonly in the tens of thousands of acres. Further, corporations often purchased the land, so that the area suffered the problems typical of absentee landlordism, and speculation pushed values up along the Red River.

Billings was aware of these problems. To discourage speculation, purchase terms were changed, with land east of the Missouri River set at $4.00 an acre (with a rebate of 25 percent for land cultivated within two years) and at $2.60 an acre west of the river, where Billings continued to hope that the future lay with the small homesteader. The railroad was consciously lax about enforcing contracts, since it wanted settlers to remain, and in combination with market forces the Land Department worked to break up the largest farms so successfully that by 1880 the average farm was only 326 acres, arguably too small for the region.[6]

On the West Coast there was a modest land boom for coal, timber, and the rich agricultural meadows in the Cowlitz and Chehalis valleys. Early in 1874 Nathaniel Jacobs was removed as assistant land commissioner and George Hibberd, who had performed well as superintendent of emigration, replaced him. In October the Pacific district office was moved to Tacoma, sharing quarters with the Tacoma Land Company. Land sales soared: between 1872 and 1874 only 17,000 acres had been disposed of, while in 1875 over 144,000 acres were sold.

Still, Billings, Power, and Hibberd could not do enough. Most land purchases had been made with bonds, which produced little operating capital. In 1874 the company leased and sold its surplus rolling stock, dismissed more staff, sold its investments, and continued to fall behind. In January 1875, it passed up its option to buy back the stock of the Oregon Steam Navigation Company. In April, the company was ordered into receivership.

Not all bondholders had wanted land, of course, and a suit against the Northern Pacific in August 1874 brought a court order prohibiting further bond issues and instructions to the trustees to sell lands for the bondholders' benefit. A pending foreclosure suit in January 1875 precipitated the final crisis, and a bondholders committee consisting of Frederick Billings and five other directors recommended receivership. On May 20 bankruptcy proceedings were heard before the U.S. Circuit Court in New York, with Judge Nathaniel Shipman presiding.

Shipman was one of the most learned judges on the federal bench. A friend of railroads, he realized they demanded innovation in the application of Anglo-American common law. He appointed President Cass as receiver, ordering him to operate the property and apply all proceeds to reducing creditor claims, and Cass resigned the presidency, with Vice-President Charles B. Wright, who was a conservative and prudent Quaker, succeeding him. A reorganization committee, composed of the same six directors as on the bondholders committee, was instructed to come up with a plan for salvaging the company.

As chairman of the executive committee, Billings was senior member of the reorganization committee, and felt it was his responsibility to design the plan for restoring the Northern Pacific to solvency. In doing so he won renewed respect throughout the financial community and put himself in line for the presidency. There were, he later explained when defending the proposal before Congress, three parties whose interests had to be considered: the bond-holders, of whom there were eleven thousand, those who held stock, and the "proprietary" or original interests. If the reorganization plan were subjected to prolonged litigation, the larger goals of harmonizing these interests, ridding the road of its debt, and beginning the next stage of construction would not be realized.

Billings proposed to divide $100,000,000 of capital stock into 51 percent preferred and 49 percent common stock. The bondholders would receive $30,000,000 in preferred stock, with $1,400 in stock for each $1,000 bond. (The original 7.30 gold bonds were set at 8 percent payable in currency. Two years' interest had already accrued, and the preferred stockholders would be given three years' advance, that is, five years' interest at 8 percent being 40 percent, hence the figure of $1,400.) This would absorb $42,000,000 in pre-ferred stock, leaving the company with $9,000,000 for general purposes. The common stockholders would exchange stock share for share, could not vote until July 1, 1878, and were not to receive dividends until the preferred stock-holders had received theirs. After deducting both preferred and common stock, the remainder of the capital stock would be distributed among the proprietary interest. Thus each investor received something: the bondholders the most, the common stockholders the least.

There was a good bit of public opposition to the plan, largely on the hostile ground that the company had forfeited its rights and could not continue to function. Billings hired George Gray, a leading figure in corporation law, to help the Northern Pacific's general counsel, F. A. Lane, argue the case, and Gray held that Congress, by authorizing the company to make a mortgage, including a franchise to incorporate, had intended that in the event of foreclo-sure any purchasers would still be the Northern Pacific Company—that "the right to be, the entity, the spiritual life" of the company went with the sale as surely as material property did. Gray's "somewhat novel" argument[7] pre-vailed, and soon after he replaced Lane as general counsel.

There followed two weeks of great tension. On July 26, 1875, the court order of May was approved, so that the sale might proceed on August 2, and on July 30 the purchasing committee reported that two-thirds of the bond-holders accepted the plan. Billings assembled the committee on July 31 and discussed the details of the sale; there was, they concluded, a problem. Thus on August 2, Billings, the committee, Jay Cooke, and others met on the steps of the New York courthouse, where the Northern Pacific sale was to take place, and in a lengthy confrontation with a variety of interested and often hostile individuals, in unseasonal cold and with strong winds blowing the gentlemen's hats down the street, discussed a delay so that application might be made to the court for approval of modifications "desired by several parties."

The delay was granted, though as Billings confided to his diary, August 2 was "a very worrying day": he relieved it by going out to buy a shot pouch and powder flask for Parmly. On August 3 he worked with Gray, drafting and honing, and the next day he and director George Stark presented themselves at the Farmers' Loan and Trust Company, which agreed to be custodian as necessary. Billings then left immediately for Newport to confer with another director, William B. Ogden, and the two men, joined by Lane and Gray, labored further over the terms of sale. On August 6 Billings and Gray appeared before Judge Shipman on a motion to amend his decree. Shipman, recognizing that thousands of people would get nothing unless the Northern Pacific were given the opportunity to renew construction, accepted the amendments.[8]

At noon on August 12 the sale of the reorganized company was declared open, no creditor or court having posed any last-minute objection. Once again, on the courthouse steps, in a chilling rain, the group assembled and all the property of the Northern Pacific, except for patented and certified lands, but including the right to be a corporation, was auctioned off to Johnston Livingston, a member of the bondholders committee, for $100,000 (it was the only bid). By this step the bondholders became "the body politic and corporate" still known as the Northern Pacific Railroad Company.*

Billings walked to his hotel in the rain that afternoon to have dinner with Samuel Bowles. "This is," he wrote proudly if somewhat blandly, "a great triumph of the Plan for wh. I have worked so diligently." The next day he went home to Woodstock, knowing that he more than anyone else was responsible for refloating the railroad. Reorganization was, the Portland *Oregonian* reported when the facts were more fully known, "a brilliant stroke of financial genius."[9]

The company had been given a second chance. On August 25 Judge Shipman confirmed the sale, and the Farmers Loan and Trust Company was made custodian of securities. By the end of September, powers of attorney representing five-sixths of the bonds were received and converted to preferred stock. The debt was extinguished. The Northern Pacific was in possession of 575 miles of railroad, ten million acres of land, and had the unencumbered right to earn some thirty million acres more by completing the line from the Missouri River to Washington. At the board meeting of September 1877, Billings would have the pleasure of making a motion possible only because of his plan: that construction on the Missouri River and Columbia River divisions of the railroad begin at once. This was, as he told his diary, a "very harmonious and agreeable" moment.[10]

When the board of directors met in December 1875, with reorganization

---

* Often railroad companies altered their names slightly at reorganization, typically from Railroad to Railway Company or some other minor adjustment.

triumphantly vindicated, Billings was honored by a tribute from the purchasing committee and a resolution that authorized placing an engraving of his portrait on the company's stock certificates. Though he protested that the portrait would give him too much prominence, he accepted the compliment, declaring the reconstruction of the Northern Pacific to be "one of the most gratifying events of my life." Characteristically, he concluded his response by charging his fellow directors with carrying the great project forward in such a way that they would always have occasion to rejoice, "never to be ashamed."[11]

21

"My Life Work Was Done": President

THE TRIUMPH OF RECONSTRUCTION put Frederick Billings in line to be president of the Northern Pacific, since no one expected Wright to hold the office for long. He had other interests, he was quite rich, and he lacked the conviviality the job demanded, now that the prospect of wine and roses appeared once again. The company had suffered greatly from the public's tendency to believe that its debt precipitated the great Panic of 1873. Railroads generally were not in good public repute yet and the expected Republican standard-bearer in the election of 1876, James G. Blaine of Maine, had lost the nomination because of a railway-related scandal. The nation was in a tempered reform mood—the Civil Rights Act of 1875 had guaranteed blacks equal rights in public places, Secretary of War William Belknap had been removed from office for corruption, and civil service reform was being pressed. A railroad would be wise to put at its head a man known for his integrity and constancy.

This was also a good time to run a railroad, for though public confidence was returning only slowly, a growing conviction arose that the nation must complete the network of lines that had been envisioned, not only by the Northern Pacific but throughout the nation. Grangerism, that series of anti-railroad reform bills and other manifestations of farmers' discontent with political and economic life in the Midwest, had passed its peak; indeed, late in 1875 Minnesota, one of the first states to establish an office for overseeing railroads, reversed many of its Granger laws.

Billings did not need to be president, certainly, and had no particular desire to be. He was quite comfortably well off—as Julia had remarked to him long before, most people would be content to live on his taxes alone[1]—and after reorganization had nothing he had to prove. He had many interests—his children, travel (he and Julia took a particularly happy trip to Charleston in 1878, to walk over the scenes of her mother's childhood), the Marsh estate, and New York's Brick Church, in which he was becoming increasingly active. In 1879, at fifty-five and not particularly well, he could have retired. But, as he wrote to a friend, "I am getting old, and I want to ride over the Northern Pacific before I die."[2]

For the new Northern Pacific, the first task was to regain public confidence. This would best be done by winning some sign of favor from Congress. Of course, Congressional assistance was no merely symbolic matter, for though the railroad had net earnings of $97,000 in 1875, $300,000 in 1876, and $480,000 by 1878—a healthy and steady climb—this was scarcely sufficient to the grander design and came, in any event, largely from the bonanza farms.[3] Now was the time for the investment in Wade and Potts and others to pay off, and indeed the two collaborated on an appeal to Congress to guarantee interest on the company's bonds, though to no effect. The lobby did win a further extension from the Senate, when in February 1876 it agreed to give the railroad until July 4, 1887 to complete construction, but the effort died in the House, as did another despite Billings's best effort at a Congressional hearing, in the next session. Billings decided the company ought not to ask anything more of Congress, especially when the U.S. attorney general gave his opinion that even though the deadline for completion would run out on July 4, 1879, the contract remained in effect unless Congress explicitly declared a forfeiture of the grant. Thus the mortgage of 1878, taken on the Missouri Division for forty years at 6 percent, was the essential step toward renewed construction: $2,500,000 was raised, half from stockholders and half from the public. This was another success for Billings, who was chairman of the purchasing committee.[4]

In July 1878, the common stockholders regained their voting rights under the reconstruction arrangement, and soon made it known that they were unhappy with President Wright. He was, in any case, in need of an operation for cataracts. In March 1879, Cass left the board, disposing of his interests. Clearly there was a vacuum of leadership. Everyone knew that at the board meeting in May 1879 a new president must be chosen. Several directors spoke to Billings, who told Julia that he was being pressed and did not want the responsibility. She told him to do what he had to do. With eighty thousand shares, he was, after all, the largest individual stockholder. He also was the most experienced board member.

An important policy difference had arisen between Wright and Billings, and Billings emerged the winner. In 1873 three well-placed businessmen, known as "the associates," realized that the Northern Pacific would not retain control over the St. Paul & Pacific Railroad in the midst of its post-September decline. The three were Donald Alexander Smith, head of Hudson's Bay Company in Canada, a Scot who dreamed of a great Canadian railroad to the Pacific (and who would become Lord Strathcona and Mount Royal as a result of his success); Norman W. Kittson, a shrewd and experienced pioneer fur trader who had put the first steamers on the Red River; and James J. Hill, a Canadian who emigrated to the United States in 1856 and was to become one of its leading railroad entrepreneurs, a man with a clear grasp of the geopolitics of Minnesota and Manitoba determined to make that knowledge pay. A fourth associate, George Stephen, Smith's cousin (in time to be Lord Mount Stephen, Great Britain's first colonial peer), and a highly regarded Canadian manufacturer and president of the Bank of Montreal, was brought into the association by Smith in 1876. These four men made a formidable combination, and in 1876 set out to acquire the SP&P and complete it to the Canadian border.[5]

In the summer of 1878 the Northern Pacific encouraged the formation of a new company, the Minneapolis, St. Cloud & Sauk Rapids Railway, to build a road along the west bank of the Mississippi River to connect Minneapolis with St. Cloud and thus to compete directly with the SP&P. The associates felt they were in control of the situation, but realized that a compromise with the Northern Pacific had to come, even though they held the trump card: possession of terminal arrangements in St. Paul. Wright was demanding joint ownership of various SP&P properties. While Stephen and Hill refused, they agreed to bargain about some of the SP&P's land in the Twin Cities, since they knew Wright badly wanted an entry point in St. Paul on the east side of the Mississippi. Billings, on the other hand, opposed such an arrangement, arguing that the Northern Pacific needed every bit of its capital to build its line to the West Coast, and that nothing should be allowed to divert it from its primary goal. Minneapolis businessmen backed Billings, as did Vice-President Stark, Potts in Montana, and Hill behind the scenes, since he wanted to keep the Northern Pacific out of St. Paul. On November 8, acting on behalf of the associates, Stephen made an offer that the NP accepted ten days later: disputed land-grant properties would be settled in court, the NP would not build a line up the west bank of the Mississippi, and, in exchange, the NP would be granted running rights over the line from Sauk Rapids to St. Paul, use of the passenger depot there, and, as an outright gift, ten acres for a freight warehouse and tracks at the capital. This "Great Treaty of Peace" favored Billings, not Wright, and suggested that the Vermonter was likely to be able to deal with Hill effectively in any future business arrangements.[6]

Wright now pressed for a bill in Congress that would, in Billings's eyes, have damaged the Northern Pacific charter by admitting a forfeiture clause. He would rather have no bill, no Congressional aid at all, than allow a compromise on the railroad's rights, especially with respect to land grants, which he continued to regard as the key to success. Billings rallied Ainsworth, the Pacific director, to his side, and asked Power to provide frequent reports on the wheat crop and the grasshopper blight, once again serious across Minnesota and eastern Dakota, for he knew at the next board meeting he had to bring in a strong report about the value of the land already under cultivation.

Power obliged with detailed accounts, and as the harvest began was able to send the good news that grasshoppers had not done as much damage as feared and, indeed, had fallen heavily only on the lands served by the St. Paul & Pacific.[7]

Throughout the fall of 1878 Billings found himself at odds with Wright, pitching in to him several times about his "half & half, water-on-both-shoulders, course." In his direct manner, Billings told Wright that he must not be timid in pressing the Northern Pacific's case, that he must not try to walk both sides of the street with James J. Hill, and must not let Congress, under the guise of assisting the railroad, undermine the promise of the lands it was to gain in Montana and on the West Coast. In the Northwest, he said, the railroad should work for the election of any candidate who would support its goals and not remain blindly loyal to the Republican Party on the assumption that it would invariably be helpful. There was, Billings declared, "only just one way to carry forward this enterprise—and that way is the way of courage, frankness and no compromise with bummers." "[I]f we can get to the Yellowstone without a mortgage," he wrote, *the road will never stop.*"[8]

The split in the board deepened in November over construction policy. Billings proposed a scheme; Wright offered a counterproposal. The goal, Billings stubbornly insisted, was to rally everything to construction. He had not changed his mind one whit from 1870 yet realized how the board could be distracted to matters of alleged statecraft, or quick fortunes on the short run, or lured into a false optimism about the dead weight of debt. No flourishes, no purchases of competing lines, no unneeded stations, no advance contracting of rolling stock: everything was to go to construction, with land to be claimed meticulously as it came due. The great hurdle would occur somewhere west of Bismarck and short of the Yellowstone River, for settlers were unlikely to take up that land in numbers, so the line would have to leap those miles quickly and nurture its modest profits. Once onto the Yellowstone, and once land grants were fully claimed in Washington Territory, with the grace note of profits from coal and Tacoma town lots, the company might look to the left and right to see what opportunities were offered, but until then, everything was to be concentrated on the single goal.[9]

Billings was almost certainly right, and had he not taken this approach, the Northern Pacific might well have foundered once again. It is also true that by keeping his gaze so steadily westwards, along the narrow belt over which the Northern Pacific must pass, he failed to take seriously enough a threat developing along the Columbia River: the acquisition of the Oregon Railway & Navigation Company by Henry Villard. In the end Billings's policy would make him president. It would also cost him that presidency.

In March 1879 Wright let Billings know that he planned to resign by July

1, and Billings decided that he would stand for the presidency. "The Pennsylvania group" told Billings that he must step in. Billings made his commitment clear in mid-April: in order to provide operating capital to begin construction on the Missouri Division west from Bismarck, the company announced a new subscription of $2,000,000, and he became the largest subscriber by far, putting in $350,000, with three of his supporters taking another $225,000. Billings allowed himself a moment of self-congratulation when purchases reached 70 percent of the goal: "The wheel is over the centre! . . . The effect is electryfying [sic]. . . . It is especially my victory, for I really believe but for me we would have this year been building the road from Sauk Rapids to Minneapolis and doing nothing beyond the Missouri." Now the push to the Yellowstone could begin, and he felt he was the man to do it.[10]

The board met on May 24. Wright resigned and Billings was unopposed for the presidency. The board approved a bond issue of $3,200,000 for the Pend d'Oreille Division and Billings led off with another $350,000. He sent telegrams to Julia and to Ainsworth, his warm ally on the West Coast, to tell them of his selection. Julia wrote confessing that she "never was more reluctant to give you up, but I sympathize in your aspirations & rejoice in your executive ability." You can, she had told him, "do more good as President of the Northern Pacific than as President of the United States." And in her usual pert way, she did not let his triumph go completely to his head: the children were pleased with the news as well, "but their enthusiasm was divided between your new honor & a newly hatched canary."[11]

Billings tempered his optimism about completing the Northern Pacific's grand design with a hard-headed assessment of the situation he faced as president. The line would eat up capital quickly and renew its demand more than once. Getting labor out onto the treeless plains, especially if another Indian scare occurred, and holding wages down during the time when every penny counted would be difficult. Other railroads were building once again, and keeping good construction engineers would prove expensive. To promote Seattle and Portland respectively, Orange Jacobs, delegate from Washington Territory, and John H. Mitchell, senator from Oregon, were trying to get a bill through Congress that would cause the company to forfeit land grants along the projected Cascade Branch to Tacoma for failure to have met earlier deadlines and wanted to tie the company's hands by forcing it to commit its resources to the line in the far west before Montana had been breached.[12] There was also a new threat: Henry Villard's efforts, begun in April while Billings was preoccupied with bond sales and his quest for the presidency, to convert to rail the Oregon Steam Navigation Company, the Northern Pacific's essential link along the Columbia River between Wallula, which was to be the end of the line at the confluence with the Snake, and Kelso, opposite the great rock that blocked large vessels fifty miles down-river from Portland, where trackage began again. If successful, Villard would control the Pend d'Oreille Division's terminus.

As president, Frederick Billings brought several strengths to the enterprise. He knew the west as well as any director and had a well-deserved reputation for integrity. Under his leadership anti-railroad men in Congress could not feed their fires with talk of corrupt management. (Late in 1879, when an important extension bill was being shaped for Congress, he wrote a prospective director who was hopeful that something might be done to sweeten the pot that the bill would "have to pass on its merits or not at all. I do not propose to buy my way through Congress.")[13] He was the author of both the Missouri and the Pend d'Oreille Division mortgages, into which he had put $700,000 as evidence of his own confidence, he knew the intricacies of the company better than anyone, and he was spending virtually all his time on the railroad, unlike other directors and past presidents.

For the Pend d'Oreille Division Billings worked out a plan more favorable to the company, a sign that public confidence was returning. For the Missouri loan, $200 of bonds and stock produced $100 in cash; for the Pend d'Oreille, $170 in bonds and stock brought $100 in cash, and $4,500,000 in bonds were issued. As a result, construction work in the Pend d'Oreille Division, with Sprague in charge, began in October.

With both ends of the line progressing toward each other, Billings decided it was time to begin building on the Yellowstone to close the gap. He had taken the precaution of having grade constructed through Hell Gate Canyon, one of the most difficult mountain passages, in order to fend off any possible advance upon it by the new Utah Northern Company, thus making creditable the claim that work was in progress on all divisions. Billings, with the advice of General Counsel Gray, concluded that the time had been reached when a mortgage on the entire line, rather than piecemeal mortgages by divisions, would be accepted by the public and attract no undue criticism in Congress. Thus the Northern Pacific would return full circle, though this time, Billings was confident, under far more realistic conditions, since there was a market upsurge throughout 1879. On July 16 Billings applied to have NP stock placed on the list at the New York Stock Exchange.[14]

Billings returned to a well from which Jay Cooke had drawn much water— Winslow, Lanier & Company, of New York, a well-established house with deep roots in the middle west. One of its leaders, John W. Ellis, always thought the line could be of great importance, but had not felt comfortable about its management. In the summer of 1880 a New York banker, Julius Wadsworth, vice-president of the Chicago, Milwaukee and St. Paul Railroad, having placed millions in bonds through the Farmers' Loan and Trust Company of New York, where Billings was well-known and highly respected, told Ellis that the

Northern Pacific was now a sound investment. General Alfred Terry, who had been George Armstrong Custer's commanding officer at the time of the Battle of the Little Big Horn in 1876, and who knew the Northern Pacific's route very well, also told Ellis that the NP's land grant was of great value. Ellis advised Winslow, Lanier to contact Billings, and he was invited to call.

Or so the semi-official history of the Northern Pacific Railroad, written in 1883 by Eugene V. Smalley, to appear simultaneously with the completion of the line, tells us.[15] The account goes on to say that Winslow, Lanier sought to bring Drexel, Morgan in as a partner in the loan, that Pierpont Morgan's partner, J. B. Williams, had already approached them and they had decided against coming in, and that after a few weeks of consideration, Drexel, Morgan; Winslow, Lanier; and their allied houses in Philadelphia and London, following further meetings with Billings, decided to back a loan of $40,000,000. This is true enough in rough outline, though it misses some essential details, of which the most important is that the initiative probably came from Billings, not Winslow, Lanier.

Or from Morton, Bliss & Company. Founded in 1869, these partners specialized even more than most in railroads. They were capital middlemen, and before J. P. Morgan grew to dominate the railway scene, Morton, Bliss knew more about railroads than anyone else. George Bliss and Levi Morton were also pillars of the Republican Party, and Morton was viewed as the party's representative on Wall Street. Morton knew Billings as a fellow Vermonter and had become better acquainted with him during the Northern Pacific's brief Canadian negotiations. As the market began an upturn in 1878, Morton, Bliss carefully monitored the progress of various companies. They were involved in the Canadian Pacific and the St. Paul & Pacific, and thus they were well aware of the NP's position. Bliss had taken a hard look at the company in August and concluded that it had yet to demonstrate its ability to pay.[16] He appears to have felt favorably toward Billings, perhaps because at the time two men whose judgment he trusted, James Hill and George Stephen, did so, or perhaps simply because Billings came in from time to time to see Morton and to make a donation to the Republican Party. This was especially so now, for Billings felt deeply about the need to elect James Garfield as president, since only Garfield, he thought, would bring both reform and stability.

Billings called at Morton, Bliss twice in September. On September 13, after looking in on Cooke's old partner, Harris Fahnestock, at the First National Bank to discuss financing for the Northern Pacific, Billings went along directly to have a "long talk" with Morton, during which he agreed to take one hundred thousand shares in the syndicate for the Canadian Pacific. He also agreed with Morton that he should give $5,000 to the Republican cause. On October 19 Billings went to see Morton again and gave another $5,000 and "talked Canadian Pacific & No. Pacific." It cannot be entirely coincidental that Billings went directly from Morton to Drexel, Morgan, and then to the First National Bank once again to discuss the NP, and the inference is strong that it was Morton, Bliss that steered him toward Drexel, a contact that, if Billings's diary entries are correct, preceded any call on Winslow, Lanier. The issue may not be important, though given the fact that Drexel, Morgan soon became the NP's principal banker and, in the contest between Billings and Villard, supported the latter, it may be of some significance.[17]

That a Northern Pacific syndicate would take on a contract of the proposed magnitude showed how much the market had changed in the last eighteen months, for this was the largest single railroad issue until then. The syndicate consisted of seventeen firms, and by sharing risks the financiers could amass long-term capital in great volume; with a mortgage of this size, the major partners—Winslow, Lanier and Drexel, Morgan—were in a position to become "the guiding force."[18]

This was a serious mistake, though it is difficult to see what choice Billings had, since the Northern Pacific could scarcely have begun to build again without the syndicate. Despite the bankers' hard bargaining, Billings did not anticipate that they would interfere with his policies, which he sometimes pursued somewhat high-handedly, making a decision, as he admitted to an old friend, and then putting it to the board to ratify it after the fact. When the deal was struck, Billings wrote to Cheney that he was grateful the Northern Pacific had "not been on our knees to Vanderbilt . . . nor anybody! The N.P. stands on its own base, and so far as I am concerned it will stay thus!" Pensively, he told his fellow director, "we are getting old Mr. Cheney, and it is best to push the road through."[19] Soon, however, Billings found that Villard was playing on the syndicate's fears, for when Billings remained adamant about building across the Cascade Range so that the Northern Pacific would not be dependent on Villard's Oregon Railway & Navigation Company, Villard persuaded Drexel, Morgan, a firm that had harbored a deep resentment against Jay Cooke and, consequently, against the earlier Northern Pacific, that this would be folly.[20]

A loan of $40,000,000 was a great sum of money, and the principals—Billings, the Northern Pacific's general counsel, George Gray, and A. H. Barney, the treasurer, on the one side, with Ellis, Egisto Fabbri, a Morgan partner, and J. C. Bullitt of Philadelphia as lawyer on the other—met several times for day-long sessions. During the negotiations Northern Pacific stock strengthened even further, and Billings received excellent reports about the next wheat harvest. The process took four and half months, from the first discussion of the need for a consolidated loan at the NP board meeting of August 18, which was interrupted by Wright's taking ill and having to be carried out, to December 30, when the parties inscribed ten copies of the necessary legal documents. Throughout Billings appears to have been confident that solutions to all problems would be found, and though he reported more sleepless nights than restful ones, he also found more time than usual to be with his family in Woodstock, walking with Ehrick, talking with Parmly about his anticipated entry into Williams College in October, and working with the Reverend Lewis Hicks on behalf of the Congregational Church, to which Frederick was contributing a new chapel. There was much else on his plate—particularly consideration of the Superior South Shore Line and frequent meetings with Villard concerning the situation in Oregon. But with a board that was ready to back him he experienced none of the tension of reorganization or the earlier loans.[21]

On September 29, at the annual stockholders' meeting, Billings was re-elected president, and $36,000,000 in stock was voted. At the November board meeting, the directors accepted the plan for a forty-million-dollar forty-year 6 percent gold bond issue on all construction and on all lands west of the Missouri River. The mortgage, payable semiannually, would not fall due until 1921. The bonds would be issued at the rate of $25,000 per mile on each twenty-five mile section of completed road as accepted by the government (a clause that would make for difficulties later, when money was needed in advance of construction to blast and dig out the great tunnels through the mountain ranges), with the syndicate taking $10,000,000 in bonds at ninety with options to take like sums at a slightly higher rate in 1881 and in each of the next two years. In order not to conflict with another bond sale being managed by one of the syndicate partners in December, the sale was to open on January 1, 1881.[22]

"It seems like a dream," Billings recorded in his diary on November 19, the day the deal was struck, "and somehow I feel as if my life work was done. But I wd like to live to see the rails laid across the continent. . . . I am thankful to God." Two days later he left for Woodstock to celebrate Thanksgiving, sleighing with his children, distributing forty-eight turkeys to the needy at the market, and helping his daughter Laura decide against attending Smith College.[23]

From Woodstock Billings wrote to Alexander Ramsey, former governor of Minnesota, senator from 1863–1875, and now secretary of war, to tell him that he had just concluded an arrangement that would make it possible for the Northern Pacific to be completed by 1883. Billings was soliciting letters and telegrams of congratulations that he would then send to the press, a trick he had learned from Jay Cooke, for this would increase the space given to the Northern Pacific. He had, he confided, just written to the president. Ramsey replied with an eminently quotable telegram, heralding the NP as "a great national enterprise," and adding that "any man identified with its consummation may well be proud."[24]

On the morning of November 24, Billings read details of the negotiation in the Boston papers since ex-president Wright let them out well in advance of the formal signing. Worried and indignant, both because he thought Wright was trying to take away what he felt was his own triumph and because the price of box cars and steel rails might well go up before he could place orders, Billings sensed once more that his board had proved weak. But congratulations now poured in, letters from Generals William Tecumseh Sherman and Winfield Scott Hancock, from President-elect Garfield, and a "charming" letter from President Rutherford B. Hayes.[25] On December 16 the syndicate's agreement was formally ratified by the board, and so that any public announcement said what he wanted it to say, Billings wrote out "a supposed interview by a reporter" for placing in the friendly press. On December 30 copies were exchanged with Ellis in New York and, given that Parmly was in bed with the measles, Frederick passed up any celebration.

As president of the Northern Pacific, Billings did far more than negotiate the great $40,000,000 loan and, in the final months, jockey with Villard over the Columbia route. When asked what Frederick Billings's accomplishments as president had been, the state railway commissioner for Minnesota listed six. He doubled wheat acreage, established a system of grain elevators from Duluth west, reopened the foreign emigration agency, and changed the company's fiscal year from September to June 30. More important, Billings restored faith in the enterprise through his refinancing scheme and resumed construction.[26] The commissioner did not mention Billings's most important achievements: his pre-presidential organization of the company so that it might merely survive through the long wait for financial recovery, and the fact that, though Billings was not president on the day when the last spike was driven at Gold Creek, he was the one who made it possible. He had, after all, said that the line would be completed in 1883, and it would be.

The Northern Pacific was something of a reflection of its president. Then and later, it was a no-nonsense line: it had no nickname, unlike the Ma and Pa, the Soo, the Chessie, and dozens of others that entered the American vocabulary; it appears to have evoked no ballads or folk songs; it created no heroic legends. To Billings the railroad meant energy, new horizons, power, the pressure to perform, the sense of time and orderly movement a railway schedule brought. Yet he knew there was something romantic about a railroad even when he did not find the language for it. In 1879 to 1881, when he wrote about discovering the right route over the mountains, he weighed all the factors—grade, supply, resources, safety, weather—but always left a word for beauty, for the needs of the tourist, and for conservation values.

Unlike many of his contemporaries and so far as the evidence shows, when Billings left the presidency he had not resorted to bribery, had not ruthlessly bulldozed his opponents under, had not taken the railroad to the brink of disaster through speculation and imprudence. His hard work, fair-mindedness, and good spirits meant that he was liked by nearly everyone. In 1879, when the Utah & Northern appeared to be threatening the Northern Pacific's projected control over the Montana Territory, Billings had been interviewed for the Helena *Daily Herald.* The interviewer expected to encounter a man hostile to the Utah & Northern.* Instead, Billings diplomatically praised the line, somewhat disingenuously to be sure, saying that he hoped it would push along as quickly as possible, since it would call "public attention" to Idaho and western Montana, "which were not known as they deserved to be," and this could only help both the people and the NP. Billings had, the reporter said, "jocosely observed" in closing the interview, that when people came to know him better they would like him.[27] This usually was the case.

Having achieved the presidency, Billings seems to have relaxed more with his family and spent more time with the children. Julia often came to New York to stay at their apartment in the Brevoort House—they had given up their New York home and did not buy another until 1881—and made a point of being there when he returned from his trips along the line. He used his private railway car more for long weekends in Woodstock—or tried to, for it was annoyingly unreliable—and though he often was at the office before eight o'clock, he equally often left by four. He took his daughters to the theatre or to see Punch and Judy shows in the park, he enjoyed long talks with his sons and going hunting with Parmly. He charged Parmly with reorganizing the books on the shelves at the mansion, perhaps in the hope that this would make the often unruly adolescent more contemplative, and he experimented with each of the lines of New York's new elevated railway (he liked only the Harlem line and, rather than ride back, would walk home). He often remarked on the pleasures of being in the arms of his family, of having the seven children (six after Parmly left for Williams) at table together, or sixteen for Christmas dinner.

Billings could afford to relax more: construction had, at last, begun once again. In September 1879 surveyors for the Northern Pacific located a definite route from the mouth of the Snake River to the falls of the Spokane, a distance of 140 miles. Billings prepared a letter for public release when the NP was ready to declare its course, and this was now done through a friend in Washington Territory, the Reverend Charles Andrew Huntington. In his letter Billings advocated a route to Puget Sound directly across the mountains

* The Utah Northern and the Utah & Northern are different. The latter absorbed the former in April 1878. Montanans appear to have ambivalent feelings about the Northern Pacific. Even today, one may find no fewer than thirteen state historical markers to the NP implanted in North Dakota, while there are only two in Montana.

rather than along the Columbia River. This line would annex to the NP some of the richest timber land in Washington. If completed expeditiously it would also destroy Villard.[28]

Construction crews assembled at the end of September, and a boom town, which took the name of J. C. Ainsworth, sprang up where the Snake ran in to the Columbia. On October 2 Sprague arrived to lead the time-honored ritual, turning over a shovelful of flaky, grey soil, and work began. Detailed surveyors moved out ahead of graders, who moved ahead of track layers. Ties were brought down from the headwaters of the Yakima River in the Cascade Range, and by late January 1880 ten miles of track had been laid. At the same time Billings admitted that efforts to work out with Henry Villard an agreement for joint building down the Columbia under the Northern Pacific's charter had failed, and he instructed Sprague to survey a line from Ainsworth down to Wallula, only twelve road miles but strategically critical, to forestall an effort by the Oregon Railway & Navigation Company to move in on the juncture of the Snake and Columbia.[29]

In the meantime, impatient with progress on the Missouri Division, and always one to visit the source of a problem, Billings had travelled out toward the end of the line at Sweetbriar, ten miles beyond Mandan, in Dakota Territory, in June 1879. He suffered from a prolonged illness, diagnosed as diabetes, throughout most of the journey, which was in unseasonably cold weather, and returned impressed with the land and depressed by the quality of engineering. Heavy truss bridges were needed at six places in the first few sections beyond the Missouri because the Heart and Knife river crossings proved more difficult than expected, and these were not in the budget, while money was spilling out as though from a millrace at a huge cut that was far more complex than anyone anticipated. There was no nearby timber for ties, which had to be of oak and brought from afar, and snow was a major problem, with more snow fences needed. Estimates of costs were off by 50 percent. Billings went up to Winnipeg, where he discussed connecting lines and potential agreements with the mayor, American consul John Wickes Taylor, and the local member of Parliament, and dined at the new Canadian Pacific Hotel, which he liked and wanted to emulate. Between Fargo and Duluth he felt he saw signs of improved business, and became cautiously optimistic. When in St. Paul he met with several railroad men to work out the practical details of the "treaty" with the St. Paul, Minneapolis & Manitoba.[30]

The principal remaining bone of contention was the north-south route on the west side of the Red River. The Northern Pacific had agreed not to build any line closer than twenty miles west of the river, and had wanted to begin a branch at Casselton, in the heart of the bonanza farm country about twenty-five miles beyond Fargo. This was quite proper, but as the Red River bore in a slight westerly direction as it flowed north, so too would the branch line have to veer westwards. There were technical problems with this and with the route south of Fargo to Breckenridge, on the Minnesota side of the Red River. Billings found his opponents, as he had been told they would be, entirely reasonable, however, and he concluded a satisfactory agreement, feeling just a bit like a diplomat working out the boundaries between two sovereign powers in some faraway colonial empire.

When he returned Billings fired off various hot telegrams and memoranda, charging everyone to work harder and warning that "everybody will get cleaned out" if there was not more "push."[31] By the end of the construction season grade was finished a hundred miles west of Mandan, with track laid for half that, and a contract let for the remainder of the line to the Yellowstone River. There was daily service to Bismarck, with both coal and Montana cattle being loaded on at the end of track in the territory. That December General Thomas Rosser, superintendent of construction in the Missouri Division, whom Billings blamed for the engineering errors, resigned to become chief engineer of the Canadian Pacific. Early in 1880 Milnor Roberts followed, so that Billings soon had a new team, with General Adna Anderson taking over as chief engineer in Roberts's place. On November 10 the line reached the Montana border, and a silver spike brought over from Helena was driven in celebration of the event. Soon after winter put an end to construction.[32]

Now was the time to get tough, and Billings did. He smothered the persistently ill Cheney with a blanket of courtesy, though he thought he had "become a baby"; he denied a renewal of the vice-presidency to Stark, who was too reluctant to visit the field; and he brought his own men on to the board: Colonel Joseph D. Potts and Walton Ferguson. The latter would, Billings confided to Ainsworth, resign at any moment he asked him to.[33]

Billings also got rid of employees if they did not seem to bend their backs to the hoe. Thomas Doane, the chief surveyor, was slow and therefore replaced; Homer E. Sargent, hired away from the Michigan Central as general manager of the Eastern Division, was insufficiently attentive to detail and also replaced; Robert L. Belknap, who replaced Barney as treasurer, was reprimanded and almost certainly would have lost his job had Billings remained president. James Power, with whom Billings had worked so well, was let go, in part because he failed to act quickly enough in discharging an unsatisfactory employee in the Brainerd office, in part because he had given the appearance of misuse of funds to his own advantage, a charge against which he offered an able but insufficient defense. His subsequent effective work for a competitor, James J. Hill, was in some measure a result of Power's determination to exonerate himself.[34]

Indeed, Billings was even prepared to abandon Tacoma, the terminal, and the town lots, if another route to the sound proved quicker and cheaper. As he wrote to Ainsworth in October 1879, the spring would bring engineering parties out in force to survey the Cascades in hope of finding a route over the mountains so that it would not be necessary to follow the Columbia. Billings thought the Snoqualmie Pass was the best route and, if so, the railroad must run there "no matter how it affects Tacoma,"[35] a clear indication that he was not, in fact, underestimating the risk that a combined operation with Villard along the Columbia would entail.

With able men in place—Adna Anderson for Rosser, Joseph Thompson Dodge, another old friend from University of Vermont days, now a noted engineer with extensive experience, for Doane—Billings drove the surveyors, the construction foremen, and the engineers hard, with a steady drum-beat of encouraging, demanding, uplifting, subtly threatening letters. Speed and economy were now the cry, as he told Dodge: "We must get a cheap line and a safe road," he wrote, "but at the outset we must in grades and curves try to save—and trust to the future for the higher finish."[36] He wanted no labor strife, for this would slow the work: be fair, be straight, be tough, he told the contractors. He wanted no diversions, no loss of time from Indians or disgruntled settlers: pay the Crow $25,000 if necessary, trade beads and blankets for their land, be conciliatory, make a point of getting along pleasantly with them, he told the many military men, the successive secretaries of the interior, and the land agents to whom he wrote. He worried that the construction crews drank too much—which they did; the beer supply agent for the railroad revealed that he reckoned it took 2,640 ties to build a mile of line and that laborers drank 110 cases of beer, or 24 bottles each, per mile of construction, with a tie costing fifty cents and a beer the same[37]—and were too long away from their families, which they were. He worried about competitors, real and imagined, and instructed Dodge and others to communicate in cipher. He worried about the landscape, reminding the surveyors to damage any scenic resource as little as possible, for tourists would one day pass over the route. But above all he worried about keeping costs down.

To benefit from tourist traffic, to bring settlers onto the land, and to begin to make real profits, the NP would have to retain its land grants, however, and during Billings's presidency the grants were under persistent attack in Congress. Though the House of Representatives had always been restive over the "great land giveaway," it looked as though both houses might take action detrimental to the NP's interests in 1880 and pass a forfeiture bill to deprive the Northern Pacific of its grants. The House Committee on Pacific Railroads, now chaired by Congressman Robert M. McLane of Maryland, met throughout the spring, discussing the NP. A number of Congressmen wished to force the Northern Pacific to abandon its land grants on the ground that it had neither begun nor completed its line as required. Billings had anticipated this attack and while Wright was president had persuaded him to begin construction of a road to the Payallup coal mines, maintaining that this was a segment of the branch. More effectively, he continued to lean on the court opinion that forfeiture was not automatic, and that unless Congress passed a statute for forfeiture, the lands remained in railroad hands. However, some on the committee were advocating just such legislation. On the positive side, there was the possibility that the committee would recommend an extension of the deadline for completion of construction, which would render the legal question moot. Billings had preferred to let sleeping dogs lie, but Wright had stirred them up, and Billings now had an obligation to clarify the matter. The entire process made him very anxious, and during the two years of his presidency he did not have as many sleepless nights and sudden "starts" in the early mornings as during this period.

The enemies of the railroad added fuel to the House Committee's apparent hostility. In March strikes on the line led to sensational reports about mistreatment of workers; Billings assured himself that these were without foundation, though damage had already been done. Other reports said the company was handing out stock to sweeten up some of the committee members, or that it refused to hire Southerners, and Billings felt it necessary to refute both these charges as well as the rumor that railroad baron William Henry Vanderbilt was buying up stock. There was, Billings knew, a public belief that the Northern Pacific was rich, and in a sense it was, but he knew it had no money to spare. He judged forfeiture to be a real possibility, since there were many in Congress who saw the NP as a giant monopoly controlling both the Columbia River route through the Oregon Navigation Company and the Cascade route by virtue of the land grants. Villard, and residents of Oregon who hoped Villard was a possible savior from the NP octopus, particularly supported this view.

Billings went to Washington four times in 1880 to testify at the request of the committee, to meet with NP's lawyers and talk with various congressmen. While McLane, whom he had known through Peachy in California, and from whom he had fruitlessly hoped for some relief over the New Almaden mines when McLane was minister to Mexico, was "evasive and diffident," one or two committee members were encouraging. Billings alternated between believing that the committee would recommend against extension and require forfeiture to thinking that extension was all but assured, and at one point concluded that the committee was unanimously against the railroad. The situation was, he said more than once, "very ugly," and he would feel personally responsible if the railroad suffered, given his earlier opposition to Wright on the issue.[38]

On April 15, Billings rode down to Washington in the company of William A. Wheeler, vice-president of the United States. The day was hot, and Billings had slept hardly at all the night before. He judged the issue of forfeiture to be in the balance and knew that he must put forth one of his best efforts.

Walking to the Capitol from the station, he addressed the committee for an hour and a half, eloquently according to observers, and was pleased that the members, who often did other work during testimony, gave him their close attention. His spirits lifted a bit, he called on several political figures who, if a bill were reported out from the committee, would have important roles to play. On the following day he met with President Hayes for thirty minutes. When he returned to New York he immediately drafted replies to questions put to him by the committee, briefed the board on the situation, and sent the company treasurer back to Washington with his statement, which he then had printed for wide distribution.[39]

In addressing the committee, Billings argued that three sets of men had already gone to the wall for the Northern Pacific, and that the great crash of 1873 had occurred in an era "which was full of craziness" that would not recur. The huge land grant had not saved the road from bankruptcy and, even now, the grant up to Bismarck was not paying a cent of the railroad's real costs, for the lands had been brought onto the market so slowly they would pay only the interest on loans. He appealed to the congressmen's patriotism—pointing out that the Canadian government was doing so much more for its transcontinental line—and to their pocketbooks, arguing that the further the railroad progressed, the more frontier forts could be dismantled, saving the government $1,000,000 a regiment. He appealed to their sense of fairness, stating that previous congresses had given the company every reason to believe it could proceed in good faith. If the company were to be stopped, who would build the railroad, so needed for national unity? If you doubt us, he concluded, grant no extension; but do not punish us as though we were wrongdoers by taking hostile action; do nothing.

Billings now hired his old *bête noire* from his New Almaden days, Jeremiah Black, to prepare an opinion on behalf of the railroad. He had feared that Black would be called upon against the Northern Pacific, for he had seen the old lawyer as an opponent. Black handled several railroad cases, sometimes for and sometimes against the companies, and had represented John C. Frémont's railroad interests in litigation until 1875. But the NP director from Baltimore, Henry E. Johnston, urged Billings to hire Black, and Johnston needed some soothing, since he was one of the few members of the board upon whom Billings could not count automatically. Accordingly, Billings asked Black to come to see him, and offered him $500 to write an opinion on the section of the charter that could be argued to require a specific act of forfeiture from Congress. Black's statement proved to be "most satisfactory," and Billings had it run in all the morning papers and five thousand pamphlets printed from it.[40]

Black's opinion produced results: Billings was summoned to Washington once again, and on April 26 he caught the late afternoon train, riding with

David Dudley Field, the counsel to Jay Gould during the Erie Railroad litigation in 1869, and Beverly Tucker, who had worked diligently on behalf of Frémont and Billings during the New Almaden cases. Billings again met individually with members of the committee, feeling so agitated he could not sleep, and—having had to work on a Sunday—depressed that business was allowed to profane the Sabbath. Then he was told the committee would take no action for some time.

In all his carefully kept diaries, Billings seems not to have fallen into so deep a depression as during the House Committee's inquiries. In May when he went to Woodstock to view the foundation for his family's cemetery monument, he began to brood. He could take little more of this pressure, he wrote; he wished he could cast all his cares onto the land, that he could simply remain in Vermont. On Sunday he listened to a strong sermon on total abstinence and decided that he would not drink anything again, even though when depressed he felt the need for something. He walked down to the cemetery on River Road by himself and stood looking at the foundation—it needed a bit more elevation, he thought—and concluded that his time on earth would not be long. That evening he dug out old letters from his sister Laura, sent to him from South Carolina nearly forty years before, thinking about what he had accomplished and about what he had failed to do. He went to bed after a "most earnest" prayer for God's blessing, and lay awake until one o'clock in the morning.[41]

But all came right in the end. The House Committee did not recommend forfeiture. The essence of Black's argument was accepted. Beyond Billings's expectation, Congress voted the Northern Pacific an extension of eight years, the Senate—with an old friend of the railroad, William Windom of Minnesota as chair of the committee at the crucial moment—being unexpectedly generous. Samuel Bowles congratulated Billings for having achieved his goal "with perfect honesty."[42]

There was one problem to which Billings applied himself consistently from the time he was first appointed chairman of the Land Committee until the end of his presidency in 1881: the need to extinguish Indian titles along the route of the Northern Pacific as the government had promised in the original charter. As rails advanced through Minnesota and into Dakota, Billings had begun to remind successive secretaries of the interior, beginning with Columbus Delano, President Grant's second secretary, of the need for action. There was the pressing issue of Wahpeton and Sisseton claims. There was the matter of the boundaries of the Fort Berthold Reserve for the Arickaree, Gros Ventre, and Mandans, which as drawn in 1869 lay across the proposed Northern Pacific route for a hundred miles. Looming ahead, far out on the plains beyond Bismarck, were the Sioux and Crow, and further on, the Flatheads. On the West

Coast were the Puyallups. If there was any one aspect of his presidency for which Billings might, with hindsight, be held to have acted without the perfect honesty Samuel Bowles attributed to him, it concerned Indian rights. But in the end, though he wished to break the rules, he did not.

The situation on the plains was such that little progress in the matter occurred during the Cooke years, after which, with construction suspended, many felt the question could be put on the shelf. Not Billings, however, and he continued to nudge Delano's successors, Zachariah Chandler, who took over as the interior's head in 1875, and Carl Schurz, whom Billings had known when Schurz was senator from Missouri, and who was President Hayes's secretary of the interior from 1877 to 1881. Only with the reform-minded Schurz in office was Billings able to bring the most serious problem, that of the Crow land, close to resolution.

In 1851 the Crow controlled a quarter of Montana, including much of the Yellowstone system; by 1868, as the iron horse began to champ at its bit over by distant Duluth, they had been reduced to five million acres—still "a lot of territory." In 1882, however, they lost a quarter of their reservation, and with that the Northern Pacific and its president, by then Henry Villard, were home free, since in 1880 Secretary Schurz, following a siege by Billings, had restored to the public domain the large area of western Dakota through which the railroad passed. Further west the story of the Flatheads was even sadder, as they were removed from their traditional home in the Bitteroot Valley, settled on a reservation on Flathead Lake and the Jocko River, and then stripped of part of that land.[43]

Throughout this process, Billings's general attitude is best described as careful of the law and indifferent to the Indians. He did not want to take land unilaterally, for he insisted as a matter of principle that clear title was essential to so durable an enterprise as a railroad line. He also wanted the tribes, or those who represented them in Washington, to have due process. But he wanted that process to move rapidly, for he shared the general view that nomadic Indians, restless, likely to forage (or, in the more negative turn of phrase, pillage) along the line, were a brake on progress, and when weeks dragged into months he was tempted to shortcuts. He wanted the Indians Christianized and turned into settled dwellers with assigned allotments. He was not so naive as many of the government policymakers far away in Washington, who argued that the Indians must become strong-armed yeomen farmers in the Jeffersonian tradition, so that they would enjoy the stability of "pride of individual ownership" in a bit of property, and he knew that dispirited Indians sitting about at an agency house were not a tourist attraction to stir the blood. Still, he generally thought them an irrelevance who, once dealt with fairly and through the processes of the law, would have no role to play in the emergence of the new and progressive Dakota or Montana.

Billings read the annual reports of the commissioner of Indian affairs and the popular early ethnologies of Henry Rowe Schoolcraft, but he appears to have taken many of his views on native Americans from his fellow Vermonter, Charles Andrew Huntington, the man with whom he trusted his statement in 1879 concerning the primacy of the Cascade route. Huntington had gone to the University of Vermont with Frederick, graduating in 1842. They had, however, lost contact with each other until 1865, when Billings was on his hurried trip to the Pacific Northwest.

As Huntington tells it,[44] he was a lay preacher in Olympia, where he had gone to be chief clerk to his brother-in-law, the superintendent of Indian affairs for Washington Territory, and was well into a rattling good sermon when he noticed a middle-aged man seated in the rear of the church giving him his undivided attention. The man was, of course, Frederick Billings, who passed his card to Huntington, inviting him to call at his hotel, where the two men renewed their friendship. Huntington explained that "after our graduation [Frederick] went to California and got very rich. I went to Illinois and got very poor." Learning that Huntington was preparing to return to the East, where he had left his wife and eight children, Billings persuaded him that he must remain in Olympia to raise up the Indian civilization. He would, Frederick promised, pay for Huntington's family to come to join him whenever he wished, and two years later he fulfilled that pledge and thereafter remained a firm friend, sending money as needed, as Julia continued to do after his death.

Remaining in the territory, Huntington became agent for the Makah Indians and took charge of the Neah Bay reservation, from which he appears to have kept Billings informed about Indian habits, the problems of native land ownership, and the corruption of the War and Interior departments. Both thought the government quite wrong to confirm Indian tribes in a proprietary interest in lands, as an act of 1853 had done, and while both wished to see the Indian progress, and saw that individuals did so, neither believed that the generality of Native Americans would ever learn the ways of plow, train, and commerce.

Even so, Billings was shocked at the corruption in the Office of Indian Affairs, and hailed the housecleaning on which Secretary Chandler, followed by Schurz, embarked. Though he knew he was not especially well informed on the intricacies of Indian policy, and certainly not of inter-tribal politics, he did not join those who railed against the Army for not trying to put down every alleged Indian scare, some of which were the product of fevered imaginations or deliberate attempts on the part of Montana settlers and politicians to get the U.S. Cavalry in to shoot a few Indians or roust them about the countryside for a bit.

The trade-off for Billings, who actually favored Secretary Schurz's more restrained and deliberative approach to Indian claims over Montana's governor

Potts's persistent demands to throw their lands open to settlement, was the governor's strong advocacy of the railroad in the face of vocal opposition from many Montanans, largely Democrats, who feared the monopoly over transport that completion of the line would give to the NP. Montanans were divided into two factions, one eager to see the east–west Northern Pacific built across the territory at all speed, the other, even before the collapse of 1873, supporting the shorter line that could be brought up from Utah. The question was not simply one of speed and route, of course, since a line along the Yellowstone River would create competitive communities there to such towns as Helena or Deer Lodge in the west, already flourishing and eager to be the seat of government. Potts remained loyal to the NP route, at least until 1879, when he appeared to be moving toward the Utah & Northern, as a division of the Union Pacific.[45]

Certainly the Northern Pacific benefitted from Potts's support and from the presence of the U.S. Cavalry, charged with helping the line to advance, and neither friendly (Crow) nor hostile (Sioux) Indians were allowed to slow its course. Many thought that the projected route of the Northern Pacific, which was to be built up the Yellowstone Valley, violated Sioux territory, but few appear to have debated the point. While Billings solicited opinions on the definitions of the unceded Sioux lands, the treaty in which the boundaries had been set out used such vague language that no one could really give an accurate opinion. A useful opinion was a different matter, and there were plenty to say that "east of the Bighorn Mountains and north of the North Platte River" was far too loose to define rights with respect to the Yellowstone.

In such light, the Northern Pacific, with the support of the army, and with Billings impatiently calling for quicker results, got on with its surveys. Each summer from 1871 through 1873 teams had set out accompanied by substantial military escorts from Forts Abraham Lincoln or Ransom, and though the Sioux made threatening gestures, realizing full well that the railroad would mean the destruction of the buffalo, they did not attack in force until August 1873, when near the mouth of Tongue River, above present-day Miles City, Montana, Sitting Bull's Sioux served notice that they intended to make war, attacking Lieutenant Colonel George Armstrong Custer's Seventh Cavalry. Custer scattered the Sioux, and the survey expedition continued its work west to the Bighorns and north to the Musselshell. The collapse of Jay Cooke & Company three days before the expedition returned from the field ended Northern Pacific threats to Indian lands for over six years, by which time the Sioux nation had been defeated.

In the Pacific Northwest, the Northern Pacific had succeeded in holding up Puyallup occupation of land they regarded as theirs, while the coal fields around Carbonado and Wilkeson were exploited or sold. Billings wanted to see the Puyallups amalgamate with another tribe—never a realistic proposition—but this led to nothing. Finally, the government issued the necessary patents, and in 1885 the tribe would move onto its designated land, after which the railroad could deal directly with the tribal council in order to contract for coal.[46]

As chairman of the land committee, Billings added his voice to those who objected in October 1875 to President Grant's withdrawal of land north of the Yellowstone and east of Big Timber Creek from settlement so that it might remain in Crow control. Such objections were quickly effective, and Grant revoked his executive order the following March. As president, Billings recognized that previously undesirable land was becoming more appealing, as ranching, farming, and mining expanded, and he worried about renewed Indian restlessness in the face of what he believed were inevitable new pressures from white settlers.

Did the Northern Pacific help the army to subdue the Indian, did it bring prosperity and peace to the northern plains? Or were the costs to society, white as well as Indian, far larger than anyone ever conceived? The wholesale killing of the buffalo destroyed the Indian's mobility, eliminated his primary food supply, and forced him onto reservations. The dole, niggardly as it was, cost 400 percent of what the government had expected. Was the reservation policy, or indeed the slaughter of the buffalo, cost effective? Here, on the cold, high plains of Dakota and Montana, in the years the railroad, mile by mile, was laid down over once rich Indian hunting grounds, is fertile soil for economic historians to ask questions about social costs.

Certainly no one connected with the Northern Pacific asked such questions. This was not a time for much discussion of public redress for private suffering, and there was no discussion of how government, or large corporations, might be expected to mitigate the human costs of economic change. The public, and certainly the business community, assumed that inequality, where not related to race, arose from personal frailties, not from the social environment, and that entire groups of people, such as Native Americans, were destined to be swept aside by the engine of progress. Quite apart from assumptions of social Darwinism, few would have believed that several thousand persons—even white persons—should be allowed to stand in the way of benefits that were presumed to bring fundamental change to the lives of many thousands more.[47]

Exasperated by the many mistakes that he felt had been made on the Missouri Division, and deeply anxious to keep construction moving into Montana, in 1880 Billings instructed Dodge to carry his survey across the vast Crow reserve. When Schurz learned of this in September, he sent a telegram to Billings, urging him to withdraw his survey crews. Billings replied that he would do so, while also instructing Dodge to find some way to get on with the survey locally. This he did until Chief Spotted Horse intervened and, in December, Billings belatedly stopped the survey. Adna Anderson appointed

Major F. D. Pease agent to procure a right-of-way across the Crow lands and provided $3,000—a goodly sum, for this represented fifteen months of Pease's salary—for presents for the Crows. Pease sent samples of beads and blankets back to New York, where Anderson did his best to have them duplicated, finally having the blankets made in Minneapolis. Otter skins were added to the package and distributed freely in the spring of 1881, apparently at a meeting with Crow leaders at which Pease exceeded his authority and made commitments that were improper. The survey was resumed.[48]

In May 1881, Billings made one last attempt to clear the Crow titles. He had, on advice of counsel, interpreted a recent Supreme Court decision and an order from the Department of the Interior to mean that the company had a right-of-way by virtue of its charter and that the government would compensate the Crows. Billings instructed Anderson that, whatever the outcome might be, the NP must get along with the Indians, since an open clash was not in anyone's interest. Through an intermediary Billings informed the new secretary of the Interior, Samuel J. Kirkwood, that construction across the Crow Reservation would begin within sixty days, clearly anticipating a favorable response, since Kirkwood had been a railroad president himself and was believed to be anti-Indian. In a trice came back an abrupt reply to the effect that, as Schurz had permitted survey teams to complete their tasks, the department had gone to the extent of its authority, and that construction could not begin until Crow titles were extinguished. This required an agreement between the U.S. government and the Indians and ratification by Congress; even prior to negotiations, the NP would have to supply a map of its survey. This was done late in June, and negotiations were concluded on August 22. By this time Billings was no longer president.[49]

The judgment that Billings's presidency was one of great, if not perfect, honesty is fundamentally correct, and certainly so within the ethical climate of the time. Billings skirted close to the edge when dealing with the problem of extinguishing Indian titles, and like all the directors he realized substantial personal income through other businesses entered into by virtue of the Northern Pacific. But he did not speculate in land in any sustained way, he was consistent and straightforward with Congress and the public, did not pay out silver in exchange for votes, and curbed the type of misleading publicity that had been commonplace in Jay Cooke's time. (Indeed, he was so "absolutely ashamed" of one promotional folder that described the Northwest in terms that were crude, ungrammatical, and untrue, he ordered the destruction of all ten thousand copies.) He wanted honest administration: land agents who connived with speculators were to be dismissed at once. He wanted settlers on the land: to satisfy the Grangers, to take away political hostility, to bring stability to the frontier—and so he ordered that the price be kept at $2.50 an acre. He wanted schools and churches in all the settlements, for these marked "the right kind of property." He wanted, he said, to concentrate on "the steady, honest building of the road."[50]

To this end, Billings sought to choose upright men to handle the construction of the line, and cautioned them against self-enrichment. One of his last acts as president of the Northern Pacific was in March 1881 when he employed Herman Haupt to be general manager of the Eastern Division of the Northern Pacific. Haupt was a distinguished military engineer with a long railroad background, and Billings negotiated with him carefully, for he had several other job offers. Billings wanted him so badly, he was prepared to offer a $25,000 annual salary, though of course he kept this from Haupt. When Haupt dictated his terms Billings could not have been more pleased: Haupt wished $15,000 and a five-year contract. Very quickly Billings alerted Haupt to the probability of an engineers' petition for higher pay and told him to be firm and reasonable, which Haupt was in an able circular the following October. Billings preferred such "firm and reasonable approaches" to all problems, and he assumed others would reply in kind. He felt he had not received reasonable responses from Villard when they counted most, during their contest over the Cascade route, and for years afterwards, though friendly to his successor, he would tell the press that Villard had ultimately failed the Northern Pacific—which he would in fact do in 1883—because he lacked experience in practical business matters and was neither "firm nor reasonable."[51]

22

Villard and the Blind Pool

H ISTORY FINDS IN THE VILLARD-BILLINGS CONTEST the stuff of high
drama. Most accounts tell the story as though the two men were locked
in battle; the old-line conservative Vermonter, filled with rectitude, a bit inattentive to the risk that Villard posed for him, against Villard, the upstart
former journalist and immigrant from Germany, ruthlessly preventing the
locomotives of the Northern Pacific from passing beyond the point where the
Snake met the Columbia because he controlled not only the steamers and local
politics but also the Columbia route along which the NP might have built.

Yet, in fact, for the most part Billings and Villard negotiated with respect
and in friendship, and though Billings was taken by surprise at a crucial moment in that negotiation, no prolonged battle occurred. Billings had achieved
what he most wished to achieve: reorganization, successive loans on the divisions by which the line marched forward, and then the great loan on the
entirety of the land, which assured the Northern Pacific's completion. Certainly he would have liked to have been president when the railroad was finished, but given his unwillingness to buy his way through Congress and his
insistence on financial probity, he cannot have wanted to be president if it
meant doing as Villard felt he had to do once he had taken control of the
railroad, since Villard's management brought the triumph of a finished line in
1883 and the collapse of that line into bankruptcy, receivership, and yet another reorganization only five months later.

Virtually all accounts of the contest between Billings and Villard also suggest that when Billings gave up the presidency in June 1881, he retired into
the mountains of Vermont, content to expand his estate and watch over his
investments. Nothing could be further from the truth. He remained active in
Northern Pacific affairs, was a member of the executive committee, and, when
Villard resigned in January 1884, was widely mentioned as a candidate for the
presidency once again. Further, in April 1881, while still president of the
Northern Pacific, Billings purchased the Arnold house, a large and somewhat

run-down structure at 279 Madison Avenue. He and Julia took up residence there that fall, so he was quite active in the New York office and until his final illness in 1889 remained one of only two long-term NP directors who almost never missed a board meeting.[1]

The colorful past and rather flamboyant personality of Henry Villard has led many writers to take him at his word, so that most histories of American railroads in the early 1880s praise his role, both at the Northern Pacific and in American business generally, very much in the terms he used for himself. They consequently diminish the presence of Frederick Billings. One need not inflate the one to deflate the other, for both were important.[2]

Villard is something of an enigma—no full biography of him exists. His elegantly written memoirs are full of misleading statements and accounts of events that did not occur as he described them. The colorful biography of Villard provided by Eugene V. Smalley in his history of the Northern Pacific is understandably an admiring portrait, both because it was written for the celebration of the completion of the line in 1883 and also because, as Villard later admitted, he dictated those portions of Smalley's book, apparently as it was going to press.[3] Almost all accounts of Villard's background in Germany, his remarkably dramatic life as a journalist, and his activities outside business, though told in dozens of books, appear to derive from two basic sources: his own memoirs and Smalley's chapters. Yet the chapters in the memoirs are a close rendering of those in Smalley: that is, having dictated his version of his life so that it was committed to print by Smalley, Villard then simply used that account, with little further reflection, sixteen years later to deal with his Northern Pacific presidency. It was reasonable that he did so, for his health was failing and he believed he had little time left. However, this means, in fact, that most accounts prove to come from a single source, Villard himself. Seldom has a figure of historical prominence so effectively created his own biography.[4]

Henry Villard was tough, extremely intelligent, very able, and apparently enormously attractive: people rallied to him, believed in him, wanted to follow his leadership. He was courteous, even courtly, and he had winning ways, which with fellow businessmen, and certainly with Billings, consisted of an apparently straightforward air of solicitousness, a brusque sense that one must be up and doing because great matters beckoned, and a clear-headed grasp of the fundamentals of a problem. He could also be ruthless and sly: his papers even today contain a ticking time bomb, for in them one finds notarized copies of letters Billings had sent to James B. Power in 1878 concerning the purchase of Dakota land, letters Villard apparently intended to use against Billings during their contest of wills in the spring of 1881. Villard obtained these letters from C. F. Kindred of Brainerd, for there in the file, dated April 23, 1881, is a short note from Kindred transmitting the letters with the explanation that they showed "pretty conclusively . . . that [Billings] had something to do with the lands of the Co."—a point Billings was contesting at the time. Kindred had been summarily dismissed by Billings the previous December for "personal speculation and favoritism . . . [and] absolute personal dishonesty." Villard was neither the first nor the last person to seek an advantage over a competitor by invoking a disgruntled former employee of that competitor, though one wonders at the person who retains such letters long after they have served their purpose.[5]

Villard was born, he tells us, in 1835 as Ferdinand Heinrich Gustav Hilgard in Speyer, on the Rhine. His father was in the civil service and the family moved often. Educated at a gymnasium, then at a French lyceum, and then apparently at universities in Munich and Wurzburg, Villard was well-read and quick-witted. In 1853 he came to the United States, and eventually lived with relatives in Illinois, where he read law in a German attorney's office. He did not pursue the law, however, and recognizing that a legal practice or journalism confined to German would cut him off from the mainstream, began a peripatetic career as a writer, covering the Frémont campaign in 1856, the Lincoln–Douglas debates in 1858 (where he met Abraham Lincoln, taking refuge with him in a box car to get out of the rain), and traveling by stagecoach to the Rockies, where he sent back the first detailed reports of the Colorado gold discoveries and became well acquainted with Horace Greeley. In 1860 Villard covered the Republican National Convention and rode on Lincoln's train from Springfield to New York.

His Civil War triumphs were legion. He was at Bull Run, and rode back through the fleeing Union troops to give the bad news to the president. He was with Sherman, Burnside, Buell, and Rosecrans. He was on the admiral's flagship as a Union fleet bombarded Charleston. He was at Shiloh, Perryville, Chickamauga, and the Wilderness. Late in 1863, with two partners, he began the first newspaper correspondence bureau to send dispatches from Washington to a combination of papers not linked by the Associated Press, taking as his clients, among others, the Springfield *Republican* and the Chicago *Tribune*. Shortly after the end of the war he married the daughter of the widely heralded abolitionist, William Lloyd Garrison, and became a foreign correspondent for the New York *Tribune* in Europe.

While Villard was in Europe, living at Wiesbaden, he cultivated the leading German bankers, and when American railroad companies defaulted on their interest payments to German bondholders after the crash of 1873, he was asked to work with various groups organized to protect the bondholders. In April 1874 Villard returned to the States with a formal commission to represent these groups, and with the specific assignments of concluding a contract with the Oregon & California Railroad Company, one of the defaulting lines, and discussing the default of the Oregon Central Railroad. Villard found that he liked Oregon, and in 1875, when the bondholders forced the Oregon & California Railroad to give up its properties to them, Villard became president of it and of the Oregon Steamship Company, whose mortgage the railroad held.[6]

The following year Villard was appointed a receiver for the bankrupt Kansas Pacific Railway Company, which was in debt to German bondholders, and soon found himself in conflict with Jay Gould. He contested Gould in the courts and won a settlement for the bondholders that, in his words, "raised him to a position of influence in American financial circles, while it added greatly to his reputation abroad." He had, in a very short time, become a figure to be reckoned with in American business, especially since he had German support, which for the first time showed a serious interest in American investment opportunities.[7]

Villard was filled by now with the kind of grand dream that seized many men in the railroad age: to play a major role in developing a transcontinental railroad. To do this, he began with the assumption that he must build a line along the Columbia River to bring the Union Pacific from Utah to Portland. Surveys of both banks made it clear that construction would be very costly and quite difficult. Further, he knew the Northern Pacific would oppose him since it intended to construct its line down the Columbia from the Snake River, or over the Cascades directly to Puget Sound, or possibly both.

Villard determined to close those options down so that the Northern Pacific would have to deal with him. The Oregon Steam Navigation Company, of which J. C. Ainsworth, member of the Northern Pacific's board and Billings's West Coast confidant, was general manager, ran three fleets of steamboats on the Columbia; one fleet each was on the upper, middle, and lower river, connected by short portage railroads around the falls and rapids. The Northern Pacific had once owned the steamship company but had given it up during the crisis of 1873.[8]

Villard purchased the navigation company on behalf of himself and a group of associates. He then offered the Union Pacific half his interest in the new enterprise, to be called the Oregon Railway & Navigation Company, which was to combine Ainsworth's operation with the Oregon Steamship Company. The Union Pacific declined, and Villard and his associates acted alone, incorporating the new company on June 13, 1879. Villard became president and Ainsworth a director, and the company made it known that it intended to build a railroad along the south bank of the Columbia River.

Frederick Billings had given the Northern Pacific its second chance, and he was emotionally committed to seeing the line through to the Pacific Coast. Indeed, this has been the dream from the beginning, including use of the south bank of the Columbia River, for when Thomas Canfield had gone to the coast as general agent for the Northern Pacific in 1869 he reported in favor of the river's south bank. The north bank had one obvious advantage: as Washington was still a territory, the line could claim twice the amount of land it would be granted if the line were run through Oregon, a state. This had not persuaded Canfield, for he knew that construction on the north bank would be far more difficult and expensive, and also predicted that a south bank line would control the traffic of a large and fertile area, which the north bank of the river, given that mountains ran directly down to it, would not. Furthermore, Canfield recognized the need for preemptive annexation; if the Northern Pacific did not build along the south bank someone else eventually would. This would make Portland a rival to Puget Sound, which the NP did not want. The Portland business community, on the other hand, for the same reason favored nearly anyone except the Northern Pacific. However, to keep options open Canfield had recommended that legislation ought not to stipulate either river route, and the Northern Pacific directors had, for the most part, remained silent on the point so as not to stir up opposition in Portland or awaken potential rivals.[9]

This strategy had been sound when no rival existed and the Columbia was in the hands of a friendly power, in this case Ainsworth and the Oregon Steam Navigation Company. Villard had nullified the policy, and for a time, or so Billings felt, had led Ainsworth into an act of treason. As Ainsworth shuttled back and forth between Villard and Billings, attempting to mediate between the two, Billings came to suspect his motives (they were, in fact, clear and reasonable: to get the best deal for Portland possible), and only as Billings stepped down from the presidency did he warm to Ainsworth again.

The Oregon Railway & Navigation Company now held control of the Columbia River and, to some extent, of the south bank, since it possessed the short portage lines there. Late in 1879 Villard approached the Northern Pacific, suggesting that they combine in advertising the Pacific Northwest, and very shortly after proposed an agreement for sharing routes and traffic. Billings was preoccupied with meetings, especially with General Nelson Miles, concerning Indian titles in Dakota and Montana, and with Julia, who was quite ill. The tenor of his diary and his early letters to Villard shows that he did not take him as seriously as subsequent events proved he ought to have: Villard was full of "cheek," a somewhat dismissive judgment.[10] Further, Billings was refusing to work with Villard over the line Villard intended to construct (with Northern Pacific cooperation, he hoped) from the Snake River to Celilo, midway down the Columbia, ostensibly because the rails Villard had purchased were English and the Northern Pacific charter required that all iron be American. Billings saw this as an important point of principle; Villard may have regarded it as an effort to slow him down. In Wright's judgment Billings was uncivil to Villard, and in January 1880, Billings wrote to apologize: "let us have no personal differences," he urged, agreeing that they should unite forces in building a railroad down the river. In the meantime Billings had, of course, committed himself publicly to the Cascade route[11] via the release of his Huntington letter.

Villard felt the Northern Pacific was dragging its feet. Billings asked Colonel Potts to negotiate, and go to the Pacific Coast to look at the situation on the spot, but Potts seemed unavailable at the times Villard suggested and was unwilling to set out as soon as Villard hoped. On the day Gray met with Villard's lawyer, Billings instructed Sprague to locate and grade the line between Ainsworth and Walulla, clearly a preemptive strike, and when Villard sent two urgent telegrams four days later, after his agents observed the NP crews at work, asking Billings to stop construction, Billings ignored the first and replied to the second that he would not, naming a date for a meeting five days hence, a meeting he then cancelled. On March 1, however, the principals met in New York, and in a friendly two-and-a-half-hour meeting, Billings and Villard agreed that they ought to find a means for uniting. On the same day Billings received word from Milnor Roberts that the NP line should go over the Cascades via Snoqualmie Pass.[12]

Billings knew the NP did not have the funds for building two lines west of the confluence of the Snake and the Columbia, and clearly hoped to get a decision as to which was the more likely and begin construction while holding Villard in negotiations. Further, Billings was adamant about the stretch of rail line from Ainsworth down to Wallula: it must be filled by rail not boat, and only NP rail. "The Northern Pacific has always insisted it would build to Wallula even if from Wallula down the two companies should build together under Northern Pacific Charter. I do not see how negotiations can affect this gap."[13]

This was a long way from what Villard had wanted. He had proposed that his company lease the Pend d'Oreille Division, effectively annexing the Palouse country of eastern Washington to Oregon. Having lost on this bid, a traffic contract was an acceptable second best, but he still wanted a rail line, and through Potts he offered to build a road along the north bank of the Columbia to The Dalles, from which the NP would use his line, thus guaranteeing him the Portland terminal. Though Billings most likely would have rejected this proposal too, it was withdrawn before the Northern Pacific reacted. Negotiations proceeded at a somewhat desultory pace until October 20, when a traffic agreement was signed.

This agreement was a compromise, and despite public declarations of pleasure, neither Billings nor Villard could have been happy with it. The Northern Pacific accepted the Oregon Railway & Navigation Company's right-of-way along the south bank of the Columbia and agreed that the OR&NC could build south of the Snake, while the NP would control the Palouse north of the river. OR&NC track had been narrow gauge, since cargo had to break bulk when it reached the Columbia in any case; now it was to be to the same gauge as the Northern Pacific. Villard would have the right to build a single line into the Palouse north of the Snake in return for building from its mouth to Portland in three years, granting the NP the right to run its trains over the new road, and giving the NP an advantage on the calculation of traffic mileage in any future interchange. Billings had given up the NP's right to build on either bank of the Columbia, which he had always insisted the NP had by charter, and Villard had failed to block the NP from building down to Portland along the north bank, so neither felt safe.[14]

Then came the Northern Pacific's $40,000,000 loan through the syndicate formed by Winslow, Lanier and Drexel, Morgan. This had been kept as secret as possible, and Villard learned of it (apparently from Morgan) only shortly before the deal was struck. Without money any Northern Pacific threat to build along the north bank of the Columbia was purely theoretical; now it was very real. Billings was believed to want to build both over the Cascades and along the Columbia. Villard had proposed a $10,000,000 syndicate plan to the NP himself, and now understood why it had been politely rejected. He was faced with a choice: to sell the Oregon Railway & Navigation Company to either the Northern Pacific or the Union Pacific and retire from the field or, far more difficult, attempt to acquire a controlling interest in the Northern Pacific. In a daring maneuver he chose the harder task.

Quietly Villard began buying up all the Northern Pacific stock he could, until he had nearly $20,000,000. He also called on Drexel, Morgan, which was nervous over Billings's growing insistence that the Northern Pacific must build both routes to the coast. One of the members of the syndicate, C. J. Woeris-hoeffer & Company, broke early, apparently coming to Villard's support and, when he reached the point in February that his own resources were insufficient, helping him. At this stage Villard created his famous "blind pool," one of the most audacious ventures of his daring takeover bid.

In February Villard sent a confidential circular to fifty-five friends asking them to subscribe to a fund of $8,000,000 to which he would also be making a large contribution. The fund would create a new company, for which he would give an accounting on May 15, 1881; until then no one, including the subscribers, were to know what it was they were blindly purchasing. They would be trusting Villard without information or security. The state of the market, Drexel, Morgan's desire to see Villard victorious, and Villard's personal reputation for daring and successful ventures inspired confidence and the pool was fully subscribed within a day. Villard then postponed his accounting until June 24, and in the interval he was able to call up an additional $4,000,000. With these new funds Villard had the money needed to complete his purchases of a majority interest in the Northern Pacific.[15]

The blind pool occupies a major place in the lore of Wall Street. It was, one historian writes, "Napoleonic" in its boldness. Another calls it "one of the notable achievements in the annals of railway finance." Villard felt it gave him "a most distasteful notoriety," but bankers hailed it as a "stroke of genius." It was all of these, and Villard is due full credit for realizing just how attractive the air of mystery surrounding the pool would be. But had it not been for Charles Woerishoeffer, the German plunger, and for men like George Pullman, who had long wanted to get more of his cars onto the Northern Pacific and who subscribed $500,000, and had it not been for Drexel, Morgan's ultimate control over other members of the syndicate that had financed the Northern Pacific's loan and now controlled a friendly press—that is, had it not been for the financiers into whose power the Northern Pacific had, of necessity, passed—Villard most likely would not have succeeded. Looking back, historians see in the success of the blind pool the first major evidence that the railroads, whose builders were once movers and shakers themselves, would soon to be in the hands of even larger forces.[16]

On the face of it this is correct. Indeed, well before the Northern Pacific's 1880 sale of $40,000,000 in bonds began, Villard was fully informed on the terms the Winslow, Lanier et al. syndicate had imposed on the Northern Pacific, including a provision that no money was to be spent on construction west of Ainsworth until the line from Montana had reached that point. While this stipulation had given him breathing space, he had predicted that he and the syndicate would gain control of the NP so that it would be consolidated with his company long before the line was completed across Idaho. Drexel, Morgan and other members of the syndicate were, he wrote, "very friendly towards me," and the firm subscribed heavily to his "great unknown," the blind pool, since Morgan, unlike most subscribers, must have known its purpose. Clearly Billings's deliberate, rather stately progression across the continent, his insistence on an orderly procedure and meticulous observation of all the NP charter's provisions, made the syndicate anxious about the prospects of early completion. They preferred the more daring approach they knew Villard would take. Thus perceptions by the syndicate of the differences in personalities of the two men also played an important part in the outcome.[17]

Throughout this short and intensive period the directors of the Northern Pacific were unaware of the danger. Billings was more distracted than at any time in the presidency. His daughter Mary was very sick, and his youngest daughter, Elizabeth, was believed to be dying. Frederick was suffering from diabetes and self-diagnosed malaria, going to bed every night at 7:30 after difficult days at the office, trying nostrums handed him by friends and through chance encounters. He was so dissatisfied with his life, he wrote, and so discouraged with the time he had to give to business when he wanted to be with his daughters in Woodstock, he thought of giving up the presidency.[18]

Villard could not be certain that he would succeed when he began his silent campaign, and so he took steps to block the Northern Pacific at Wallula. He ordered his agents to study the Northern Pacific's maps of the north bank of the Columbia—maps the NP had given him during their earlier negotiations—and begin buying up parcels of land which, in their judgment, would give his company control over the essential strips and thereby forestall construction. Villard also offered to purchase all of Billings's stock, preferred and common, and at the same time solicited any information he could obtain that might prove damaging to Billings. Drexel, Morgan and other members of the syndicate that effectively controlled the Northern Pacific began to make it known publicly that they thought it was time for the Wright–Billings–Cheney "old line" leadership, with its emphasis on deliberate construction of the line, its inattentiveness (in the syndicate's eyes) to enticing ancillary opportunities, and Billings's stubborn desire to maintain the integrity of the NP's charter by building (or at least threatening to build) both routes to the coast, to make way for different leadership.

Though Billings had not detected Villard's larger scheme, he was aware that the well had been poisoned at Drexel, Morgan. In February he wrote to Joseph Potts, in Montana, blaming Villard for convincing the house that the Northern Pacific was unwise to build over the Cascades. Villard had so affected Drexel, Morgan, "they seem to think we are guilty of absolute folly in not falling in with Mr. Villard's notions." Further, Villard had convinced the house that Billings's insistence on building over the mountains was influenced by his private interests in the Puyallup coal lands rather than in finding the cheapest, shortest, and quickest route.[19]

On February 21 Villard told Billings that he and his associates had acquired control over the Northern Pacific. George Pullman told him much the same via another director. Billings did not believe them—there was not enough stock on the street, he thought, to give Villard control. Still, he was worried, for he understood that Villard also had the support of William Endicott, a major Boston investor of great wealth. Then on March 11 Villard again boldly offered to buy all of Billings's stock, said by the press to be worth $10,000,000, in part because he no doubt wanted it, but possibly also because such a bid would suggest that he had not yet acquired enough for control and might temporarily disarm the now alerted president.[20]

On March 14 Billings replied at length to Villard, declining to sell his stock. To do so, he said, would be construed as an act of disloyalty to those who had supported him for so long. Villard had said that he hoped to take control of the Oregon & California and perhaps even of the Union Pacific; Billings thought this scheme "too big" and did not want to be part of it. While he did want to work together, he insisted, to go beyond the intended cooperative arrangements with the OR&NC would jeopardize the entire operation. He appealed to Villard to be reasonable: the NP should build over the Cascades "not for the purpose of hurting you, but because the Northern Pacific, chartered to go to the Sound by two routes, should have a through line by at least one, and because the Cascade Range route is the shorter one; will be a necessity for half its length or more, for local purposes; and lies through coal, iron and timber fields furnishing always freight back to the interior." Not to build would lead to an unfavorable report from the government auditor. Villard had his own route along the river, Billings argued, and with this he could develop eastern Oregon and Washington jointly with the Northern Pacific. There was enough business for both. To back away now would lead Congress to rise up and give them trouble. "Why, virtually, put a pistol to the breast of the Northern Pacific at Ainsworth or Wallula and say, thus far and no further! It is of this I complain." Surely Villard could not expect a railroad that possessed the right to build by two routes to the Pacific to let the OR&NC build one and someone else take the other and never reach the coast itself? "This is what will happen if this Company postpones work on the route to the Sound."[21]

With this frank letter Billings had, as was his custom, revealed all his anxieties, and appealed for equal frankness in return. In refusing to sell his stock, he was, he felt, simply doing what he must: "my duty without reference to pecuniary or personal matters." Gray and Wilkeson, still the secretary, praised him: "Not one man in a thousand wd have declined 5 to 6 millions from a sense of duty."[22]

The Northern Pacific had one last counterattack it could make against Villard's initiative. The company's treasurer held, as trustee, $18,000,000 in reserved stock, which was to be issued to pay for sections of the road as they were finished. Used strategically, this stock would make it difficult for Villard to complete his plan, for there would be more stock, dispersed and presumably in the hands of company loyalists, that he would need to acquire before he held or regained a controlling interest. At the board meeting on March 17 Billings read aloud his letter to Villard, spoke at length and with great intensity, insisting that the Northern Pacific must never be subordinated to the OR&NC, and carried the day unanimously.[23] The next day the executive committee ordered the issue of 180,000 reserved shares, with the certificates distributed between 240 individuals. Billings had been a holdout on this strategy, for without this stock the line could not be built, but he was persuaded by the other directors that unless it were used, they would have no line to build. In the end he went along, perhaps hoping that Villard had bought his stock on margin, so that the issue of so much new stock would lower the value of Villard's holdings and lending banks might recall their funds.[24]

When Billings declined to sell his stock—the single largest bloc—Villard realized the board might take such a step and he was ready. On March 25 he obtained an injunction from the Superior Court in New York City to restrain any issue of the reserved shares on the ground that the stock issue was illegal. Now Ellis of Winslow, Lanier and John C. Bullitt of Drexel, Morgan, presumed backers of the new issue, also objected.[25] Billings was devastated. He had thought Villard thoroughly discomfited; now there would be litigation which, in New York, "was uncertain," and there was bound to be much malicious newspaper gossip.[26] This was a declaration of war. Billings went to bed early, feeling ill, for he now understood just how alienated the syndicate felt, and the next day, when he read attacks on him and on the Northern Pacific's "conservative" board, he understood that he had lost the backing of the key financiers. The board employed Judge Shipman, who had been so helpful at an earlier time, to represent the proprietary interest holders with respect to the injunction. Jeremiah Black was engaged once again, for though Villard made noises suggesting compromise, Billings was so angry over personal slurs in the press, mostly to the effect that he was secretly hoping for Villard's success, he insisted that the suit must be tried.

On the day of the hearing on the injunction, Billings backed off a little, concluding that he must sell some stock, for the litigation might well go on for some time with an unanticipated outcome, and he therefore should "get into a safe condition for funds," since he had only a few days earlier agreed to the purchase of the Arnold house. The week was as miserable a time as he had ever known: Mary's illness was diagnosed as a very serious combination of pneumonia and Bright's disease and it was feared she was at death's door. Bleak news of Elizabeth's further deterioration arrived. His old friend and mentor O. P. Chandler was dying, and Billings was making frequent depressing calls on him. The Arnold house, to which he obviously had committed funds at precisely the wrong moment, proved to be in a shocking state, requiring more attention. On the day the hearing was renewed, Billings came down with a malaria attack. Had he seen the reports in the St. Paul press—that the "Billings dynasty" was still going its "serene and undisturbed" way, and that he was "impregnable to assault or capture"—he would not have recognized himself.[27]

The news did not improve. Black advised that the Northern Pacific's case was not strong. Villard began a second suit intended to get at the NP's books. Billings concluded that Villard was going to win, and decided that, with the public abuse being heaped upon him by newspapers friendly to Villard, his stock would fall and that he must sell, though he still held on, giving up only his Erie stock. Patently this was not enough, and Billings concluded that he must sell enough that Villard would call off his suits. "It is a dreadful calamity, all this trouble. I feel it deeply," Billings wrote. The next day, April 20, he collapsed on the floor and had to be put to bed.[28]

What Billings did next has had widely differing interpretations, in the press at the time and by historians since. He allowed Villard, secretly, to acquire his common stock, at substantial profit for himself. When this became known to the press and his fellow board members, many denounced Billings as a traitor, as the person who, having fought Villard for so long, had now sold out to him for personal gain, making possible Villard's acquisition of the Northern Pacific.

This was not precisely so. A group was working on a compromise position: Ellis, for Winslow, Lanier, and Potts and Ainsworth, who now was interested in developing Redondo Beach in California, for the Northern Pacific. Billings was convinced Villard would win since he believed the syndicate had abandoned the company. He also thought that Villard already possessed enough stock to assert his controlling interest at the next board meeting. On May 19, as an indication that other parties agreed this was the likely case, Villard was granted two places on the Northern Pacific board, and two Billings stalwarts stepped down. The following day Villard's suit was dismissed though the issues raised by the injunction were not yet settled.

On May 21 Billings agreed to sell to Villard ninety thousand of his shares for $4,500,000, which according to the press represented a net profit of $1,925,000.[29] To Ainsworth, who had been working toward a truce, convinced that if he could not do so the NP would stop in its tracks, Billings sold $9,000,000 of common stock, receiving half a million more than the stock was worth on the market. Ainsworth then disposed of the stock to Villard at the end of May.[30] If Villard did not in fact already hold a controlling interest, the sale of Billings's ninety thousand shares had the effect of nearly cancelling out the advantage the Northern Pacific had gained from selling its $18,000,000 in reserve shares. If he still did not have control—and there was general belief that he did—he certainly acquired it through Ainsworth. The *New York Times* reported on March 14 that he had purchased $25,000,000 in Northern Pacific stock, which made Villard and his backers by far the greatest shareholders. While one cannot know today precisely how many shares Villard did hold on May 21, from the available evidence and the contemporary press it seems likely that he did in fact have effective control of the line without either sale. At least three board members knew of Billings's intention to sell, and they appear to have raised no objection. When, on June 8, the alleged terms of the agreement worked out between Villard and Billings were printed in the Portland *Oregonian*, Billings believed Villard or Ainsworth were responsible for the leak. The *Oregonian* was delighted, for with Villard in control Portland was likely to replace Tacoma as the terminus for the Northern Pacific.

On the day the story broke in the press, to much negative comment, Parmly arrived home with the sorry news that he had been expelled from Williams. Frederick was deeply upset, urging him to return and beg for reinstatement while preparing alternate applications to Yale and Amherst. The following day Villard advised Billings to resign his directorship as well as the presidency.

"My life seems crushed out of me. . . . The N.P. is no longer to be my life work," he recorded in his diary, and with a sense of the deepest gloom wrote to his brothers and sisters to tell them what they would receive in his will.

At the board meeting on June 9, 1881, Billings resigned as president, to be succeeded by A. H. Barney, the New York banker and long-time treasurer. While a few members stood by him, Wright led an attack on his leadership, and the board declined to pass the customary motions of approval and good wishes. Smarting from being so "meanly treated," impressed with Villard's politeness and frequent courtesies—Villard was, he wrote, "a better friend than many of my associates"—Billings warmed quickly to his vanquisher. In his usual direct way he confronted Barney, upon whom he had depended as a faithful friend, when he learned that the treasurer had called him "able but not faithful" for selling his shares to Villard. Over the next several weeks Billings brooded over how his associates had treated him, while Villard took the trouble to come to see him, seek his advice, and write in kind terms.[31]

Villard and Billings worked together in a successful effort to forestall the Utah & Northern Railroad. Counsel Gray had searched the statutes and concluded that the competitor had no right to locate in Montana at all and should be treated as a trespasser. Anderson had hoped to bring construction down Clarke's Fork secretly, and Dodge proposed to occupy the Deer Lodge route to Helena and Fort Benton, thus shutting the Utah & Northern out. But the line was still far from Deer Lodge, and construction could not remain secret for long, so three weeks after Villard succeeded Billings, Anderson instructed Dodge to obtain a Montana territorial charter for the Rocky Mountain Railroad Company, with Villard, Billings, J. B. Williams, and Barney as the directors. It would run the Butte Coach, he told Dodge, which would connect Butte to Helena. The next day Anderson had second thoughts, advised Dodge to suppress the names of the NP directors and recruit only "friendly outsiders" from Montana who could be counted on to resign their interests to the four shadow directors when called upon to do so. However, the Utah & Northern backed away, content to end its challenge at Butte, and the new company turned its attention to the building of a rail line from Livingston up the Yellowstone River valley to the National Park.[32]

For the rest of the year Frederick was in a state of gloom, from which the recovery of his daughters, Parmly's admission to Amherst, and the occupation of the Arnold house in October did not fully draw him. The president on whose behalf he had contributed, James Garfield, was shot on July 2 and lingered until mid-September. Not only was Billings's hope for political reform crushed, but the crucial confirmation of titles to Indian lands, on which he worked with Villard in the weeks after his resignation, was put on hold, for Vice-President Chester Arthur, believing at first that Garfield would recover, was unwilling to take any action. As a result Billings was, a little to his surprise, lost "with nothing to do." He had imagined, when president, that he wanted to be free to spend his time in church work—Reverend Hicks had accepted a call to Wethersfield, Connecticut, and Billings helped interview candidates to replace him—and his several philanthropies, and with Julia and the children. Now he remained in Woodstock, taking long, often solitary walks. He worked on his estate. He took an interest in genealogy. He thought about how to use his money (of which there was a great deal now) wisely for the family and in "doing good."

The recovery of good spirits was slow, but by the end of the summer Billings was taking a close interest in the Northern Pacific once again. Though he had rather hoped to see its name changed, which would put distance between his presidency and the new enterprise—to Villard he suggested Northern Transcontinental Company—in time he was happy this had not happened. He began to make investments once more, in Scranton Steel and Montana land. He spent time with old friends from California as they came through, with Oscar Shafter's widow and others. Then on September 16, as the newly elected president, Villard appointed Billings chairman of the executive committee, and he was, once more, back in his element, reconciled with most members of the board, working effectively with vice-president Thomas F. Oakes, whom Villard left in control. It was not like being president—"I was very proud of the office" he wrote in his year-end summary—but he was busy and working with the new regime, which was as determined as he had been to see the Northern Pacific through to the coast.[33]

23

The Last Spike

VILLARD SOON FOUND HIMSELF in difficulties because of the terms of the $40,000,000 loan. Bonds could be issued only as sections of twenty-five miles were completed, examined, and accepted. Because Vice-President Arthur was unwilling to order the commissioners to begin examinations while Garfield was still alive, many months passed and the railroad completed 275 miles without an inspection. Thus Villard had to pay out large sums in advance of taking in new funds. The Oregon and Transcontinental Company, Villard's new creation which now sheltered the NP and the OR&NC, therefore had to meet these demands, so that it was increasingly dragged down. Further, Villard felt compelled to take a lease in perpetuity on the Oregon & California Railroad Company lest he be invaded from the south, and he decided to develop terminals in Minneapolis, St. Paul, and Portland, and to consider bridging the Willamette River at Portland. Billings had plowed most surplus capital back into improving the railroad, while Villard chose to disburse surplus earnings as preferred dividends. Where Billings had proceeded cautiously, Villard decided to forge ahead.

Costs soared, reaching $2,000,000 a month. As president, Billings contracted for an iron bridge across the Missouri River at Bismarck and for construction along the Yellowstone, and he committed the Northern Pacific to large orders of equipment, doubling the number of stock cars the company owned and adding substantially to the fleet of box and flat cars. Now Villard repaid George Pullman for his support, placing a large order with the Pullman Palace Car Company to equip and jointly operate passenger cars for fifteen years. Expansion required a new roundhouse and new construction at Brainerd; further, two major tunnels were to be constructed in Montana, the first at the Bozeman Pass, where a detailed survey had revealed that the line could not remain within the stipulated maximum grade without tunnelling. A cave-in on July 4, 1882 wiped out five months of work, forcing the company to

THE NORTHERN PACIFIC RAILROAD
CIRCA 1883

Northern Pacific line and its branches (including associated lines in Oregon)
Uncompleted Northern Pacific lines
Connecting other railroads

build a switchback line over the pass to move traffic to other construction points while continuing the tunnel.

In August 1882 President Arthur at last appointed six commissioners to inspect the track, and when approved, nearly $7,500,000 in bonds were issued. At the end of the year the Northern Pacific also had over $7,000,000 in earnings, enough to sustain perhaps seven months of construction, assuming nothing else went wrong. Villard pressed for completion of the line by August 1883 before snow could delay work into yet another year. Crews labored under great pressure, and, as they narrowed the gap, a sense of competition developed between the east and west teams. On what would be the last day, August 23, both crews began at five o'clock in the morning, racing toward a red flag set down in the middle of the final ten-mile section; the west reached the flag while the east was still three-quarters of a mile away, and at three o'clock that afternoon the rails met.

Villard's grand celebration was extensive, dramatic, and in retrospect quite foolish.[1] On September 3 the official excursion, with 350 guests, began in St. Paul. There were four trains, the first with Villard, his wife and four children, and a party from Germany. A second train was made up of British notables, and included the Billings family and Samuel E. Kilner, Frederick's secretary, in their private car, the *Northern Adirondack*. The third train brought ambassadors, senators, governors, and a variety of railroad notables, and the fourth hauled a bevy of journalists and members of management. As a result, the press often was not on hand at the stops where Villard and others spoke, for the crowds were large and would not wait for the last train, which sometimes ran an hour or more behind the others. The people especially wanted to see former President Grant and Indian fighter Bill Sheridan, and since Grant was riding in Billings's car part of the way this brought Billings more attention than he wanted.

The excursionists were a stellar group. The great British historian and diplomat James Bryce had advised on the choice of the other British guests.[2] They included the British minister, Lionel Sackville-West, Lord Justice Sir Charles Bowen, the earl and countess of Onslow, the governor of the Bank of England, the chief aide to the Prince of Wales, the secretary to the Board of Trade, and several members of Parliament. Distinguished German scientists drank Bavarian champagne with Villard in his private car, where they were joined by well-known German writers, diplomats, and bank directors. Villard chose William Maxwell Evarts, for ten years president of the New York City Bar Association and more recently secretary of state under President Hayes, to be the official orator. (This was a pleasant touch for Billings, who had often visited Evarts at his home in Windsor, Vermont.) President Arthur belatedly designated secretary of the interior Henry M. Teller to represent him. Also along were the U.S. attorney general, the governors of Wisconsin, Minnesota,

Dakota, Montana, Washington, and Oregon, August Belmont, Marshall Field, James J. Hill, George M. Pullman, Alexander Ramsey, Edwin Godkin, editor of the New York *Evening Post,* Joseph Pulitzer of the New York *World,* and Carl Schurz, not as secretary of the interior but as editor-in-chief of the New York *Evening Post,* which Henry Villard had purchased in 1881.

At St. Paul every building within sight was covered with flags and bunting, a grand arch passed over the street by which "Villard and Guests" arrived, and lines of small girls dressed in white threw flowers down from high stands onto the entourage. A parade of seven hundred wagons took four hours in St. Paul and, because the Twin Cities could not agree on a program, seven hours more in Minneapolis. President Arthur, who had just returned from a private tour of Yellowstone National Park, spoke at the opening banquet, which did not end until midnight, when the trains left immediately for the west.[3]

In each town along the way, the celebration cars were greeted by large crowds and more decorations. At Fargo, a pagoda of cereals was erected, and the speakers were required to crawl through the display of wheat to mount the rostrum. The first train would arrive, Villard would speak, and his section would move on; the second train would pull up, Grant and Evarts would speak, and their train would hurry to catch up with the first. The fourth train would arrive and journalists from twenty-three newspapers scurried among the crowd to get reactions for their colorful stories without hearing the speeches on which they were soliciting opinions, and then dashed along after the other trains. Villard and Grant could command the crowds, and Evarts generally could hope to hold them because he was a noted orator, but anyone who spoke after Evarts usually was drowned out by the restive throngs.

The journey was both exhausting and exhilarating. In Bismarck the dignitaries helped lay the cornerstone for the new capitol building, and the territorial governor, Nehemiah G. Ordway, spoke at great length, followed by Villard, Billings, Schurz, and Grant. Chief Sitting Bull was present with a number of his followers, and Villard introduced him to the audience, persuading him to say a few words in Lakota. (The chief is said to have denounced all land thieves, but the army officer who was translating converted his words into the traditional language of English hospitality. Whatever it was he said, Sitting Bull received a standing ovation.) West of Bismarck the third train was held up for seven hours with a broken axle, which caused an engine to overturn and required laying a quarter-mile track around the wreck. That night a coupling on the second train gave way in the Badlands. ("If this is God's Country, it must be because he is still at work on it," one passenger observed.)

Frederick Billings's best moment came at the town named after him, which was reached at nine o'clock in the morning of September 8 on a "very fine day." Standing on the balcony of the railway hotel, Billings told his audience that theirs was "a great town" settled in "a great valley," which irrigation would make most marvelously productive. The Bull Mountains—in which he had been investing quietly—would produce coal (at this point he gestured toward a huge piece of coal, weighing hundreds of pounds, on exhibit nearby). Trade would now grow along the railroad. Billings possessed "all the elements for great material prosperity." He was honored, he said, that the town had taken his name. If his humor and ever-present puns fell rather flat—he was very tired—the people appeared to enjoy seeing the person after whom their town was named. In a second talk, he declared that he would always be a friend of the community, as he and his descendants proved to be.[4]

Until he had been to Billings, Montana, the community that bore his name had been an abstraction to Frederick. Now he recognized that he had, somewhat by chance, been paid a distinct compliment. This was, of course, an honor bestowed on many Americans in the nineteenth century as the frontier moved westward. Though he had been honored before, in the still-tiny village of Billings, Missouri, and briefly at the former Depot Springs, in Washington Territory, which had changed its name to Billings and then to Cheney, he sensed that this community on the Yellowstone would become prominent, and he would help it with gifts, with efforts to see that branch lines began there rather than at some point a few miles east or west, and through investments. In doing so he would also help himself.

Why Billings? In the fall of 1881, as railway construction workers had neared the settlement of Coulson on the Clark's Fork bottom, the Northern Pacific's general contractor, Heman F. Clark, realized that it would be better for the railroad if any permanent town were to be on flatter ground and chose the spot where a town plat was to be located. A name was needed, and he suggested Billings in honor of the immediate past president of the line, largely because Billings was helping to organize the Minnesota & Montana Land & Improvement Company, which was to develop the site. There were no other suggestions, and Frederick was asked for his permission, which he readily gave.[5]

The record of railroads and their related land improvement companies, especially when town sites were involved, obviously differs from company to company and according to time and place. However, site strategies were aggressive only when competition was anticipated, and once the Northern Pacific was a few miles past Fargo, it did not expect an early rival. The railroad wanted towns to grow where they would benefit the railroad company, and used their power to see to it that this happened. Once a line was built, they watched carefully any move to locate a town elsewhere, for this would be an incentive for a new railroad to come in or would require a preemptive strike by the railroad already present, either in the form of a branch or spur line to the new town site or by lowering rates at the local elevator or stock pens to be certain farmers or ranchers would continue to bring their produce to them and not to a potential competitor. The railroad surveyors were looking for much the same kind of typography as site planners wanted: level land, well-drained, with water nearby. A farmer would prefer to haul a load of grain ten flat miles rather than drive a four-horse wagon five miles over a steep incline, for one was less likely to own four horses than have the time to cover the longer distance. Site was all-important, and all railroad companies were ruthless in taking control of the land best for their purposes.

The Northern Pacific used at least three systems for laying out town sites and involving local capital. The railroad might hold an undivided half interest in all lots. It might own fully half the lots, laid out in an alternate arrangement, thus controlling to some degree the uses of the lots not owned, or, preferably, putting the unowned lots into the hands of a friendly syndicate. Or the company might own land in scattered blocks chosen because of their situation regarding water or some other desirable feature. By building irrigation ditches, once undesirable land could be made desirable. At Billings the Northern Pacific followed all three practices at different times.[6]

When the town plat was filed, it established the usual pattern for a railroad town along the Northern Pacific, with business lots fronting on the tracks, a gridiron of squares extending back from these choice locations, and a station and other railway buildings arrayed along one side.* In effect, the train would arrive by running down or at least immediately parallel to Main Street, an adaptation to flat land of the pattern long before imposed on many small communities in Vermont. Thus Billings or any town along the Northern Pacific was not only the product of the railroad, it also looked as the railroad wished it to look, and a name connected with the railroad could not have been more appropriate.

The Minnesota & Montana Land & Improvement Company was incorporated in March 1882, five months before the first train reached Billings, with Clark, Billings, and Thomas Oakes as the first three directors. Five thousand lots sold quickly and the initial price soon increased 600 percent. Clark left the Northern Pacific to become president of the improvement company, until he was discharged in 1883 for speculating in land and raising the price so high on a body of colonists from Wisconsin that they chose to move on. Without such tactics the company did well, for Billings quickly developed as the center of the stock-raising industry and a major wool market for Montana. By the mid-1880s it was as true a cowboy town as one could find, a creator of cowboy lore and the prosperous and growing city that Frederick had predicted in his

*In time this pattern would lead to the phrase "the other side of the tracks," the American equivalent of "below the salt" or "beyond the pale." Grade crossings were few, and walking over the track was dangerous, so prime businesses and the better residential area grew up on one side of the tracks, back of and near the station.

speech. Until the devastating winter of 1886–1887, the worst ever seen on the high plains and the beginning of the end for the cattle kingdom, Billings was a boom town.

With Clark gone the improvement company needed a supervisor, and when Billings learned that Benjamin F. Shuart, the local Congregational minister, was leaving the pulpit, he hired him. Shuart had arrived in 1882, and on his behalf a townsman had the inspiration to write to Billings to ask for a gift to build a church. Julia responded with $5,000 of her own money and the improvement company donated a lot. When the builder's plans were ready, they were submitted to Frederick "for his approval," and he offered to meet the full cost, which was an additional $8,000. (He also gave $4,000 for a school, repeating his pattern of philanthropy from San Francisco.) Billings was impressed with the way Shuart accounted for the gift, and when Shuart left the church to expand a farm he owned near Hesper, west of Billings, he offered to report quietly to Frederick on land values and good purchases, in effect acting as an agent for Billings's investments. In time Billings asked the minister to be agent for both the land company and the Northern Pacific Coal Company, which was developing a mine in the Bull Mountains to produce the high grade coal Herman Haupt required to reduce the risk of fires that would destroy grazing and timber land.[7]

Frederick Billings also developed a close working relationship with Samuel T. Hauser, arguably the most powerful man in Montana until J. J. Hill arrived on the scene.[8] Hauser was a Kentuckian who had come to Fort Benton in 1862 after working as a construction inspector for the Pacific Railway Company. He knew trains, trestles, and Montana's need for transport, and believed in the political future of the vast region, having been one of a delegation of three who had gone to Washington in 1863 to lobby Congress to create a territorial government. In on the ground floor, he was able to influence legislation from the time the first representatives were chosen to attend territorial meetings, and though he proved to be a poor business manager, he was a superb promoter, investing heavily in mining properties and silver smelters. In 1868 he had founded the First National Bank of Helena, and by 1880 Hauser dominated banking throughout the territory. In 1885 he was appointed governor.

Just as Benjamin F. Potts had been the right man to bring to one's side politically, Hauser was the essential economic ally in Montana, and the Northern Pacific was quick to seek a mutually supportive relationship. Unlike Potts, however, Hauser initially favored the Utah & Northern Railroad; by 1881, when it was apparent that the NP was going to win, Hauser shifted his loyalties, later helping to finance branch lines from Livingston toward Yellowstone National Park and from Helena to the mines at Wickes.

Billings invested in Hauser's ventures both on behalf of the railroad and privately, buying shares in twenty-one mines. He also bought into Hauser's most ambitious project, the Helena and Livingston smelting works. Together with Hauser, Conrad Kohrs, Granville Stuart, and others, Billings took a one-ninth interest in the Pioneer Cattle Company, which by 1886 was running nearly nineteen thousand head of cattle. The two men maintained their business relationship, and well after Billings's death his nephew, Edward Goldsmith Bailey, who at Frederick's suggestion had worked for a season at Hauser's DHS ranch on the Fort Maginnis range, continued to watch over his investments. Frederick provided his son Parmly, together with Bailey, the funds needed to open a private bank in Billings in 1886, but when Parmly died two years later he sold his interest in both the bank and the land and improvement company. He had by then broken with Shuart, who had practically accused Parmly of lying to his father about the terms of a land acquisition—the evidence tends to support Parmly—and he was focusing more on the prospects of Yellowstone National Park.

Billings did not linger in his namesake during the great excursion of 1883. Neither Edward nor Parmly, both away on a cattle drive, had been able to be present to hear his speech. They also missed the great event—the driving of the last spike.

But for Billings the town, Billings the man was no small matter, and it was right that it should bear his name. In 1887 he was by far the largest individual landholder, in assessed value, in the county. Having taken control of the Minnesota & Montana Land & Improvement Company in 1884, and having charged Parmly and Edward with the local management of the company's holdings from 1885, he controlled even more property. The Northern Pacific held substantial land on which Frederick made many of the key decisions. These properties together accounted for nearly 20 percent of all assessments in Yellowstone County. Further, Parmly—together with I. D. O'Donnell—had been instrumental in the development of the community's essential irrigation ditch (and O'Donnell in the beginnings of scientific agriculture at his Hesper Farm), and the Bailey and Billings Bank would, in 1891, become the Bank of Billings. The modern city of Billings would have every reason to honor Frederick with his statue in 1983.[9]

That grand moment came at Gold Creek. At eleven thirty on the morning of September 8, Villard's lead train pulled up for the ceremony about a mile and a half beyond Garrison. F. Jay Haynes, the Northern Pacific's photographer, was there to record it, and the emigré artist Henry Farny was along to illustrate the event for *Century Magazine*.[10] A large pavilion had been constructed in the midst of the wilderness where lunch was served while waiting for the later trains. Billings's train was delayed, for coming down the Mullan Pass in the dark his car had broken away, to be stopped abruptly and dramatically, with the shattering of much crystal, by its air brakes. Clouds came down to obscure the hills as a delegation that had travelled up from San Francisco and Portland brought their train forward to touch engines with the lead train from the east.

Villard spoke briefly and well; Grant had a few words to say, by which time the rather liquorish crowd was becoming unruly. Only those in the front could hear Evarts, who was exhausted and spoke at length. Teller spoke. Billings spoke. Neither could be heard, which was too bad, for Frederick had worked over his speech with his customary care, and unlike his somewhat flat remarks at his namesake city, his speech was witty and wise—if rather unsuitable to the restive audience, starting as it did with a flourish to Agamemnon. He reminded his few listeners that he had been faithful to an idea that he had initially heard from Asa Whitney, the person credited with having been the first to project a railroad across the continent, when Whitney addressed the Vermont legislature four decades earlier. Now, he said, the railroad had brought peace with the Indians, new states to the Union, new power to the nation, and a new bond of unity to the people. Billings ended with a warm tribute to Jay Cooke; he did not mention Villard.

But the speeches were not over—there were eighteen in all—and the crowd grew distinctly restive. Sir James Hannen, a British judge, made a few remarks on behalf of the British guests. The governors of the states through which the Northern Pacific ran all spoke. Darkness was rapidly approaching and the spike had still not been struck. Those in charge at the site had forgotten to rope off the place where the track would meet, and the crowd milled about so that a force of soldiers was needed to clear the spot. The crowd jeered, swore and sang, trampled on their hats, and tore up their shirts, drinking from the many bottles going through the assembly. At last the gilded iron spike was brought forward and H. C. Davis, who as superintendent of construction in 1870 had driven the same spike, struck the opening blow. Eight people, including Villard and his wife, Grant, Frederick, and Julia, each had a go at the hammer. Cannons boomed, locomotive whistles tooted, the multitude cheered, the two trains nearly afloat in bunting met toe-to-toe on the newly laid track, and a military brass band that had accompanied the excursion from Fort Keogh played "God Save the Queen," "Der Deutsche Rhein," and "Yankee Doodle."[11]

Villard and the bulk of the party continued to Portland, though many guests returned to the East after the ceremonies were over. Given the near disaster on the Mullan pass, the five trains (with the Portland section added to the convoy) set out at thirty-minute intervals so as to remain safely apart through the mountains. At the Snake River crossing the entire excursion, sixty cars in all, was ferried over, four cars at a time, on the transfer boat *Frederick Billings*.[12] After a rest in Portland, the party continued to Tacoma, Seattle, and a final grand reception. A hard-core group of fifty returned to see

Yellowstone National Park. All the festivities had cost Villard between $250,000 and $300,000.

Some reporters concluded that the celebration excursion was a fiasco, deeply damaging to Villard, and others thought it did not contribute in any significant way to his subsequent downfall. There were reports of foreign investors rushing off the train at various station stops across Dakota, sending telegrams to Europe ordering divestment, because the land they were passing through seemed to them so barren. Certainly a flurry of telegrams were sent, especially from Portland, though more likely because a sharp fall in the stock market had begun just as the excursion set out, and the financiers were in constant fear of heavy losses.

Villard now hurried back to the East Coast to try to halt the collapse of his empire. He had spent too much too quickly: the grand celebration, money for a hotel in Portland (into which he had put the company's stock without authorization), a steamer before it was needed, another hotel in Helena. Still, all were peccadillos compared to the basic problem: expenses of building had doubled and the decline in the stock market, begun in 1882, had steepened as Villard plunged ahead. He faced a deficit of $10,000,000. Villard tried to raise the value of the railroad stocks by arranging to have the Oregon and Transcontinental acquire outstanding shares while using his personal holdings in the Oregon Improvement Company to take a large loan. His erstwhile supporter William Endicott wrote, "I cannot quite make up my mind whether it is you or [P. T.] Barnum that has the greatest show on earth." Word leaked out about the deficit and that Villard proposed another bond issue; shareholders began to sell. At the September Annual Meeting the Northern Pacific announced a shortfall of nearly $8,000,000; the next day a rush began.[13]

The collapse was sudden. In October Drexel, Morgan and Winslow, Lanier suggested a new bond issue that would have put Villard wholly in their hands. The railroad resorted to extreme economies, even removing lamps from switch stands. Herman Haupt resigned in anger. Construction stopped and public confidence plummeted. To his diary Billings confided that he could not stand Villards' management: "if the public knew exact situation his life wd be in danger."[14] On the night of December 16 Endicott, Fabbri, and a representative of the Farmers' Loan & Trust Company awoke Villard to tell him that he and the company were at the point of bankruptcy. A syndicate would help him, however, if he would give up the presidencies of the O&T and the OR&NC. On December 17 Villard resigned from both and on January 4, 1884, abandoned the presidency of the Northern Pacific as well.

Billings refused to believe that Villard was personally impoverished, and sought him out in his lavish new home on Madison Avenue where for three weeks Villard had refused all interviews. There the two men, one exhausted and on the verge of hysteria, the other in a state of high exasperation and anxiety, shouted at each other amid the new paint and varnish. The press said that Villard had acted dishonorably, while others backed him as he cried that he had exhausted all his personal resources in his attempt to stave off the final collapse of his three companies. Some of his assets were known to be in his wife's name—ought he to be expected to call those into play as well? Villard was able to convince Billings that he had done all that he could, and they parted with appropriate Victorian sentiment, more in sorrow than in anger. Wounded by the publicity, the Villards went to Europe for a time, to reemerge triumphant once again in 1887, returning to the Northern Pacific board with the help of German investors.[15]

For two weeks in January 1884 the Northern Pacific board met to talk their way through the disaster and choose a new president. A committee of six, including Billings, was appointed to find a successor to Villard. Some members wanted Billings to return to the presidency, and, though tempted, he knew by now that his health truly would not permit him to do so. Villard told him he must not, that the burden was too great. Still Billings wavered: he enjoyed the show of support that old friends gave him, removing the last sting of that angry board meeting at which he resigned thirty-one months before. He was moved when reporters came to his office to say that the public wanted him to take over. As the largest stockholder Charles B. Wright was a possible president and Billings was so opposed to him that he might have accepted if necessary, but Wright declined to be considered because of his health. Billings felt he could safely leave the post—and the problem of dealing with the syndicate—to others, and with that, Robert Harris, an able man whom Billings had brought onto the board and hoped to see Wright make vice-president years before, was chosen. Billings was pleased, especially since Harris also favored the Cascade line. Harris also had Wright's support and the confidence of the syndicate, and on September 16 the board confirmed that he was their choice.[16]

Billings continued to serve on the executive committee and be a faithful attendant at the board meetings, though he was involved in many other enterprises. In July 1883 he joined in organizing the Northern Pacific Express Company with offices in Portland and San Francisco; in Montana there was the Bull Mountain mine, the Rocky Mountain Railroad Company of Montana, the Montana Improvement Company, and the Helena and Livingston Smelting and Reduction Company.[17] Billings was a director of the American Exchange Bank and of the Farmers' Loan & Trust, which held several key Northern Pacific mortgages. He had joined with John Jacob Astor, Pierre Lorillard, and other New York capitalists in backing the Yedras Mining Company in Mexico, and he was now on the board of several smaller railroads: the Woodstock, the Vermont Valley, the Connecticut River, the Connecticut & Passumpsic Rivers, the Rutland, and the Delaware and Hudson Canal Company, which operated two short rail lines in Vermont.[18] He was also involved in the American Steel Barge Company, organized to build and operate vessels on the Great Lakes.[19] He was valued for his knowledge of railroads, was solicited often for advice, and with the help of the faithful Kilner was meticulous about his holdings. He was seldom in the press any more, but when his name came up, it usually was as an example of someone who conducted business "without concealment and without misrepresentation—with economy—in a plain straightforward business way."[20]

Billings was never swept away by the boom psychology of the time and recognized in Villard—in his talk of creating a vast monopoly over transportation from Puget Sound to the Golden Gate, in taking over the Union Pacific, perhaps in commanding western Canada—a fine case of the disease. Because Billings insisted on pursuing a sober path, those who craved high adventure thought him dull. Because he lacked Villard's exceptional self-confidence, some thought him hesitant, unwilling to take risks. Often the person who pays his bills and meets his deadlines is, when standing next to one who dreams greatly, a somewhat dour figure. There is a saying among authors that you either talk your book or you write it. There were those who built railroads and those who talked railroads. Billings belonged to the first group. He had never, one historian concluded, allowed emotion or rhetoric to overcome reason.[21]

There was a debit side. Some argued that Billings's conservative leadership had kept the Northern Pacific from getting off to a quick start when the market improved in 1880. He had failed to anticipate the dangers in allowing Drexel, Morgan and Winslow, Lanier and other members of the syndicate such control and did not adjust soon enough when he realized he was losing the syndicate's confidence. Others pointed out that he was slow to comprehend Villard's grand design, and his insistence upon the NP's reaching Puget Sound on its own rails has been judged quite foolish by some historians, an act of sheer pride, since the NP could have had full running rights on the OR&NC in perpetuity.

There remain the charges that Frederick Billings betrayed his friends when he sold ninety thousand of his shares to Villard and that he benefited from speculating in land. The first charge is not well framed and involves many value judgments. Billings firmly believed that Villard already had control of the company when he sold his shares, which he was convinced would decline quickly and precipitously in value if Villard's takeover was a subject of sustained public commentary. He also thought that Villard might well prove to be good for the company—not as good as he believed his own leadership to be, but not the disaster it became—and that in any event, once the syndicate was determined to put Villard in, there was no resisting them. There was no betrayal, simply a judgment made that friends and associates might legitimately disagree with and did.

As to speculating on land, Billings consistently—and more than almost any other railroad president—resisted sales that would benefit monopolists and speculators. The bonanza farms did so to some extent, but the trade-off seemed necessary, to produce a quick turnaround in public perceptions about the productivity of the allegedly barren Dakota prairies. He purchased some land in the name of his sons—a widespread and perfectly legal activity—and on one occasion publicly stated that he owned no town lots in Billings, Montana—which was narrowly true, though he did hold a substantial interest in the Minnesota & Montana Land & Improvement Company that did own lots in Billings. Some of these lots most certainly were sold for speculative profits. He realized profits of 60 percent or more on lots in Tacoma, doubled his money in Dakota, and for that matter his New York townhouse trebled in value by the time of his death. (He lost money in Superior.) Such gains were well within the norms of the time and none could be used to mark him as a real estate speculator.

Billings continued to enjoy the railroad he helped build. In 1884 and again in 1885 he and Julia traveled across the continent on the NP, using their new private Pullman Palace Car *Glacier,* to see the changes the railroad had brought along the line. He felt pleased that he had joined in a great national dream. He had been an empire builder, had helped in the winning of the West, and was a captain of industry. He had never thought the issues were simple, though "no railroads, no settlers" had always struck him as a command to build for the benefits of both the individual and society. It never occurred to him that any of these popular phrases would, one day, be used negatively.

In 1883, when the Northern Pacific Railroad was completed, it was the only one of the transcontinental lines to have reached its goal under the title in its original charter. So far as Frederick Billings was concerned, he could take pride in his twenty years with the Northern Pacific. They had also, he was convinced, been a service to the nation.

24

Reform Republican

W HEN FREDERICK BILLINGS TOLD Jay Cooke in the spring of 1872 that he wanted to resign as chairman of the Land Committee, Cooke guessed that Billings was once again seized of the notion that he ought to run for office. Though Billings had toyed with the idea of political office on several occasions in California, he had resisted such thoughts after returning permanently to the East—that is, until 1872, when he concluded that he had a real chance of becoming governor of Vermont.

In political terms Vermont was, throughout Billings's life, remarkably stable. With one exception, all governors between 1836 and 1854 were Whigs, and from 1854 until 1962 all were Republicans. From the founding of the party until 1964, Vermont was solidly Republican in presidential elections. In many states dissident or reform-minded Republicans formed splinter groups or bolted to the Democratic Party, but in Vermont there were few identifiable half-breeds or mugwumps; nearly all reform Republicans remained loyal to the party and worked for change within. Because the state and the potential spoils of office were small, little corruption and few scandals existed to give any candidate leverage.

Such change as there was in Vermont politics largely came from the constricting of the state's political influence as the new states to the west grew. In 1823, when Billings was born, Vermont elected five representatives to the House; in 1890, at his death, there were only two Congressional districts. At the beginning, therefore, Woodstock—or at least Windsor County—might reasonably expect to elect a congressman, and in all elections save two it did so from 1824 until 1854. Only twice again in the century did a congressman come from Windsor County and never again from Woodstock. During Billings's post-California years, Vermont had only two senators—George F. Edmunds, from 1866 to 1891, and Justin S. Morrill, from 1867 until 1898. Further, the two senators were not, in fact, elected at large, for there was the unwritten though quite clear Mountain Rule, which pre-dated the Republican

Party, that one would come from east and the other from west of the mountains. The Mountain Rule also applied to the governorship, which with rigid exactitude rotated from one to the other side of the divide until 1940. Thus a man in Billings's position knew the state's political leaders intimately, and he knew which elections were, by rule, denied him.[1]

The political scene in Vermont was marked by a sense of independence, especially from national trends. People took part in the political process well above national norms, knew their political leaders, accepted their idiosyncracies, punished their malfeasances. They supported public education and, within the context of what they understood, protection of the environment, so that these issues were less divisive than in many states. They believed in religion—anyone's religion, so long as one subscribed to some church—in reform, and in the stability of the social order. Vermonters did not like show, and while they were proud of the success of so many who had gone away to make their fortunes, they expected those who returned to behave themselves and make no unseemly public display. A very few leading figures built or remodeled large mansions, but these were modest compared with those built elsewhere in the country by other millionaires. Wealth did not mean that one could automatically achieve any office sought. (True, there were exceptions, notably the Proctor family of marble quarry fame, which established a modest dynasty and elected four of its members to the governorship, and the Smith family of St. Albans, which used its influence in railroading to achieve substantial political power from the 1860s to 1890s.) Vermonters had a clear sense of their own identity, stemming from the time when Vermont was an independent republic and had not immediately entered into union with the original thirteen states, a sense enhanced by isolation and strengthened by experience.

Most issues of significance in Vermont politics were, as in most states, local in their expression. Vermonters wanted protective tariffs and a hard money policy for the nation's treasury. What they wanted protected was relatively constant: sheep, dairy products, marble, and some manufacturing. Morrill had an unassailable power base because he was sound on money, had framed the enormously popular land grant act that supported public education in agriculture, and, as a congressman, had sponsored the protective tariff of 1861. To win election one also had to favor temperance, at times perhaps even prohibition. Generally, political views on the west side of the mountains were a little more liberal than on the east, the one looking at times to New York State, the other to Massachusetts.

Billings met all these criteria, but there were some he did not meet. He was a gentleman farmer, not a dirt farmer, and while his contributions to the development of the dairy industry in Vermont, which was taking off rapidly in the 1870s, were important, he would not have been particularly attractive as a candidate had the Vermont State Grange been well organized (nor was there a vigorous state railroad commission that might have stirred up trouble against a company man). What did tell against him was that he was not a joiner or a Civil War veteran, nor had he any legislative experience. Some men used the agricultural societies, the Masons, or even the Morgan horse or Merino sheep breeding clubs as a political base. Redfield Proctor would build his political strength on the basis of veterans, with at least one well placed in every community on the west side to help him. Billings had no such nucleus with which to begin, and he developed none.[2]

Nonetheless, given his success, to the Republican Party in Vermont Billings was a sound choice for one of the few significant offices: governor (when the position fell east of the mountains). He looked every inch the part, for over the years his waist had thickened, his beard had turned steely gray, and his piercing brown eyes looked almost haughtily over the countryside. This was a time when the electorate liked a little stateliness, not to say pomposity, in their governors. The responsibilities were not onerous, and one could carry on a business, even a major one; indeed, the public and the press recognized that the position was somewhat honorific, to be awarded to a figure distinguished for his service to the state, and then passed along to someone else also deserving. Thus the tradition that one not serve more than two years—which prior to the constitutional change of 1870 meant two terms, the earlier terms being for a single year—was felt to be a sound one, and many assumed that, now that two-year terms were the law, governors would accept only one. This assumption would lead to some confusion and probably cost Billings his best shot at the office in 1872.

In 1869 Billings had tried a little self-promotion with President-elect Grant, orchestrating a petition from eleven well-placed residents of Washington Territory asking that the expected West Coast Cabinet position be given to him. Billings framed the essential language himself and worked with his classmate, Charles A. Huntington, to get the document put before the right people. B. F. Dennison, associate justice of the territorial court and a family acquaintance who knew Oel Billings well, joined in, and the Washington group sought the aid of Peter Washburn, Vermont's Governor. The petition was put into the hands of Alan Flanders, Washington Territory's delegate to Congress, whom Billings had known at school in New Hampshire and Flanders presented the petition to Grant. Stephen Field wrote from the capital that it ought to happen.[3]

But illness and death continued to plague Billings's hopes for high office, though this time it was someone else's illness. Washburn died at the crucial juncture. Grant had far too many political debts to pay to let a Cabinet post go and no one from the West Coast was appointed. It had been a long shot, for Billings was not from Washington—"Your being in N.Y. the last 2 years is nothing in the case" Huntington had written, apparently in response to

Billings's worry that the men in Olympia would react as those in Sacramento had done four years earlier—and it was now clear to him that whatever office he might attain must come through his native state.

In 1872 the public was well aware of many scandals attached to the Grant administration (though there were more revelations to come). The spoilsmen were all too evidently in power. Reform Republicans were deserting the party for independent movements or, in New York, to the Democrats. Carl Schurz and others had begun a stop-Grant movement, alarmed that the Republican party organization was bent upon a second term. A Liberal Republican Convention was called for Cincinnati, where Horace Greeley was chosen as the presidential candidate. Soon after he received the endorsement of the Democratic Convention as well.

In May the Vermont Republican state convention assembled in Bellows Falls to choose delegates to the national party convention in Philadelphia.[4] Windsor County sent 24 of the 170 delegates; Billings was one. As expected, he was chosen president of the state convention, and in an opening speech he struck all the right chords. He was earnest, eloquent, highminded. His was "not the speech of a politician," said the Burlington *Free Press*, and it was "so frank and outspoken, so full of ideas, so strong . . . in its devotion to republican principles" that the audience broke again and again into hearty applause. In truth, he was effective precisely because, as he wrote to George G. Benedict, the editor of the *Free Press*, "the natural ambition wh. almost every man has for political place seems to have gone out of me."[5]

Papers friendly to Billings now touted him for the governorship. The Governor, John W. Stewart, was expected to adhere to the two-year precedent and not run for another term which, though only a second, would result in four years as governor. Three names were mentioned as Stewart's likely successor: Julius Converse of Woodstock, Horace Fairbanks of St. Johnsbury, and Billings. All were from the east side of the mountains. Converse was, at seventy-four, perhaps a bit too old; further, he had been lieutenant-governor, a sufficient honor.

However, it soon became apparent that, unless he could be assured a seat in Congress, Stewart did wish a second term. Several newspapers on the west side deserted Billings in favor of their local man, while east siders began a campaign against Stewart, accusing him of vanity and power-brokering for seeking two more years in the face of precedent. Public opinion focused more on "the governor question" than on the candidates, and by the time the Republican Union state convention met in Montpelier, twelve of the state's leading papers, including three from the west side, had come out firmly for one term.

In the last days before the convention the mudslingers went to work. Stewart was linked to patent right swindles in Addison County; Billings was attacked as a nonresident who had been away too long to understand Vermont's needs. The Woodstock *Vermont Standard* endorsed Converse, the Woodstock *Post* backed Billings, and efforts to keep two men from the same town from running against each other broke down when Converse reversed himself on what Billings's supporters thought had been a commitment to stand down until the next election.

On June 26 over four hundred delegates assembled to select the Republican candidate for governor. Stewart's name was placed in nomination by James M. Slade and Charles P. Marsh of Woodstock nominated Billings, reminding the assembly that the governorship should return to the east side. However, Billings did not control a majority of the delegation from his own county, and for them Major William Rounds of Chester nominated Julius Converse. On the first ballot, with 203 votes necessary to win, Billings received 181, Stewart 125, and Converse 99. Clearly by plan, Slade arose, announced that someone must surrender, and withdrew Stewart's name, asking those who had supported the governor to cast their votes for Converse. Excitement swept the hall, with cries of "Billings" and "Converse" going up, and reporters nosing out the scent of a deal. New tellers were chosen. Two delegates left the hall. With 201 votes now necessary for nomination, a second ballot was taken; Converse received 202 and Billings 197, with four cast for others. By a majority of one Converse was selected.

Billings now rose to make what many who had heard him before said was the finest speech of his career. The object, he knew, was to accept defeat gracefully, close ranks, and look to the future. Converse was seventy-four, Billings forty-nine. Converse was childless, Billings had a family who might play a role in state or national politics one day. Further, Billings's natural inclination was toward harmony, and he deeply feared a divisive outbreak of Greeleyism if the people of Vermont came to believe that the Republican Party had not behaved honorably. Nothing that had been done to thwart him was illegal, though some of it was very clever. As he usually did, Billings voted for party unity and the future.

He had, in fact, prepared an acceptance speech. In it he meant to reach out to heal any wounds his victory might leave, to praise Vermont for the honor it paid him, and to close by charging all, whether in or out of office, to be good citizens. He now rose to his feet and wove these same themes into an act of reconciliation. He declared that he would not have run at all but for a misunderstanding about the intentions of another candidate—presumably Stewart, though he carefully left open the possibility that he meant Converse, thus bowing toward both camps—and at some length he praised the delegates, the nominees, the citizens of Vermont, and the party.[6]

At the end applause was prolonged and tumultuous, and Slade stood, extended his hand to Billings, and to the cheers of the crowd the two men shook hands. "It was," one paper reported, "a scene never to be forgotten and Mr. Billings left the convention and Montpelier really the victor of the day." His was "the moral triumph of the Convention," the Burlington *Free Press* commented the next day, based on "the most extraordinary and magnetic" speech the editor had ever heard. (Its editor had reason to extend the olive branch to Billings, for Benedict had created the problem, by vacillating privately with Stewart on whether he should run. In February Stewart had told Benedict he intended to run again and yet the editor had doubted this so strongly, he appears to have said just the opposite to Billings.) Had Billings spoken before the vote, several papers commented, he would have had the nomination with ease. If Frederick Billings had not come fully home until that day, had not been accepted as a returned Vermonter, he left the convention hall able, according to the presiding officer of the convention, to count on the delegates "in future for anything he wanted."[7]

But Billings never redeemed this promissory note. The governorship returned by custom to the west side in 1874, and in 1876 Billings was fully engaged in other activities, including national party matters. That year Horace Fairbanks of St. Johnsbury was elected from the east side, and thus by inexorable rhythm Billings's next chance, if he wished to create one, might come in 1880.

Late in 1879 the Vermont press began its usual campaign of rumor, innuendo, and occasional fact. Six or seven names were prominently mentioned for the 1880 gubernatorial race, including ex-governors John Stewart and Redfield Proctor, who was said to be so popular he might challenge the one-term precedent. Another round of public commentary about one-term governors brought up the alleged misunderstanding about Stewart's candidacy in 1872, forcing Billings's supporters to replay that election. The Burlington *Free Press*, in the guise of a letter, painted an unflattering portrait of Billings: that he had "laid about him right and left" after the 1872 convention, striking "below the belt" at friends and foes alike (there is no evidence of this), that he had "spent or invested very little of his great fortune among us," that few people knew him, that his father had been a "genial, social fellow" who had been committed to the Woodstock jail as an insolvent debtor (this was technically true, though to Billings a low blow, and he felt no shame about it), and that his success in California had come from speculation after a start given him by "Beza [sic] Simmons, a wealthy shipmaster," who had taken him to San Francisco as his clerk.[8] This tissue of half-truths, confusions (it was Frederick's brother Frank who was the clerk), and snide comments about his family background were deeply wounding and reminded him of just how unpleasant politics could be. Preoccupied with the Northern Pacific, Billings chose not to pursue the nomination.

Nationally the Republican scene was grim. President Hayes stated that he would not run again, and the party was divided into three factions, one hoping to bring Grant and the spoilsmen back for an unprecedented third term, another supporting James G. Blaine, former speaker of the House and Hayes's chief rival for the nomination in 1876, and a third promoting Senator John Sherman of Ohio, Hayes's secretary of the Treasury. Grant's handlers had kept him out of the country on a world tour for the last two years, and Billings knew that he could not support a man who had proved so ineffective in rooting out scandals from his administration. Blaine had been accused of corruption relating to railroads, and though he had defended himself, many had not believed him; the president of the Northern Pacific could ill afford to appear to be backing a man some thought beholden to men like himself. Sherman, however, was a dry old stick without charm and, from Vermont's perspective, not quite right on the monetary issue. However, Billings believed that Vermont's Senator George F. Edmunds, whom he knew well, might have a chance, and so he joined the state delegation at the national nominating convention in Chicago.[9]

The convention was loud, crowded, and unruly. On June 6 Billings, who was chosen to place Edmunds's name in nomination, was the last to speak and was very tired. When he rose at eleven o'clock in the evening excitement in the hall was such that he could scarcely be heard, and he kept his speech short. On June 7 the convention moved inexorably toward deadlock as balloting proceeded. To win, a candidate needed 378 votes. Grant had 304 delegates on the first ballot, Blaine followed with 284, Sherman with 93. Seventy-five votes were split among several favorite sons, with thirty-four largely mugwump votes going to Edmunds. The deadlock continued through thirty-three ballots on the 6th, throughout the 7th, and into the 8th. On the twenty-ninth ballot there was a movement toward Sherman. On the thirty-fourth ballot the Wisconsin delegation, which had supported Blaine, suddenly moved to James A. Garfield of Ohio, the former House minority leader who had made an excellent and subtle nominating speech for Sherman. On the thirty-sixth ballot word swept across the floor that Sherman had released the Ohio delegation to his fellow Ohioan and a stampede from the Blaine column began at once. The Vermont delegation was released and joined the movement. Grant held firm at 306, and Garfield was carried to 399 and victory. Vermont-born Chester A. Arthur was nominated as Garfield's running mate.[10]

Billings liked Garfield and felt he had been a friend to the railroads, and his support for him was more energetic than for any other presidential candidate. That fall Billings gave generously to Garfield's campaign and on election day he rose early to be the first to cast his ballot. The next night he marched at the head of a torchlight parade, spoke once more with great feeling in the Woodstock town hall—one account reported that he was so moved, he threw a new silk hat into the air and jumped on it—and joined in "the revels." Perhaps; his diary simply says that he "did self honor" in a "brilliant & orderly celebration."[11]

Billings continued to be an active presence in the Republican Party in Vermont, though he never sought office again, even when friends and supporters tried to push him into doing so. He consistently espoused a Reform Republican position. Late in 1883 the Vermont press began its usual dithering about who might or might not run, and the Montpelier *Watchman & Journal* let it be known that Billings's name was being tossed about. The Democratic press, apparently fearful that he would take this heavy breathing seriously, implied that it was "understood" that Billings was a tax dodger, as he had failed to make out a legally required inventory of his taxable property under a new listing law. Billings had just made a major new gift to the University of Vermont to expand its library, and he was angered by this charge; he was reminded again of how much he disliked public exposure. His explanation—that most of his property was outside the state and thus not subject to the listing law, and that it was technically true that he had not supplied a list, but rather had agreed to a list the tax assessor had supplied to him—did not convince everyone, and Billings well knew that those who read headlines often do not read small print.

In the end the press ran Billings when he would not run himself. Again Billings "decidedly refused" to run. He could not refuse, said one paper; he surely would, said another; he was the only man who could deny the office to George Nichols, who was in J. Gregory Smith's pocket, said a third; he was so popular and so able, even the Vermont Central clique could not block him this time, declared a fourth. Throughout this rodomontade Billings remained serenely quiet, tending to his estate and giving thought to whether he ought not, after all, take over the presidency of the Northern Pacific once again. At the state convention on June 18, Samuel Pingree was nominated for governor and, in the fall, was elected.

By 1884 Billings was clearly identified with the reform group in the national Republican Party as well. In Vermont this was less divisive than it would have been had he taken such a position in 1872, for the state's native-son candidate, again Senator Edmunds, was the figure around whom the "third section," as the Independents were called, might rally. Those who had opposed Grant in 1872 and had voted for Horace Greeley, and those who had helped to block Grant and Blaine in 1880, were willing to support Arthur if they could not do better, but in Edmunds they felt they could do much better. The Vermont delegation, with Billings serving as the third delegate, were solidly for him. Nonetheless, and as expected, the nomination went to Blaine, and in November the Democratic candidate, New York's Governor Grover Cleveland, was elected President.[12]

After 1884 Billings played no significant role in national or state politics.

He kept his hand in, of course: in May 1885 he invited former President Grant and his wife to accompany him to the West Coast in his private car, but Grant was dying from cancer and was trying desperately to complete his memoirs so that his wife would have an income after his death. Though Billings sent a contribution to Benjamin Harrison in 1886 and wished him well in his campaign two years later,[13] the political fire was thoroughly banked. There were ritualistic mutterings in the press about the possibility of his running for governor in 1889, but this was scarcely creditable, for he had no interest and was quite ill. Though once he had had some modest political influence, mediated through others to be sure—Greeley, Bowles, Evarts, Edmunds, even Frémont earlier—now he had very little. Nor did he care; his interest now was almost wholly in matters of conservation.

# 25

Commerce and Conservation

I N THE LAST SIX YEARS OF HIS LIFE Frederick Billings devoted his time
    principally to issues of conservation, forest management, and scientific
agriculture. These were not new concerns, though before they had not been
front and center. They were basic to his perception of Vermont's needs and
to his ideas about Vermont's role in the nation's ethical economy. He believed
Vermonters had drawn a special strength from the land and that as a Vermonter he had a responsibility to that land. As a businessman and Reform
Republican, he felt that the interests of conservation and commerce were not
antithetical. There could be—indeed there must be—an alliance between the
machine and the garden if either activity were to flourish. This alliance was
best expressed by his attitudes toward the two early national parks, Yosemite
and Yellowstone, and the use the Northern Pacific Railroad hoped to make of
the latter.

As a child and as a student at the University of Vermont, Frederick Billings
had shown an intense interest in landscape. At the university he studied with
Zadock Thompson, the Windsor County boy who expanded his horizons beyond Bridgewater to encompass all Vermont's natural and human history,
who had taught Billings astronomy, who had prepared the gazetteers Billings
so enjoyed. Billings's letters home from California were strongly visual, filled
with a sense of place, of how the land rose before him or the sand dunes fell
away below. He enjoyed nothing more, his letters suggest, than getting away
from people to hunt, ride, and read. He had not been in California for more
than a year before he began commenting on the need to preserve the state's
natural wonders in the face of a rapidly growing and highly diverse population
motivated primarily by the desire for material gain.

Though Billings was interested in nature, he did not approach it scientifically, and except for reading the works of such popular giants of the time as
von Humboldt and Agassiz, he appears not to have been particularly well
informed, at least until the 1860s. Rather, he responded to the picturesque

and the grand, to the kind of landscapes depicted by the great painters of the West such as Albert Bierstadt, and to the more bucolic though no less romantic painters of the East and in particular of the Hudson River School, such as Thomas Cole, Frederic E. Church, Asher B. Durand, and John F. Kensett. When his interest turned to photography, he showed the same concerns, preferring the sweeping landscape and the panoramic view to the detailed study of an individual plant. He was not a naturalist.

He was, however, a conservationist by the lights of his day. Of course, mid-nineteenth-century conservation arose in part from patriotic sentiments and from the desire to hold to the values of the past, and less often from any well-reasoned argument about the need to conserve natural resources. The language of present-day conservation, of ecology and biospheres, did not exist. Many of Billings's friends would have thought of themselves as preservationists, vaguely disturbed when the scenes of their childhood were transformed though hardly at all concerned when the childhood scenes of their political enemies or business competitors were swept up by an expanding city, a wood-hungry railroad, or the need to produce more grain. Billings went rather beyond his friends, though not until 1864 when he read and was deeply influenced by the first American book on ecology, George Perkins Marsh's *Man and Nature*, could Billings be said to have had a philosophy.

Billings marked Marsh's book, singling out an early passage that suggests a transition to more systematic thought; that he marked it would suggest he already had begun to sense that love of landscape (the nineteenth-century equivalent of what, in the automobile age, is often dismissed as "a windshield experience" with nature) was not enough. "Only for the sense of landscape beauty did unaided nature make provision," Marsh wrote. "Indeed, the very commonness of this source of refined enjoyment seems to have deprived it of half its value." Nature would need help.[1]

Marsh's book reminded him of his childhood in Royalton and Woodstock, bringing what he recalled into sharper perspective. Royalton once had been surrounded by a vast forest cover and excellent soil, but the early settlers had quickly cleared the land and destroyed the supply of big game and fish: deer and bear were scarce around Royalton even by the time the Billings family moved out. To early settlers the environment was the enemy, and they wanted to be rid of the trees behind which—or so they largely imagined—hid Indians, bears, and wolves. By the time Billings had returned to Vermont, the entire township of Royalton was cleared and pastured right over the tops of the hills, game was gone, and stream pollution from tanneries (or the odd distillery) had put waste into the brooks.[2]

When they could afford it, Oel and Sophia Billings enjoyed "visiting with nature," as the nineteenth-century phrase had it. They especially liked their visits to Franconia Notch, staying at the Profile House, which when enlarged after the Civil War became the outstanding summer resort of New England. It did not occur to them that commerce might ultimately destroy the landscape they were going to see, that a hotel able to accommodate five hundred guests within the notch itself and a railroad line that brought tourists directly into the notch might one day do irremediable damage. Frederick, on the other hand, worried in distant California about the next century, for he had seen the erosive effects of unprecedented population growth and uncontrolled exploitation of the environment. Some hurried to see the natural wonders, fatalistically convinced they would never survive, while a small minority, of which Billings was one, thought that quick action could protect the most appealing landscapes.[3]

One must not attribute too much to Frederick Billings, however. He too did not grasp how quickly the coast and the great trees might disappear. He joined most commentators on California who, like the English traveller Frank Marryat in 1855, believed that the redwoods literally were inexhaustible, less because they did not see how rapidly the forests were being cut down—after all, in 1849 lumber valued at $8,450 could be taken out of a single tree[4] — than because they could not envision the day when the technology of the timber industry would be so effective that entire forests could be clear-cut in a matter of days. Many were frightened by a true forest primeval and preferred the poetry to the fact, and Billings was among those who, when he rode down to San Mateo to spend a weekend with his friend John Maynard, favored clearing out trees to provide for a "more entrancing" view of the redwood stands.[5]

Billings's inclination toward saving landscapes was reinforced by like-minded friends. When Horace Bushnell visited the Big Trees, he was deeply moved, calling the Mariposa Grove "the park of the Lord." That man could cut down such grand creations of the Lord surpassed all contempt, he wrote. Later, as the Yosemite Valley was made more widely known by Clarence King, Josiah Dwight Whitney, and John Muir, Billings read their books but does not appear to have sought them out in support of his goals.[6]

By the 1860s there was a standard tour from San Francisco to the mountains on which Billings had gone often, either alone or with his friends or wife. The traveler visited Bear Valley and then went up the Merced to the Yo Semite. After a few days in the valley, one moved via the Toulumne on to the Calaveras Big Trees and, if time was abundant, all the way to Silver Mountain (Virginia City) and Lake Tahoe, to return to the bay area by way of Placerville, the recently discovered alabaster cave near Folsom, Mare Island, and the geysers of the Napa Valley. One also stopped at Colonel Haraszthy's vineyard near Sonoma and, most likely, climbed Mount Diablo, back of Oakland. Already these sites were showing signs of overcrowding—the alabaster cave was visited by over four thousand people within seventy days after its discovery in 1860 and it had to be closed because of abuse in the 1890s—and Billings was advocating their protection. The Yosemite Valley and the Mariposa Big Trees might well be one large park; the Calaveras Big Trees were isolated, divided into two groves, and visited by only a fifth of the travelers who went to Yosemite, and they were less likely to require immediate attention.

The least of these stops on the grand tour, in Billings's mind, was also the most instructive: the alabaster cave. A trip to it was comfortable yet adventurous, and the entire excursion marked a perfect union between nature and rail travel. One rode on California's first passenger line, the Sacramento Valley Railroad, of which he was briefly president, from Sacramento to Folsom. From Folsom clean stage coaches run by pleasant entrepreneurs carried visitors ten miles further to the cave via "instructive" mining camps. At the destination there was a first-rate hotel. On the walk to the cave's entrance one could view a lime-kiln, from which a large portion of the lime consumed in San Francisco was manufactured, and if one stayed overnight, one might also see the kiln in satanic full blast, a dramatic symbol of the age. Guidance at the cave itself was expert and polite. In short, the entire formula for a successful wedding between commerce and conservation was set out at this one site: an agreeable ride, low cost, efficient and attentive service, good food and lodging, the *frisson* of an encounter with the industrial age (and the future), and the uplift and visual pleasure of a natural landmark.[7]

California's natural beauty contrasted with the essentially historical nature of the popular European grand tour. Despite the pleasure one might take in visiting the English Lake District and savoring the antiquity of Europe, one quickly grew tired of all the "old tapestries" and saw the value of a "new bright country, where there is so little history." People who took more interest in visiting a dungeon where some poor creature had been ill treated than in seeing natural beauty were odd indeed; Britain was "a place of sepulchres," a Golgotha, a constricted land of prudence, while California was a place of wild beauty, challenge, and even adventurous folly, where all Nature's processes could be seen plain.[8]

The urge to preserve nature's wonders in California was an amalgam, then, of several factors: a New World patriotism, a frontier delight in the unknown, a fascination with the picturesque, a tendency to humanize nature into landscapes that could be internalized by art and photography, a shrewd judgment about commerce and tourism, nostalgia for the scenes of one's childhood sojourn, and a desire to understand God through his handiwork. Commerce could serve the cause of conservation by bringing visitors to a site worthy of preservation, thus building a political constituency and creating a source of funds to meet the costs of purchase and protection. Billings saw no contradiction between these goals, and in the uncrowded late 1860s there was none.

The Yosemite Valley was first entered by white men in March 1851 and the Mariposa and Calaveras trees were made known to the public in 1852.* Billings first visited Yosemite in 1852 and the Calaveras grove in 1856.[9] He and his friends heard of these wonders in San Francisco as a citizen of the twenty-first century might hear of the discoveries of inhabited planets, as though time were in some way renewed. Thomas Starr King delivered a sermon on Yosemite in 1860 that taught the religious lesson with didactic clarity: "Who is insensible to the wonders of nature is indifferent to the patient and continuous art of God." An Emersonian concept of the manifestation of God's power as revealed through nature—the dimunition of even the giant trees by the cliffs high above the valley, the transformation of man's sense of time by observing how a waterfall, known to be moving at great force, appeared to be motionless from across the valley floor, the autumnal gentleness and spring rebirth of the meadows (which were far more prominant then than now), the savagery of the scene above which hawk and eagle soared—all these King noted with Transcendentalist awe. Yosemite was scripture in stone, an expression of the Divine Mind. Billings agreed with every word of it, for all the elements of the amalgam were part of his nature, and he wanted to see the valley preserved for posterity. One means to that end was photography.[10]

Born in Oneonta, New York, in 1829, Carleton E. Watkins,[11] pioneer photographer, arrived in San Francisco on May 5, 1851, the day following the destructive fire that swept away so great a portion of the city. He travelled via the Chagres route in the company of Collis P. Huntington, the future railroad financier, and until Huntington's Sacramento store was burnt out in 1852, Watkins worked for him. Watkins then moved to San Francisco and took employment as a clerk to George Murray, a bookseller and stationer with a room in the Montgomery Block. There Watkins surely met the partners in Halleck, Peachy, and Billings, since they purchased their stationery from Murray. Watkins became acquainted with one of the city's many daguerreotypists, Robert H. Vance, who in 1856 asked him to look after a photographic gallery in San Jose when the operator there quit suddenly. At some point that summer Watkins took his first photograph at New Almaden.[12]

In 1858 Vance sent his partner, Charles Leander Weed, to the American River to take photographs of river mining, and the following year to Yosemite, where Weed made the first known photographs of this cathedral of rocks. He made both large-format negatives and stereographs, which were displayed

---

* There was some vague awareness of the Calaveras trees as early as 1833, and they were, in fact, "discovered" in October 1841, though by individuals who did not grasp the significance of what they had seen. John Marshall Wooster visited the grove in June 1850 and discussed it later that year in a Stockton paper. In 1852 Alexander T. Dowd, who is usually credited with the discovery of the grove, first wrote of it.

in San Francisco in August 1859, where Billings, Watkins, and the entire city viewed them. Watkins, in the meantime, was beginning to take landscape photographs to be used as court evidence; the first clear instance, in 1858, was for a dispute over the Guadalupe quicksilver mine near San Jose.[13] Almost certainly Billings saw this photograph since HPB was involved in the Guadalupe case. Watkins's biographer, Peter E. Palmquist, believes it likely that Watkins worked for and with another photographer, James M. Ford, between 1856 and 1858, and quite possibly earlier, and that Watkins may have taken some of the photographs that appeared in a remarkable album on San Francisco, produced in 1856 by George Robinson Fardon. Many people in San Francisco, Billings among them, wanted views of the important buildings and landscapes of the city to send to friends and family on the East Coast. Palmquist thinks it probable that many photographs of the activities of the Vigilance Committee were taken by Ford and Watkins, since their gallery overlooked Portsmouth Square, scene of vigilante meetings. If so, Billings would have known of this work as well. Billings visited photography studios fairly often, as his diary reveals, and while he seldom supplies names, the premises of Ford, Watkins, and Fardon were quite near his office. Billings appears to have thought about commissioning photographs of the New Almaden Mines as early as 1857, and while he seems not to have done so—for no photographs from that time are extant—he showed a steady interest in the products of the photographers.

In 1859 or 1860 Trenor Park, or Frederick Billings, or both, commissioned Watkins to make a careful series of outdoor photographs of John Charles Frémont's Las Mariposas, in preparation for their efforts to interest investors in the quartz mines. Accordingly, Watkins made at least forty-nine different pictures at twenty-seven different locations. These were sent on to Park, and by Park to Billings and Frémont, as all three men bent their efforts to salvaging the seriously deteriorating financial situation at the mines.[14]

The commission to journey to Mariposa first brought Watkins to Yosemite, or at the least to the Mariposa Big Trees. Inspired, Watkins returned in 1861 to prepare a set of photographs of Yosemite, taking his giant camera and wet-plate apparatus into the distant valley by mule and horseback. Quite possibly Trenor Park financed this initial trip to Yosemite: a glass stereograph survives from the series showing Park and his family at a rustic table in the valley in July. Watkins's photographs of Yosemite's natural wonders—the Sentinel, North Dome, El Capitan, Vernal Falls, Bridal Veil, the Mariposa Grove, and others (thirty in all)—were printed as an album in 1863. Billings obtained a set, which he sent through a friend to Professor Louis Agassiz at Harvard, urging him to see for himself how remarkable the Yosemite Valley was. Watkins's work put the professor into "a photographic fever," and he gave his support to a proposed reserve. When placed on the desk of key congressmen in Washington in 1864, Watkins's series was credited with convincing Congress that it must declare the Yosemite a reserve to be held for future generations. Though Watkins would make finer photographs on subsequent trips, especially in 1865, none would have the immediate public impact of those he took in 1861.[15]

In 1863, impressed by the Mariposa and Yosemite series, Billings and Halleck commissioned Watkins to do a set of views of New Almaden, to use in conjunction with the proceedings before the U.S. Supreme Court. Billings clearly admired the Yosemite photographs, which he sent back to the East; he preferred the unsullied natural scenes Watkins's camera had caught in the mountain valleys to his Mariposa panoramas showing acre after acre of barren land, cut down to tree stumps (which reminded Billings of tombstones), and he hoped that Watkins could capture New Almaden in its occasional scenic beauty as well as its industrial potential. The result was a new series of fine views, twenty-four mammoth prints that Billings supplied to the court and fifty stereographs which he kept for his own use.[16]

Frémont, Park, and Billings were Watkins's patrons for the Mariposa photographs, as Halleck and Billings were for the New Almaden series, and they put him well on the road to the official patronage he continually sought. However, it was the Yosemite series that established Watkins's fame. Billings continued to be interested in Watkins's work thereafter, purchasing subsequent photographs of Yosemite, acquiring his portfolio of pictures of Washington and Oregon, and in 1885 of Yellowstone, buying any book in which Watkins's California views were used.[17]

As we have seen, Las Mariposas also brought Billings together with Frederick Law Olmsted, the father of American landscape gardening and the nineteenth century's leading creator of urban parks. Olmsted's work appealed to those Americans (and Billings was one) who disliked bare and barren hillsides left in a state of nature. If barrenness was a result of the forces of nature, such scenes were felt to need enhancing: there was as yet little appreciation of the desert, and the description "a hillside of scrub oak" was dismissive rather than purely analytic. If the hillside were barren from the workings of man, the preservationist, hardly mindful that reconstruction drew upon different psychological needs and that one could not preserve—except in its state of devastation—an already ravaged landscape, wished to redeem the hillside with plantings. If the bare and barren hillside were to be within a city, it cried out to be protected from developers as a remodeled green space, a city park on which landscape gardening could work its marvels of transformation and redemption.

Before the 1850s there was little legislation for creating and small inclination for establishing city parks, either for recreation or visual effect, for the industrial revolution had not yet produced the well-to-do leisured class that had the time for recreation nor the congested city that required consciously created green spaces. Recreation in the city had been little more than walking in cemeteries and zoos. As more Americans traveled in Europe, where they saw the great public spaces of the world's capitals, they sensed their need to create for themselves statements about their increasingly refined sensitivities and facilities by which they might refresh their spirits on Sunday strolls. This was not contradictory of the spread-eagle nationalism of the time, which believed American democracy a better form of government and American natural wonders more splendid than any in Europe: one would simply create better parks. In the 1850s large municipal parks were begun in Hartford, Boston, New York, and (in 1855) San Francisco. Henry David Thoreau declared the need to "go out and re-ally ourselves with nature every day," and clearly one could not hope to do this at Yosemite. One must do it at one's doorstep, at Bear Valley, in San Francisco, or later in Woodstock. Further, and not cynically, men like Olmsted knew that good parks would enhance property values for wide circles within their vicinity.

While Mariposa was as dreary as he had expected, if not quite so tempestuous, Olmsted enjoyed the beauty of Yosemite nearby and gave thought as to how to tame the valley floor so that it might remain beautiful yet not intimidate as "the wilderness & wild" did. Depressed by the dry, stunted vegetation around the mines, Olmsted escaped to the fall colors of the Yosemite Valley, and two years later he was put in charge of a major study of the potential of the valley, a study that, though suppressed, spelled out for the first time a philosophy for what would eventually become a great system of national parks.

By the time President Lincoln signed the Yosemite bill on June 30, 1864, the Yosemite photographs and paintings of Weed, Watkins, and Bierstadt had won a wide following.[18] The bill, the work of Senator John Conness, transferred twenty thousand acres of federal land to the state, to be held "inalienable for all time." This was the nation's first federally created state park, and one of the very few ever to result from direct federal action.* The creation of a park in faraway California was a statement about national unity, continental status, and hopes for an optimistic future in the midst of a devastating civil war: Yosemite was a monument to union, democracy, and long-term goals for the nation, the product of a great national need.

Yosemite was also the product of people, of course. Conness, Starr King, Watkins, Whitney, Governor Frederick Low, Olmsted himself—these were the key figures. Compared to them Frederick Billings had played only a minor role, that of a layer-on of hands, a conduit through which Watkins, King, and

---

* The majority of the National Recreational Demonstration Areas, administered by the National Park Service in the 1930s, became state parks, especially in Missouri, Tennessee, Pennsylvania, and Virginia.

Olmsted were affixed to the Yosemite story. Very likely, King and Watkins would have gone to Yosemite in any event, given the valley's fame, but perhaps not precisely when they did and as they did.

In September Governor Low issued a proclamation announcing the federal grant and appointing an eight-member board of comissioners to manage the area. Olmsted was made chairman of the commission and instructed to handle all matters relating to leases for concessionaires and "improvement" of the grant. Practically speaking, this was California's first state park commission, and Yosemite would remain in the hands of the state until 1906, when it was returned to the federal government to be managed as a national park.

The person who wrote to Conness about introducing his bill was Israel Ward Raymond, San Francisco shipping magnate, who had a recreational resort—there already were two hotels in the valley—in mind. Raymond sent Watkins's portfolio to Conness, who in turn apparently showed the photographs to Lincoln and arranged to have some displayed in the office of the Senate's Sergeant-at-Arms. Billings, Olmsted, Jessie Frémont, Galen Clark—who had been a long-time resident and unofficial manager at Mariposa Grove—also wrote to Conness. Billings apparently pointed out that the owners of private property inside the boundaries of the grant would need to be bought out or removed as they had been at Rincon Point, and the commissioners so notified the largest land holder, James Mason Hutchings, publisher of *Hutchings' California Magazine*, which had publicized the natural wonders of the state until 1861, and now an innkeeper in the valley. Billings wanted to be appointed to Governor Low's commission, but he did not press the issue, and soon after he left California.

As president of the Northern Pacific, Billings had his best opportunity to promote an alliance between tourism and conservation, about which he and very few others in the late 1870s or early 1880s saw any contradiction. The natural wonders of the West seemed so extensive, so self-renewing and self-cleansing, that only those with the foresight of a John Muir were worried that tourists might destroy the very scenes they came to see. While railroad entrepreneurs, Billings among them, envisaged the day when special passenger trains would be run to the western wonders, they never imagined the time when thousands of visitors would turn into millions. There were, of course, the "progressive degredationists," who did not expect anything natural to survive indefinitely and thought that nothing could be done to prevent the general collapse of society, the natural order, or the wilderness, just as there were those who believed that God would provide and, at the last trump, restore His handiwork, so that man ought to make use of it in the meantime. Since both arguments were heard in the 1980s, one should not be surprised to encounter them a century before. Billings had no patience with the latter view, and only on his most depressed days did he incline even slightly toward the former. Rather, he believed that an alliance between the engine of commerce and the aesthetic of conservation could be achieved to the benefit of the majority of Americans.

In his last year as the NP's president, Billings initiated three projects that reflected his views. Some directors may have regarded any one of them as a frill, but since very little money was involved, they appear not to have said so, and in the context of the overall enterprise these projects were quite minor matters. Nor had any of these initiatives gone far before the end of Billings's presidency, so that ultimate credit for them belongs to others. Still, the projects tell a good bit about the man. One was a program to plant trees across the plains; the second was Billings's very personal desire to protect the few historic sites along the Northern Pacific's line; the third was his intention to create guide books for the traveller to induce tourist traffic. The second and third might also help create, or appeal to, a constituency that valued the natural and historic resources through which the line passed.

Billings had seen the need for trees when he was chairman of the Land Committee. Jay Cooke spoke of it, as did successive land commissioners, and the question of trees as snow brakes was a constant concern until Herman Haupt made a decision to build far more snow fences. Everyone understood that settlers in northeastern Dakota needed timber; so did the railroad, since one of the greatest problems encountered while laying track across Dakota was the lack of forest cover. The Land Department worked to encourage farmers to plant trees and eventually organized a tree planting department to prove that forest trees would grow on the treeless prairie. Beginning in the spring of 1882, this department, over which Billings kept a watchful eye, ran a test program. Across the Coteau du Missouri, at Tappan (now spelled Tappen), midway between Jamestown and Bismarck, and at Steele, further west at the high point on the line between Duluth and Bismarck, over four hundred thousand trees and cuttings (largely white willow, cottonwood, and box elder) were planted, while another hundred thousand were added near Tower City, a few miles west of the bonanza farms. Haupt would add windbreaks along the deeper railway cuts. The NP also gave wide publicity to the federal Timber Culture Acts of 1873 and 1878. When Billings passed over the route in 1883, he remarked happily upon the trees.

Billings's second initiative bore little fruit. He had wanted to have the occasional historical marker placed along the route, though in fact, by the canons of the time, there was remarkably little to be seen that was considered historical—certainly not the old frontier forts, soon to be abandoned, nor the scenes of obscure battles with the Indians (with the notable exception of the Little Big Horn). His only small victory came from agitating to have a screen placed over that part of Pompey's Pillar—a mass of yellow sandstone rising some four hundred feet about the surrounding plain twenty-five miles north-east of Billings—where Captain William Clark, of the famed Lewis and Clark expedition, had carved his name. Though the rock was across the Yellowstone River from the rail line, stops frequently were made to allow passengers to view it, and vandals had already damaged some of its Indian carvings. Clark's name and the date he had placed there, July 25, 1806, were finally protected by the railroad at Billings's urging in the 1880s.

Billings also encouraged the first Northern Pacific guide books and photographs to attract tourists and inform them of the significance of what they saw. The primary credit was not his, however, for the idea originated with the photographer, F. Jay Haynes, who had established himself in Moorhead, Minnesota in September 1876 and had taken photographs of the bonanza farms. Haynes sent some of his pictures of the Dalrymple Farm to NP officials in St. Paul; he was soon commissioned to prepare photographs for promotional purposes, and in October the Northern Pacific's division superintendent hired him to take pictures over the entire route from Brainerd to Bismarck. In 1877 Haynes's work was used to illustrate publicity materials, and he was appointed clerk in the general manager's office in St. Paul. Throughout Billings's presidency Haynes styled himself the "official photographer" of the railroad. Both the Land Department, where Billings had enthusiastically endorsed the idea of promotional photography to allay the fear that the railroad passed through a "Great American Desert," and the office of the general passenger agent, in the hands of the able and imaginative Charles S. Fee, supported Haynes's work. Fee in particular promoted romantic photographs of sights along the railway route—the idea of "sight seeing," of travelling to specific places to see the precise spot that had appeared in a photograph, was in its infancy—and advised Haynes on what was needed. In 1881 Haynes went to Yellowstone National Park and began to prepare illustrations for the Northern Pacific's projected new promotional brochures.

Haynes's views (sometimes converted to woodcuts and after 1881 used as photographic murals in the metropolitan stations) soon became the standard images of the Dakota badlands, then called Pyramid Park, of the Yellowstone Valley, and of the national park. He was, of course, present for the ceremonies that accompanied the completion of the Northern Pacific line in 1883, and in 1884 he opened a studio at Mammoth Hot Springs. His work commanded respect, for whatever it may have lacked in artistic sensibility it made up for in historical accuracy and attention to detail.[19]

Pleased with Haynes's work and convinced that as soon as the railroad reached the upper Yellowstone there would be a flurry of NP passenger traffic for the national park, Billings and Fee decided to commission a guidebook. Billings read the available travel accounts on the region but could find no von Humboldt or Agassiz in the literature. However, one guide book was already available for the Union Pacific, which expected to tap the national park from the west side, had commissioned Robert E. Strahorn, who had prepared a resource guide to Montana and the park for the territorial legislature in 1879, to transform his material into something suitable for travellers, which he did in 1881.[20] Billings was annoyed, for not only would the guide benefit the Utah & Northern, it would open up the wrong side of the national park. Further, he did not think it was very good, for it lacked the instructive and edifying content he wanted in a guide.

Billings was still thinking about the projected guide when Villard displaced him, and selected Henry J. Winser to be chief of the NP's bureau of information. Winser was British, having been born in Bermuda, and he wrote in a controlled, crisp style that conveyed both information and color in little space. As a reporter for the New York Times during the Civil War, and then as city editor, he had been well-known in journalistic circles. He had dropped out of sight somewhat, for until Villard hired him to write the book Billings and Fee had projected, he was serving as United States consul at Sonneberg in Germany. Winser wrote a guide to the entire Northern Pacific route from Minnesota to Tacoma, with a substantial portion devoted to the national park. The Yellowstone material was hurried out ahead of the larger guide to counter Strahorn, and the rather romantic overview of The Great Northwest was issued in conjunction with the completion of the line in September 1883. It set the standard for railroad guides for years to come.[21]

The Northern Pacific felt a proprietary interest in Yellowstone National Park, and Billings, Fee, and Villard believed that by making the park accessible to the travelling public, they would enhance the reputation and perception of social value of the railroad. Few predicted substantial profits from passenger rail fares, at least until a branch line was completed to the park's north entrance, but they did hope to open up lodges within the park to provide for the comfort of visitors and expected these hotels to turn a modest profit within a few years. While the national park may have had its origins around a campfire by the Madison River, where members of the Henry D. Washburn expedition talked into the night of September 19, 1870, it was Jay Cooke and the Northern Pacific Railroad in the person of A. B. Nettleton who most effectively urged Congress to create such a park, and it was Billings as president of the Northern Pacific Railroad who reminded his chief engineer and surveyors, as construction moved up the Yellowstone Valley, that they should try not to damage the values of the land through which they would carry the line, since the time might come when the Northern Pacific would make as much money from taking visitors to see the wonders of the West as it would make from the mines hidden in the mountains.

Indeed, the creation of the national park itself owed much to Cooke. Three months before Washburn led his expedition into Colter's Hell, Cooke had met with Nathaniel P. Langford, publicist of the expedition and a chief promoter of the idea of a national park, and had stressed how the railroad might gain funding in part because of the proposed reserve. In turn, Langford praised the railroad, its management, their intentions, and the scenic wonders the line would open up to those who could afford to travel on it, and he worked with the land committee to assure that the company received the greatest amount of land along the line as possible. When Professor Ferdinand V. Hayden petitioned Congress to finance a survey of Yellowstone's thermal basins in 1871, Nettleton had asked Hayden to include Thomas Moran, not yet the famed landscape artist he would become, so that he might prepare a canvas of the region. Cooke loaned Moran the money, and the result was his best-known oil painting, "The Grand Canyon of the Yellowstone," which stirred the nation as Watkins's photographs had done a few years before for Yosemite. On October 28, 1871, Nettleton proposed to Hayden that the Great Geyser Basin should be preserved as a national park just as "that far inferior wonder," the Yosemite Valley, had been reserved. On March 1, 1872, President Ulysses S. Grant signed the Yellowstone Park Act.[22]

Billings and Fee played their parts as well, providing the rationale for the link between the railroad and the park. Fee wrote that the falls of the Yellowstone must never drive the looms of a cotton factory, that in such a place adornment would be desecration and improvement ruin, while Billings noted, rather more prosaically, that God would not forgive those who destroyed his greatest creations.[23] The thought was not at all original, of course, but Billings believed it no less deeply for its derivation from the conventional religious sentiments of the day, from his reading of the new, rather mystical writer on nature, John Muir, who was making Yosemite better known, and from his correspondence with James Davie Butler, who had been Muir's teacher.[24]

Butler, the old family friend who had married Anna Bates and gone west to Madison, Wisconsin, where he became a professor at the university and vice-president of the state historical society, was in regular touch with Julia and more intermittently with Frederick. He wrote of his idea that there needed to be a great national forest preserve in the Montana Territory, perhaps somewhere up in the region of the Flathead River, which was to be opened up, it was thought (though a rival would beat them to it), by the Northern Pacific. Joined by one of the bonanza farmers, I. W. Barnum, who lectured on the wonders of the Yellowstone River, Butler, with Billings's endorsement, was among those who promoted what became, in 1910, Glacier National Park, a new preserve that was in good measure the product of a railroad, though of the Great Northern and Louis W. Hill, Sr. rather than the rival to the south.[25]

To link commerce and conservation is, today, controversial.[26] Pro-railroad publicists then and some historians since have argued that the Northern Pacific and Union Pacific brought benefits to Yellowstone National Park, as the Great

Northern did to Glacier National Park, and that railway hotels built in parks brought to them the influential type of tourists who encouraged their congressmen to support legislation to protect the wildlife and natural features of the reserves. Others, especially in recent years, and writing from a different perspective on the environment, argue that once the NP's branch line was constructed to Cinnabar in 1883, the railroad intended to violate the park. This view is supported by the fact that a number of NP advocates, hangers-on, and general boosters in Bozeman called for the branch line to extend directly to the geyser basins within the park, rather than stopping at the park boundary, and that they frequently attacked Philetus W. Norris, the park's superintendent from 1877 to 1882. However, Norris appeared to have gone over to the Virginia City men, who were supporting Jay Gould and the Union Pacific, which spoke of building a branch up the Madison River into the geyser basins. Bozeman's opposition to Norris seems to have turned on fear of the Utah & Northern Railroad more than on any desire to build into the park so long as no competitor was allowed to do so.

In the early years of the national parks, very few people opposed building hotels within them. The parks were hard to reach, and seeing them took much time. Generally, only the affluent could afford either the cost or the leisure, so the hotels tended to be, and for their surroundings certainly were, more than comfortable and often expensive. The government was not in the business of hotel construction or management, and it fell to those who would most benefit from tourist hotels to build and promote them. The Grand Canyon National Park, and its El Tovar Hotel, enjoyed a symbiotic relationship broken only in the 1920s with the rise of automobile visitors. National parks would not have been established in Canada when they were—the first was Banff, in 1885—had the Canadian Pacific Railway not promoted the idea and opened, with almost unseemly haste, a massive tourist hotel in 1888. Indeed, areas not considered worthy of national park status in the twentieth century—for example, Mackinaw Island in Michigan, designated a national park in 1875, with its Grand Hotel, built in 1887—qualified largely because they provided a resort atmosphere within a remote and natural setting.[27]

Thus the Northern Pacific, which was, together with the Union Pacific and Great Northern, the first railroad to promote an alliance between commerce and conservation, was generally viewed favorably for promoting branch lines and hotel construction. There were individuals, largely from the East Coast, who opposed any commercialization of the few national parks, and many, especially in Montana, opposed the Northern Pacific or one of its surrogates. But they did so because they feared what would happen if a single company or railroad were granted what was, in effect, a monopoly in Yellowstone, not because they were against railroads and hotels per se.

One reason to suspect the intentions of the Rocky Mountain Railroad

Company, which proposed to build the NP's branch line up from Livingston, was that it had been shrouded in secrecy and, when chartered, had used Montanans as a cover for Northern Pacific officers who effectively controlled it. These included Billings, Villard, and as noted earlier, Samuel Hauser. Again two interpretations are possible. The dummy officers were necessary to hide the line's real intentions from the Utah & Northern, and the company had been founded to connect Butte to Helena, not Livingston to Mammoth Hot Springs, though this became a goal soon after incorporation. Viewed negatively, the line hoped to encroach on the park's boundaries, and after completion up to Cinnabar in the summer of 1883—with service inaugurated on September 16—the board hoped to extend rails nearly to Mammoth's terraces, the first major tourist attraction. This would require a tunnel, which took years to finance and build, with the floating of other companies and other names, but in the end the promoters got their way, and the railroad reached the national park.[28]

In 1881 Billings invited Rufus Hatch, a Wall Street financier who had won and lost two fortunes, to tour the West and recommend how best to promote tourism within the Northern Pacific's sphere. Through Julia, I. W. Barnum warned Frederick that Hatch was a threat, being far more interested in short-term exploitation than long-term values, and advised that the financier did not share in "the religion of the beautiful." Hatch, with Northern Pacific support, certainly violated twentieth-century concepts of good park management, though at the time there was no agreement on such simple matters as whether those resident in the park could take game for their personal food supply or not. They did, and Northern Pacific officers dined off such takings, until the law made clear that even concessionaires could be found guilty of poaching. One could argue—indeed, Hatch did so—that railroad hotels and the system of coach roads laid out in the park helped to protect it, since individual camp sites were destructive and the passage of thousands of individual horsemen, coaches, or even hikers across the land on unmarked paths and roads would have been more destructive.

Still, there clearly was a "railroad gang" that hoped to get their hands on the park so that they might destroy it through "improvements." In 1883 the NP's divisional superintendent, Carroll T. Hobart, authorized surveys within the park boundaries for possible grading for rail lines. The secretary of the interior promptly issued an order forbidding the building of any railroads within the park. In June Hobart joined with Henry F. Douglas, post trader at Fort Yates in Dakota Territory, to form an enterprise which, with Hatch as president, was known as the Yellowstone National Park Improvement Company. Minnesota law, which now governed the Northern Pacific's charter, would not permit the railroad to construct hotels, and so the two companies were technically independent of each other. However, the vice-president of the railroad,

Thomas F. Oakes, made it clear to Hobart that the Northern Pacific would "respect and carry out any proper measures you may recommend that will make the Park a success & contribute to the welfare of this Co. & your associates." Hatch and Hobart wanted a hotel at Mammoth Hot Springs, and they hoped to see a narrow gauge railroad built from Cinnabar across the northern part of the park to the mines at Cooke City, outside the park's northeastern border.[29]

The Improvement Company sought out additional investors, including Billings, who could, Hobart hoped, help him deal with the new secretary of the interior, Henry Teller, who was pro-railroad and development minded. The company promised to invest $500,000 in the park to build hotels, liveries, and stores and Hobart undertook to send in twenty sawmills to cut lumber within the park, so that a stockpile would be ready for hotel construction in the spring. Douglas predicted that Yellowstone would become "a national resort" with "visitation . . . on a scale beyond anything this country has ever witnessed." He promised to "look after the interests of the public," to construct first-class stage routes, and too provide "entertainment" for fifty thousand visitors. This was, however, all wind, for even Teller knew better than to permit railroads into the park, and with Senator George G. Vest of Missouri as a persistent defender of the park's integrity, the Improvement Company was given no opportunity to improve in so wholesale a fashion. In 1885, stymied at every turn, the company was declared bankrupt by a Wyoming court. Billings had dropped out the year before.[30]

Ownership passed to a reorganized group, which called themselves the Yellowstone Park Association. The Association's leading figure was Charles Gibson, a wealthy lawyer from St. Louis with a record of philanthropy, especially in public parks. As a visitor to the park in 1885, he had found conditions deplorable, the half-finished hotel at Mammoth a disgrace, its privy bowls overflowing, and visitors suffering from diarrhea. The reserve was worth all the nation's city parks put together, Gibson wrote to the secretary of the interior, now Lucius Q. C. Lamar, and he proposed to improve matters.

Billings, who was aware of how easily a tourist clientele could be lost from watching the decline of tourism to Vermont's resort hotels, wanted to be certain that any hotels built in the parks would be "of quality," and he still hoped the railroad might build them, since he thought the company would be able to insist on high standards of construction. However, the new president of the Northern Pacific, Robert Harris, advised Billings that he thought the NP ought to play no role in ownership nor even show any leadership in the matter. Thus, on October 3, 1885, Gibson joined with Billings and five independent investors to request a lease so that acceptable hotels might be constructed near the geyser basins, at Mammoth Hot Springs, the Grand Canyon of the Yellowstone, and the outlet of Yellowstone Lake. Ultimately Lamar approved construction of the four hotels, and the Yellowstone Park Association was incorporated in April 1886 with $300,000 in capitalization.

The new association achieved modest success where the old improvement company had failed, and it would open two acceptable hotels by 1892. Billings quickly became disenchanted with the association, however, and divested himself of his shares, either because he had a change of heart on the question of "improving" the national park or because he felt the enterprise was poorly run. He was silent on whether he wished to see Yellowstone become a vast "summer resort," as some Northern Pacific publicists were urging after 1886.[31]

In 1880, as president of the Northern Pacific, Billings had endorsed an early handbook for tourists and settlers written by Thomson P. McElrath, who had travelled to the Yellowstone Valley the year before for the *New York Times*. That endorsement presumably embraced McElrath's rhapsodic hope that the day would soon come when "thousands of visitors desirous of feasting their minds with the contemplation of Nature in her sublimest aspects, or seeking relief from the stereotyped monotony of summer life at eastern watering places" would use the railroad to discover the "romantic mystery" of the Yellowstone country. This was, some might feel, Hatch's view in a coating of religiosity.[32] Billings seems not to have been alert to the inherent contradiction in such statements, not worried that thousands of visitors might injure the "sublimest aspects" of nature—though human beings are part of nature too, of course—and he did not anticipate the totality of the preservation and ecological concerns of another time. Nor ought one expect him to have done, for as most students of the modern national park movement have observed, environmental ethics are "an afterthought of the twentieth century."[33]

In truth, there was as yet no wilderness ethos, and few men—Muir was one of those few—had formulated a defense for that very wildness against which the Puritans and succeeding generations of immigrants, pioneers, frontiersmen, and settlers had fought. When Billings's old friend, General William Tecumseh Sherman, led an inspection team into Yellowstone National Park in 1883, he thought the park reservation was too great, his criterion for size being the area that a superintendent could effectively police, not what ecology might dictate, and he concluded that beyond the preservation of the park's natural curiosities (by which he meant essentially the geyser basins), there was "no good reason for any reservation whatever; the rest of the park should be thrown open to settlement like any other public lands."[34] Certainly when Congress created the national park in 1872, it had no conception of a system of such parks; Yosemite had been placed in the hands of the State of California precisely because there was a state to administer it, and had Wyoming been a state, Yellowstone would have been a state and not a national responsibility.

Yellowstone was viewed as a collection of curiosities, of "rare wonders," a menagerie of nature; even supporters of the park bill had seen the wisdom of justifying the park by virtue of its uselessness to civilization rather than by arguing on behalf of the positive values of wilderness, since only those of a transcendentalist persuasion, some religious divines, and a few mountain men believed wilderness per se had any value at all. The park was, one senator declared in 1883, mere "show business."[35] The Northern Pacific's brochures of 1885 showed "Alice" (as in *Alice in Wonderland*) staying at Mammoth Hot Springs Hotel, where she was "rubbing her eyes" in wonder at the majesties of Yellowstone, a "wonderland of fact not fiction." Puget Sound was depicted as a land of towering mountains and calm vistas that would refresh the soul. This drew from Richard F. Pettigrew, senator from South Dakota from 1889 until 1901, the remark that the Northern Pacific, having "patented the top of Mount Tacoma [Mount Rainier], with its perpetual snow and the rocky crags of the mountains elsewhere, which has been embraced within the forest reservation, could now swap these worthless lands" for more productive land in the best valleys.[36] To see the mountain peak itself as an attraction was well in advance of men like Senator Pettigrew.[37]

Finding the right balance between the commerce that drove America (and the majority of its senators) and conservation was difficult. In time "conservation" would be a term appropriated by foresters who favored regulated use (regulated, yes; use, definitely), with wilderness perceived as a vast timber farm, so that by default "preservation" came to be the favored term of those who wished to leave the wilderness unaltered, or as little altered as humanly possible when trying to create a constituency to defend wilderness values, themselves still in process of formulation. Billings had, in truth, done relatively little on behalf of preservation, though he certainly saw himself as a conservationist. But conservation meant sound use of a resource, not the neglect of it, and he was a good way from sharing the intense urgencies of a John Muir about what the future might bring.

An illustrative example arises from President Billings's requirement that Northern Pacific engines burn lignite coal west of Fargo. Lignite coal was cheap, plentiful, and much of it lay just below the surface of lands held by the railroad, especially west of Bismarck. More lignite lay ahead, across the Badlands and down through Montana, so that the company need not depend on better grades of coal shipped from the East or contend with traffic on the Great Lakes, with crowded port facilities in summer and ice in winter. To be sure, the use of lignite invited labor problems, for one man shovelled coal and another removed ashes simultaneously in a very hot cab. The blowing ash also started prairie grass fires that sometimes raged for miles. Still, Billings held to the conventional view of the time, that the grass would regenerate and, in the meantime, the fires would drive away the game that brought potentially dangerous Indians close to the line. Not until the railroad was moving into Montana, and ranchers began to complain that grazing land was being set alight, would Billings relent, and only then because a better grade of coal had been discovered in the Bull Mountains, only thirty miles from the line and, because of the lieu lands clause, within the land grant.[38]

What were Frederick Billings's conservation values at this time? He was deeply and sincerely interested in the ideas George Perkins Marsh had argued so forcefully in *Man and Nature*, and he wanted to honor the man and those ideas. He did not fully grasp Marsh's argument, however, until he turned his attention to his model farm in Woodstock. He vacillated between a belief in the innate goodness of mankind, which would, in the main, use God's handiwork well, and a premonition that nothing could stop the engine of progress, that it could at best be channelled to lessen or postpone damage to an environment that, while it must be cherished, would also disappear over the centuries in the pursuit of better living conditions. He had no sense of how quickly all would change: we find no word in his papers about the rapid destruction of the prairie bison. Railroads, by their nature, often eliminated what another generation would think of as scenic and aesthetic values. Those photographs of F. Jay Haynes, so enthusiastically endorsed by Northern Pacific management from 1877 on, reveal a casual destructiveness and a naive approach to nature's wonders that few would tolerate today. There, among Haynes' photos, are shocking testimonials to the destruction of wildlife, society matrons seated outside private shooting cars, guns awkwardly held across their laps, after the bagging of untold numbers of grouse, tourists carelessly walking across Yellowstone's geyser basins, standing and riding on its terraces, men climbing to the top of the Lone Star Geyser cone. Haynes would surround his studio at Mammoth Hot Springs with a fence of elkhorns, for elk were in abundance, and the symbolism of the horns they shed was then one of plenty rather than scarcity. His raw depiction of sour little towns, allegedly filled with hope, his famed 1877 views of Central City from Signal Hill in what is now South Dakota, showing a landscape blasted as through by nuclear war, or of the great stain across the sky from a smelting operation at Wickes, in Montana Territory, spoke then of what was believed to be progress.

Those photographs, and the attitudes of men like Haynes, Fee, or Billings, must be seen in the context of their times. If commerce and conservation could work together, fine; if not, commerce came first, whether for good or ill. Billings was clearly ambivalent about this reality, for as a businessman he insisted on maximizing profits. Nonetheless, he was part of the Romantic tradition, of which the desire to salvage some of the grandest wonders of America against environmental decay was a final stage. He did not believe that conservationists must scorn all businessmen nor that the thrust of all commerce must be against sound environmental practices as they were understood. It was also clear that few in the Gilded Age agreed with him.

26

Man and Nature

I N THE LAST DECADE OF HIS LIFE, Frederick Billings focused inward, on Vermont, Woodstock, and his estate. He had an estate manager, and an able private secretary in New York who managed his investments. He felt a long sleep coming on; his diaries from 1884 onwards were filled with intimations of death, concern that he must leave his mark in Vermont, must do God's will, must prepare for the day when the harvest stopped. This concern showed itself in three ways: in his work for the Vermont State Forestry Commission, his desire to develop a model farm, and his efforts to honor the memory of George Perkins Marsh.

The problem of the Vermont forests was plain for all to see. Billings had been shocked in 1866 at the denudation of the lands he had known so well as a child. By 1880 the center of the lumber industry had long shifted away from the eastern seaboard to Wisconsin, Michigan, and Minnesota, leaving great scars of untended and eroded land behind. Maine, New York, and Pennsylvania remained locally significant, but they had lost their early dominance. Vermont's contribution had always been modest; its mixed lumber went primarily to local mills and into woodenware, which by 1870 was the single greatest source of income to the state. In south and central Vermont farmers had cut birch, maple, beech, and white pine, and to the north spruce and hemlock, for their own use or for local markets. In central Vermont and along the Connecticut River much of the forest cover had fallen to the needs of the railroad.[1] Those workers who had become professionals moved to the vast northern pine regions around the Great Lakes, where about a fifth of the New Englanders hailed from Vermont. The lumber mills and tree fellers left behind were relatively inefficient and quite wasteful.[2]

Until the 1870s, the forest problem was not seen in state-wide terms. Despite a steady decline, Vermonters still thought of themselves as living in a farm state, and the 1870s saw the maximum amount of cleared land in Vermont, probably 80 percent.[3] (Not all cleared land was used for farming, of course.) Virtually everything below two thousand feet in elevation had been cleared. By the Civil War, spruce, once spurned for pine, was in demand, for the war cut off the supply of southern pine. The war's devastation also meant that the South, with many of its mills razed and bereft of local capital, could not meet the growing demand of the postwar years. (The Civil War also cut off the supply of sugar cane, making Vermont's sugar maples far more valuable.) As mills and their management became more complex, as the demand for wood products increased, and as the old Northeast faced a rapidly diminishing supply of timber, foresters began to express concern.[4] The Green Mountains were becoming "a biological wasteland, offering little for people to live upon."[5] Reforestation, planned management, and finally a policy of sustained yield—slow in coming—were the responses.[6]

The most significant voice in alerting Americans to the ecological problems they would face at the end of the Civil War was that of George Perkins Marsh. Marsh's book, *Man and Nature*, with its cogent subtitle *Physical Geography as Modified by Human Action*, published in 1864, marked the beginning of a broader public and professional awareness. Marsh had witnessed the most severe destruction of forests in America, for nowhere had forests been felled more quickly or the topsoil eroded more disastrously than in the Green Mountains. Mankind could so easily forget what an earlier generation had done to the land, could believe that it had always been thus: Tourists of modern times think of Greece as sparsely wooded, hot, dry, the land of the olive tree, but when Homer sang, Greece was under heavy forest, cover that was in some measure destroyed through the introduction of and overgrazing by the goat. As Vermont's state fish commissioner Marsh had studied restoration of a declining fish population, noting how the problem was related to the forests; cutting led to fluctuations in streamflow and water temperature and to decline in insects, reducing the food supply for fish. Industrial and urban development had polluted the streams. Badly placed wooden trestles for trains had altered migration routes and spawning beds. No one had pointed out the intimate relationship between the forests and the fisheries before. No one had argued so forcefully for a now well-understood truth: that in nature every act has a consequence.

One result of Marsh's work, and of the observable changes in the nation's ecology, was the creation of state commissions to investigate the forests. In 1876 Franklin B. Hough, a New York physician, historian, and statistician who had been deeply influenced by Marsh's book, was appointed to write a federal report on the state of the nation's forests. Among many measures, Hough recommended the planting of trees in vast quantities.[7] The Timber Culture Act of 1873, and the founding of an annual Arbor Day in Nebraska, drew attention to the need for new forest cover, and Billings had taken note as president of the Northern Pacific. As cities grew, those who looked to the future realized the need for renewed forest cover to protect watersheds, and the creation of major preserves in 1885 in the Adirondack and Catskill mountains of New York (and the first state forest board, in California) marked a new stage in public concern.

Between 1877 and 1881 Carl Schurz, as secretary of the interior, tried to limit the destructive methods then used in the lumber industry. He moved rapidly against timber trespass, seizing logs, initiating law suits, hauling offenders into court. He was not widely successful, for local juries would not convict men they knew for taking what in a frontier society was thought to be rightfully theirs, and Congress did not provide the necessary legislation. Nonetheless there were those who supported Schurz, and Billings was among them. He began experimenting with new species and new forms of planting on his estate, and he read widely in the scientific literature.[8]

In 1881 Raphael Pumpelly, a former professor of mining engineering at Harvard University, was invited by Henry Villard to direct a systematic survey of the mineral and forest resources of the Northwest on behalf of the Northern Pacific and other railroads so that they might make intelligent judgments about where to run branch lines. Pumpelly asked Charles Sprague Sargent, the director of Harvard's Arnold Arboretum, to study the forests of the vast region. From this and other research Sargent took up the cry for a large national forest preserve in Montana, urging such a course on Senator Edmunds, who introduced the first bill for a reserve between the headwaters of the Missouri and Columbia rivers. Sargent also prepared papers on the effect of forest fires, and one of these, read before the Massachusetts State Board of Agriculture in December 1882, received wide publicity. Sargent pointed out that fires were far more destructive than believed, noting that the economic loss due to fire was the least of the problems that arose from careless management. Most fires in New England, he reported, were caused by trains, brush heaps, or hunters, though the techniques of white pine logging were also responsible for many. He recommended a campaign of education and legislation to require all locomotives to install spark catchers, as well as the clearing away of dead brush through which fires spread uncontrollably. This struck close to home, for fires were frequent in the Bridgewater area, just west of Woodstock, and Billings was intrigued, writing to Sargent for further information.[9]

Early in 1883 the Vermont legislature resolved to establish a commission to assess the condition and report on the problems of Vermont's forests, which after a brief resurgence in economic value were again in decline. Upland farmers were cutting down young spruce and hemlock to sell in New York and Boston as Christmas trees, shipping them out by the trainload. As tree sugar had become increasingly important it was threatened by production in New York. Arson was increasing, as were fires attributable to careless smokers. When land was burnt over, less valuable species moved in; fireweed, blackberries, and grey birch took up the land once covered with pine. Management was essential, reforestation in some quarters was demonstrably necessary, and legislation would be required if replanted land was to result in mature forest.

Further, tourism was threatened with continued decline in Vermont because the state lacked significant "destinations"; the mountains of the far West were now well enough known to make "the Switzerland of America"—a once popular Vermont motto—seem foolish. In the 1880s more and more people began to sense that scenery had economic value and that the growing urban middle class would pay money to visit a green and forested Vermont. Rural farmland would not attract such visitors—many were only one generation off the farm themselves—while the quiet charms of relative isolation, the intense curiosities of the small town, and the opportunity to rest in a tranquil landscape might. Railroads were major forces in promoting Vermont tourism after the Civil War, well before the state became involved, and Vermont hotel and resort owners were endeavoring to keep up with popular trends by offering recreation, sports, and amenities to compete with the White Mountains and the Adirondacks. Thus a regreening of Vermont fit well into the alliance between commerce and conservation.[10]

The Vermont Forestry Commission consisted of former governor Redfield Proctor as chairman, together with Frederick Billings and Edward J. Phelps. Phelps had run for governor in 1880 as a Democrat; he shortly would be Cleveland's new ambassador to Great Britain. In 1883 he was Professor of Law at Yale University, where he lectured on matters of property, land, and resources.[11] Billings already had a reputation for reforestation, since he was introducing Austrian larch and Norwegian pine into his estate, and he was a public figure who, by virtue of his intimate relationship with railroads, could be counted on to view realistically the problems railways caused. Other individuals were added to the committee *ex officio*, though they do not appear ever to have met with it, so that the final report was the product of these three men.

Billings made the commission his special interest and apparently wrote most of its report. A questionnaire[12] was sent out to representatives, selectmen, listers, postmasters, and other assessors. The commissioners read widely, and they solicited information from other states and countries and from the Forestry Congress that convened in Montreal in August 1883. They also consulted with their counterpart commission in New Hampshire—where there were murmurings about a possible national forest in the White Mountains—and sought detailed information on western Massachusetts.

On October 31, 1884, the commission submitted its report to the Vermont legislature. Due to a mistake by the printer, very few copies were available, and Billings offered to pay to have more printed so that they could be distributed beyond the assembly, which was done in January.[13] The commissioners' findings were not encouraging: from 70 to 90 percent of native forests had been cleared in the older settled portions of the state, especially around Lake Champlain; there was no systematic wood culture; the water supply was failing year by year; springs and smaller streams were drying up. Without action, the commissioners anticipated that matters would get worse. However, they were uncertain that the state should or could intervene, since there were no public forests, timbered land being held in fee. Rather, Vermont would have to rely upon education, at the school level, through agricultural boards and colleges, and by private associations and wealthy individuals. "The experimental work of the practical farmer made with his own hand and constantly under his own watchful eye, is the best school of forestry. It educates and at the same time gives practical results." The farmer must learn that timber was a crop to be cultivated. Swamps must not be drained needlessly. Trees must be planted in school grounds, on roadsides, and by village improvement societies in public places. There should be a state Arbor Day. There should be more stringent policing and harsher punishment against arson. Migratory sawmills should no longer be exempt from taxation. A person who left his land in trees should be relieved of the obligation to fence against other people's cattle. There should be a tax on the value of wood and timber cut. Above all, the Board of Agriculture, farmers' clubs, and the state agricultural college must give more extended and professional study to the forest problems of Vermont.

The Forestry Commission report was acted upon slowly, however, and Billings did not live to see the creation of a state Forestry Department which, by World War I, put Vermont into a position of leadership in state forest practices. Nonetheless, by 1900, as a result of reports like that of 1884, wood products in Vermont were valued at ten times more than in 1850; an alliance between commerce and conservation once again brought a significant degree of public awareness. Clearly education had achieved much: in time Vermont ranked second with respect to the quality of cutting practices, exceeded only by Maine and far ahead of neighboring New Hampshire, and it was on the way to being the environmental leader that it is today.[14]

If the report did not produce any short-term results in Vermont forestry, the research Billings did for it was immediately applicable to his estate, which he wanted to make into a model farm for the breeding of animals, for forestry, and for general management. He began in 1869 with the purchase of the home of Charles Marsh, in which George Perkins Marsh had lived so happily for a time, and though it was at first a summer home it became more and more his true place of residence. The initial holding was of two hundred and fifty acres, which included the eastern slopes of Mount Tom, a large meadow between the mountain and the Ottauquechee River, and a quite solid brick house that

Billings would make over twice. In order "to control forests" he steadily added land until he had some six hundred acres in 1882; by 1900 the estate owned over two thousand.[15]

The mansion, or as the children and some family friends called it, the Castle, was an important symbol to Frederick Billings. It increasingly became a haven, the place where he was in the bosom of his family, free from the attacks of a spiteful press, both out of the limelight and yet, within the village, very much in it. He could entertain there as he wished, handsomely when the occasion arose but, since the mansion was far away from New York, without frequent obligation and by invitation. He could control his life when he was on his estate, could lay his hands directly on his work in a way that he could not do in New York. As he reshaped the land around him, remodeled the house, built roads around the slopes of Mount Tom, turned the quagmire of Pogue Hole into a sheet of attractive water, and tried to purchase the county fair grounds in the field below his house with the promise that he would maintain them, he was making his environment take the shape he required of it, rather than bowing, as at times he had to do, to a larger business environment over which, by its nature, his control was imperfect.[16]

At the farm Billings truly placed his hands on his work. He planted nearly twenty thousand seedlings imported from various parts of the world, working side by side with his men. Though he had farm managers, he knew farming: by reading and observation he made himself expert on dairying, haying, planting, grain markets, and land values. As the years passed more and more of his diary entries concerned the farm—dates of harvests, of first snow, of ice melt and river rise, of farm buildings begun and completed: carriage barn and stables, dairy barns, a manager's house and office. His plantations of white pine, set out in rows six feet apart, his stands of Norway spruce, his plans for European larch and white ash, all were examples of woodland preservation, at first for neighboring farmers and then, as the estate became more widely known, for visitors from ever further afield.[17]

Indeed, the farm's success with Norway spruce induced the International Paper Company to establish a nursery for the propagation of seedlings; thinnings from the Billings estate were turned to paper pulp, and his scientifically culled forest became an object lesson in fire control. Four days before he died, Billings ordered that the farm must be improved by new plantations of trees, and in his will he authorized Samuel Kilner to pay up to three times the value of land to acquire "the skirts of Mount Tom looking toward the village." In 1893 state foresters from throughout New England and New York would gather on his estate, where George Aitken, the superintendent at the farm, entertained them. To Billings and Aitken, the assembled group declared, was due the progress of the forest movement in Vermont, a judgment reaffirmed in

1911 when Vermont's state forester pronounced the property "the most interesting example of forestry in the state."[18]

George Aitken fully deserved his accolade, as Billings shared in it for choosing and supporting him. The first farm manager, George Weston, had served for nearly twelve years and had started up the Jersey herd, but he was less of a manager than a glorified herdsman, and Billings had kept him on a short leash. After two farm managers who served for a year each, Billings searched throughout the East to find someone who would give him the continuity Weston had. In October 1884 Aitken, a thirty-four-year-old Scotsman who was breeding Jerseys on a successful Long Island farm, agreed to take on the task of making the Billings Farm the showcase its owner desired. Aitken would remain until his death twenty-six years later, achieving all that his employer had hoped.

Under Aitken the farm flourished. A powerful man with a commanding presence—it was said that his voice could be heard six miles away—Aitken was one of the first scientific farm managers. With thirty regular full-time employees, augmented at harvest time, Aitken ran a diversified farm. The forty-acre large meadow by the river was put into corn or oats. Large greenhouses were used to ripen bananas, experiment with peaches, and make possible Vermont's most extensive grapery. Berkshire hogs and Southdown sheep were brought in, with up to three hundred sheep wintering in barns, in an effort to help find alternatives to the Merino in order to promote mutton stock. Horses were bred, too: Billings had joined the Horse Stock Improvement Company as early as 1869, and he was soon addressing state and other fairs on horse breeding, pursuing an interest in Morgans in particular, providing the financing for stud books on the horses of Windsor County, and overseeing efforts to breed Morgans to Hambletonian trotters. The enterprise was so constant and various, it became a central part of the community, and as late as 1948 it was said that nearly all male residents of Woodstock had at some time worked on the farm.[19]

Fame came not from reforestation or diversity, however, but from the Jersey cattle. Beginning in 1871, Billings systematically built a prize herd through direct importation from Jersey. Garfield Stoke Pogis became known as the best dairy bull in the land. Here, too, triumph came after Billings's death: in 1893, at the Columbian Exposition in Chicago, the farm sent all save one of Vermont's twenty-two Jerseys. When the American Cattle Club chose animals for a "battle of the breeds," two Billings entries won first prizes and the herd, in a competition against all breeds, placed third. Billings did live, however, to see Stoke Pogis Regina A.J.C.C. 48309 give two hundred ninety-nine and a half pounds of milk (which made eighteen pounds, three ounces of butter) in one week, the third largest yield of any heifer her age. In a day when farming still mattered, when people noted such statistics with the avidity of a twentieth-century urban baseball fan, and when Vermont badly needed agricultural leaders, Billings was rightly very proud of his herd, which—since he was generous to neighbors in granting access so that they might upgrade their herd blood lines—was a contribution to the community as well.[20]

Billings and Aitken were also at the forefront of quite a different kind of battle for conservation. From 1880 to 1886 the State Board of Agriculture was under heavy attack by the Dairymen's Association on the grounds of interference and by the State Agricultural College on the grounds of competition. Technically, the Forestry Commission reported to the Board of Agriculture and Billings defended the board. Because the board's primary concern was with two battles—one to achieve herds free of tuberculosis, Bang's disease, and mastitis, and the second to establish and maintain standards for "honest butter"—Billings became unpopular with some, and highly regarded by others, for taking the high road on both issues, being certain that the Billings herd delivered only TB-free milk and the best butter.[21]

Billings was also proud of Mount Tom and of the network of roads he constructed around it. The roads, he said, were "to be my monument!"[22] Wide enough for two carriages to pass, these graveled drives provided Woodstock with its largest pleasuring ground, for Mount Tom was open to the public. A month before he died Billings took his final ride on his roads and checked for the last time on the new farm manager's quarters, a handsome Queen Anne house built for Aitken.

On September 26, 1890, at Frederick Billings's deathbed, Samuel Kilner prepared a set of notes based on his final conversations with his employer, and these notes guided the future management of the estate. Billings's charge to Aitken was as it had been to the men who worked for him on the Northern Pacific: "expenditures must be reduced and the receipts increased." The tree plantations were to be looked after and new ones begun. "A great many" new trees were to be set out near Pogue Hole. In his memory the Woodstock estate was to remain, for Julia and the children, and for the people of Woodstock, a model farm.

To be sure, Billings was not Vermont's only forest conservationist nor the only man of wealth to turn his property into an experimental farm. William Seward Webb in Shelburne, Joseph Battell of Middlebury, Marshall Hapgood of Peru, and Silas L. Griffith of Darby, who had made his fortune in lumber, were all experimenting, and Webb in particular developed a farm that later surpassed Billings's in significance. Further, the Billings Farm achieved its greatest impact in the decade after Frederick's death, so that much of the credit quite properly went to Aitken. Nonetheless, it is clear that Aitken was carrying out a plan in part devised by, and in a very large measure financed by, Frederick and continued by Julia in his memory. Some of the practices at the farm were innovative, some were simply part of a well-established trend (the search for alternatives to the Merino sheep, for example), and some, as with the Jersey herd, if conventional nationally were nevertheless superlative achievements and significantly experimental for their time and place, providing guidance as dairying became Vermont's principle form of agriculture.

Often popular articles suggest that Billings's model farm helped to arrest the decline of agriculture in Vermont. It did not, and no model farm could have, for by their nature model farms seldom produce sustained profits and thus do not attract wholesale imitation. In any event, historians are not agreed on just what this decline was. We are told in many books that "agriculture was declining steadily" through this period, though generally what is meant is that farming was in decline. Certainly contemporary observers felt this to be the case, and Billings may have hoped to arrest this decline and the lessening in the culture of mutuality that farming represented by sharing knowledge of scientific agriculture with others.

In recent years the question of decline has been reopened, in some measure by the new economic history. The usual view is that economic decline coincided with the loss of population and the abandonment of the family farm; Vermonters left because they were lured to the cities or to the western states by new economic opportunities. This school also argues that more extensive and efficient western agriculture put the local farm economy into a downward spiral, and this in turn led to the closing of shops in rural communities that depended on the farmer until some towns resembled Oliver Goldsmith's "Deserted Village." In the 1970s, however, other historians challenged some of these assumptions, in particular by arguing that outward migration was, on examination, by people not needed in local agriculture; thus there was no necessary decline in the farm economy itself, with those left behind becoming more efficient and productive. While farms that had been only marginally productive were abandoned, those still in production were able to fill market needs. The result was that the ratio between man and land remained satisfactory.

More recently other scholars have held that neither of these views is sufficiently sophisticated. Contemporary observation about abandoned farms, such historians say, was flawed, for though the farmhouse may have been left to decay, the land itself was often absorbed into a larger, more effective farm unit. Towns like Woodstock, if smaller than they were in 1830, were part of a transportation network that made them outliers of more efficient units of production. The notion of a saturated economy leading to surplus labor and thus out-migration does not take into account the larger economic environment by which even a state is not a sufficiently large unit for analysis. Labor was, in fact, often still in demand—as shown by Billings's need for a substantial labor force on his estate—and outside attractions rather than domestic disincentives drew people away from the smaller towns. There was not, these historians conclude, a wholesale decline arising from the undoubted population loss; rather, it was the loss of young laborers, leaving farmers shorthanded in more isolated counties, that induced economic stagnation in some localities. Therefore each community requires its own close analysis.[23]

Often "the decline of the farm" is shorthand language for a decline in traditional values supported by the family farm as a small unit. Certainly small farms could not be self-sufficient and small farmers were at the mercy of market forces they could do little to influence. The work at the Billings Farm undoubtedly pointed the way toward an improved and rationalized agricultural economy, but it did nothing to sustain single family farms. To Vermonters with legitimate if parochial concerns about the demise of the small unit, Billings was actually an outsider with big city ideas and intentions.

These debates are too recent to provide conclusions for Woodstock. The relevance of such analyses to Billings's contributions relates to whether his experiments had the impact he desired. At this point it is not possible to say how important the Billings Farm was. What one can say is that most people believed farms were being abandoned, and anything that would diversify Vermont agriculture or give Vermont farmers a competitive edge over nearby challenges, particularly from New Hampshire, was to the good. In short, at the time Billings had every reason to believe that his efforts to lift Vermont agriculture from a narrow parochialism and once again make Vermont a leader were significant.

Billings was wrong to believe that Mount Tom and its roads would be his monument. The roads are only walking paths now, and in public hands, his estate reduced by half through sales and inheritances. The mansion was and is his monument, and it is a monument to a marriage. Built in 1805 by Nathaniel Smith and remodeled twice, once in 1869–1870, probably by William Ralph Emerson, and then in 1885 by Henry Hudson Holly, it grew forth, its roof line lifted, porches billowing out, to become a triumph of Victorian architecture and, for the time, of modern amenities.[24]

The first home remodeling, the presumptive work of Emerson[25] and an interior decorator, William J. MacPherson, with landscaping by Robert Morris Copeland, was in the fashionable Stick Style with a French roof and a broad veranda around three sides of the ground floor. This was the house the family knew best. Holly's version, in which Julia would live until her death in 1914, was more elaborate, a bit fussy here and there, yet a bold statement of how current fashions could be adapted to the countryside. Holly raised the mansard roof to create a third story and an attic, enlarged the servants' quarters, and brought in John DuFais, a designer from Tiffany & Company, to redesign the interiors. There were problems with Holly, and Frederick reacted negatively to the Tiffany stained glass windows that DuFais put in. But in the end he was pleased: he was "impressed with its dignified beauty" and pronounced the house quite fine.

For Billings the Marsh house was set into a moral landscape that represented the old Vermont virtues of thrift, good craftsmanship, and success handsomely but not vulgarly expressed. From its large veranda he looked out toward the town, taking notes on the weather, watching his children grow. From the house he took spiritual and emotional sustenance; to the house he brought art: Kensett, Cole, a George Catlin that he confessed to disliking and eventually gave away. He gravitated to Frederic Church rather late, attracted to the painter's landscapes empty of human beings. Above all he treasured his Bierstadts, which domesticated the sublime, rendering all of nature loving and beautiful. Unlike many railroad barons, Billings seems never to have purchased a George Inness or the work of any artist who brought the machine into the garden: there were, it appears, no pictures of railroads in the Woodstock house.[26]

Scientific agriculture on the estate, reforestation on Mount Tom, and the attractive mansion, though greatly changed from George Perkins Marsh's time, were also a monument to the man who, more than any other, Billings most admired. His desire to have something personal of Marsh's—a walking stick for the house in Woodstock, Marsh's own study chair brought from Florence for the house in New York City, a tall clock, a picture of Marsh at work in his Villa Forini—attests to a sense of intellectual kinship with the man he most likely met in 1846 when Frederick was serving temporarily as secretary to the State Legislative Council, for it was Billings who formally signed Marsh's certification as a state representative on November 3.[27]

Billings's greatest service to Marsh's memory came in 1882. In that year Marsh, worried about how to provide for his wife, Caroline Crane, decided to sell his library, noted for its holdings in science and philology. Brown, Harvard, and Cornell expressed an interest, but Marsh hoped to see the University of Vermont buy the collection, which he offered to sell for $10,000. The university did not have such a sum. Billings learned of Marsh's desires, probably through the university's president, Matthew Buckham. At this juncture Marsh died, and Billings wrote at once to his widow, offering to purchase the library for $15,000. The gesture was typical of him: he wished to extend charity to Caroline Marsh under circumstances she would not find embarrassing. Though the library was said to consist of thirteen thousand volumes, four thousand of those volumes already were on permanent loan to the university, and in any case, in the end the collection consisted of eleven thousand volumes. Billings knowingly overpaid for the library and paid for its transport from Italy as well.[28]

Billings then wrote to Buckham to ask whether UVM ought not to have a fireproof building capable of holding fifty thousand volumes, offering $50,000 to construct a suitable structure in which the Marsh collection could be housed. He was told that this would do little more than render fireproof the old library, and when he received the plans he dismissed them as inadequate. Determined the project should be well done, he suggested that Buckham look into the possibility of building a new library, and Buckham sensed a major opportunity. He sent photographs of several library designs to Woodstock, and in March 1883 Billings informed the university that he would finance the library for them. He pledged an initial $75,000 and suggested that Henry Hobson Richardson, then regarded as the leading architect in America, prepare the plans. While Billings had heard that Richardson was "a very extravagant man," he wanted to see what an architect of his standing would do.

The collaboration that resulted was a stormy one. Essentially Buckham and Billings either rejected or disliked nearly every design Richardson submitted to them, and yet their desire to have a building by a man whose fame was growing so steadily overcame their reservations. Though Billings promised to leave the question of design entirely to Buckham, he could not keep away, for he wanted "a beautiful building," not simply a functional one. Richardson understood this, and knowing that he could not build what Billings wanted for the sum offered, he continued to draw the university's patron in: as Billings told Buckham, "a few pennies more better be put into it, rather than anything should be wanting." In the end Billings raised his pledge to $100,000, then to $110,000, and then to an unconfirmed sum reported to be $250,000.

What Billings and the university got was a boldly handsome Richardsonian Romanesque structure of Longmeadow sandstone. Billings privately admitted that he did not care for the exterior. The interior was another matter: Georgia pine, rich ornamentation, and a special room devoted to Marsh, whose bust, the work of Franklin Simmons, was placed above his books. Billings did not attend the dedication exercises, held in conjunction with commencement in June 1885. He was, he wrote to Buckham, ever grateful for all that the university had done for him; now he had built what was, at the time, perhaps the finest university library in the east: "it is not just money that I put into the building. My love is built into it."[29]

The Billings Library was more than a generous gesture to the memory of an honored Vermonter, however, for the University of Vermont was also the State Agricultural College, and Billings viewed the purchase of Marsh's scientific library and its permanent installation as a significant contribution to agricultural education. Billings was convinced, as he reread Marsh's work, that conservation was the highest form of efficiency, and that future generations of students would learn this lesson in his library. Aside from philology, the collection reflected the proportions of *Man and Nature* as Marsh had written it: substantial material on the protection of forests, major monographs on the depletion of groundwater, and a smaller section on the encroachment of sand dunes along the margins of the seas. Most important, Billings thought, were the titles on the extinction of natural species arising from man's ill-informed actions.

In a sense Marsh represented what Billings often thought he might have been. A distinguished scientist, a person who had made a difference in a fundamental way, Marsh had practiced law, had been a member of Vermont's Executive Council and, from 1843 to 1849, a member of Congress. He had been U.S. minister to Constantinople and, after returning to practice law in Vermont once again, in 1861 had become the first U.S. minister to the Kingdom of Italy, where he remained through successive administrations until he died at Vallombrosa in 1882. His work was read and admired throughout Europe and North America, translated into many languages, cited and debated, and was the stimulus to much new research. As a young man Billings had toyed very briefly with the ministry; throughout his life he had flirted with politics; of all the roads not taken, however, he appears most to have regretted not having followed the one that, in the words of Vermont's best-known poet, "was grassy and wanted wear," the one on which he encountered so briefly men like Agassiz, Whitney, and Marsh.

27

The Best Season to Journey

B Y 1889 FREDERICK BILLINGS HAD THE ENVIRONMENT of Woodstock
pretty much as he wanted it. The last carriage roads were being put in
on Mount Tom, where reforestation was well advanced. He had provided gifts
that had restored his parents' old church in Royalton and thrown an iron
bridge across the Ottauquechee between his house and the town center. In 1880
he had funded a new chapel for the Congregational Church, a memorial to his
parents. True, he had continued to pursue the fairgrounds, and did not get
them—they would not be restored to the Marsh estate until 1933—and he
hoped for an extension of the Woodstock Railroad directly up to those fairgrounds, and thus to his house, and this too had not happened. But for the
most part Woodstock was, he felt certain, both a more attractive place and a
more economically viable community because of his efforts.[1]

Then the long silence began to fall. On October 7, 1889, his beloved Ehrick
died, ten days short of his seventeenth birthday. Billings had not really recovered from Parmly's death precisely seventeen months to the day earlier,
though Parmly's, while unexpected, had not been so great a shock as Ehrick's,
despite Ehrick's history of sickliness, since Parmly had been living in Alaska
and Montana, not Vermont. On Christmas Eve Billings was attending a meeting of the board of the Delaware & Hudson in New York City when he nearly
fell. Samuel Kilner took him home. The stroke, for such was the diagnosis by
a hastily summoned doctor, left the right side of his face drawn, his left
arm and side partially paralyzed, and his speech affected. On Christmas Day
Julia and the children brought presents to his bedroom and decorated a small
tree for the table; the same day he had two mild heart attacks. He knew that
the end was near.[2]

Though Frederick was now an invalid and Julia spent much of her time
reading to him, they maintained their interest in their philanthropies. Both
had been helping to support Dwight L. Moody's school at Mount Herman,
Massachusetts, and Moody often visited their home. Frederick, deeply im-

pressed in 1884 by the president of tiny Whitman College in Walla Walla, Washington, had been sending annual gifts. Earlier they had taken an interest in Pacific University, at Forest Grove, Oregon, where the Reverend Sidney Harper Marsh had been head of Tualatin Academy, which he turned into a college. They had continued their support for the Pacific School of Religion. Billings provided funds for memorial windows to Eleazar Wheelock, Marsh's grandfather and founder of Dartmouth College, in that college's chapel. There were gifts to Phillips Andover Academy, to Grinnell College to help it rebuild when it was destroyed by a tornado, and to Amherst. Now Frederick gave substantial endowments to Amherst, the University of Vermont, and the Mount Hermon school. He had always been a bit inclined to listen to tales of woe where religion and education were combined, Julia had told him, and as he revised his will time and again he blinked back Julia's hardy advice that there was "a good deal of the leech in some of your theological acquaintances!"[3]

One of Billings's interests in education had particularly far-reaching effects: support for a young Japanese named Sho Nemoto. Born in 1851, Nemoto was a Christian who had worked in an American mail room in the Yokohama post office, where he learned English. Nemoto was sent by missionaries to San Francisco where, in 1879, he met Billings's old friend Alfred Barstow. Realizing that the lot of an Asian was not a happy one in California at that time, Barstow wrote to Billings, asking that he help Nemoto.[4]

Billings enrolled Sho Nemoto at the University of Vermont, paid his fees, and invited him to stay with the family in Woodstock during summers. Nemoto progressed slowly but well, lecturing in Baptist and Presbyterian churches, translating American books, and graduating from the University in 1889. Sho Nemoto was a commencement speaker as Billings before him had been, but Frederick was too ill to attend. He and Julia gave Nemoto a trip around the world as a graduation present. Immediately after his return to Japan, he was made a commissioner of commerce and industry and was sent to study economic conditions in Latin America. He wrote one of the last letters Billings received, asking for advice on his forthcoming marriage and, according to Japanese custom, formally requesting Frederick's permission to marry. After Frederick's death Sho Nemoto continued to write to Julia, and, during trips to Mexico and Brazil in 1894–1895 and again in 1899, visited her. On the first visit he supplied a bronze bust of Billings, the work of a leading Japanese court artist, for the library at the University of Vermont, and he wrote and translated books on the American constitution.

Frederick had told Nemoto that he must "be useful to Japan," and he was. He proselytized for Christian missions, was an organizer for the National Liberal Party and supporter of the new Japanese constitution of 1889, worked against rising Japanese imperialism, and was elected to the Japanese Diet in 1898, serving continuously until 1924. Nemoto contributed significantly to the framing of the first free education act, and for twenty-three years introduced a bill to prohibit the use of alcohol by minors, celebrating the bill's passage at last in 1922. He also became a rich man through Tokyo real estate, having learned his lesson well in Woodstock. Until his death in 1933 Sho Nemoto remained resolutely pro-American, always mentioning Frederick Billings as the man who made possible his knowledge of American democracy. "The world," he wrote shortly before his graduation from U.V.M., gradually had become "a single room."[5]

The other persistent pattern to Billings's philanthropy had been in improving, and as he saw it, protecting Woodstock. There was a portrait of Norman Williams, hung in the county court and later in the library built by Williams's son. There was the repair of the Methodist parsonage and church. There was reconstruction of a schoolhouse. There were gifts to the cemetery association. Above all, there was the additional expansion in 1889–1890 of the Congregational church, with a new vestibule and port-cochère, a new interior—with organ and Tiffany windows—and a remodeled parsonage, much of this the work of Henry Hudson Holly. In 1887 the local press had predicted that Billings would "own the earth yet. Well, we just as lief he would own it as anybody we know of. . . . Oh! for more such men as Mr. Billings in our midst."[6]

For Billings, this land was embedded in history. He knew it all; before his stroke he had traveled the old road back to Royalton to see the house where he was born (it had burned down, however), rode up to the North Ridge or to Lone Tree Pasture noting the animal life, the signs of human trespass, the debris of the casual picnicker. He liked the idea that he could look across to the fallen walls of the farmhouse in which the great sculptor Hiram Powers was born, that daily he gazed down across the home of Jacob Collamer, whose statue stood with Ethan Allen's in the nation's capitol, representing Vermont. Billings was reading local history and novels with Vermont settings. Perhaps his most unusual gift to the town was its history: in 1881, he had persuaded his old friend Henry Swan Dana to resign as register of probate to write a full town history. For five years Billings had paid Dana's salary and met his expenses. Now the project was finished.

Billings had his own ideas about what constituted a good book, and he was not entirely happy with Dana's initial manuscript, completed in June 1886. He wanted the manuscript reorganized, he wanted more on the agricultural society, libraries and book clubs, and the Civil War, though Dana defended his sense of proportion on the ground that Woodstock had not furnished a single officer of note to the conflict. Billings wanted genealogies (Dana agreed), more pictures (agreed), and a more resounding conclusion. On this point Dana rightly dug in: "there is no such thing as concluding a history," he advised his patron; one should read all the great historians—Thucydides, Tacitus, Macaulay—as fragments.

In fact the book was quite original in its conception, approaching the history of the town as though one were taking a carriage ride down each of its streets, borrowing a literary conceit from Lydian geographer-historian Pausanias. This worked perfectly with a contemporary and knowledgeable audience. If today the book is quite unreadable, its most used part undoubtedly being its index, for its time it was a remarkable effort, clearly better than most local Vermont histories (a genre by its nature far more antiquarian than historic), worthy of both Billings's trust in Dana and Dana's dedication of the book to Billings, the "free and munificent spirit [who] has contributed so much to make Woodstock pleasant and attractive."[7]

Recording history, cataloguing one's possessions, placing plaques where idle passers-by may be instructed about an exemplary event are an old man's ways of shaping the environment, of attempting to put out a sheet anchor against change. Dana's book was explicit in this, for it was not a history as the term is used today, analytic of change, critical of any figure who appeared in its pages; rather, it was a chronicle of events, a record of material and moral progress, a series of vignettes about how the past served the present, and a compendium of all those names of which the town could be proud and a few that, in other circumstances, it would not have been had not their appearance between the covers of a printed book bestowed the authority of respect. Well after the book appeared in 1889, published by Houghton, Mifflin of Boston through Billings's old University of Vermont classmate, Henry Houghton, it was hailed by James Davie Butler for being "so accurate, so elegant, so unique, so abundantly satisfactory"; the book was, Butler declared, "a monument" that would keep Dana's name green in the hearts of his townsmen for as long as the town endured. Billings paid for the printing costs, and the book may have been his most innovative gift.[8]

On March 30, 1890 Billings's doctor told him that either he would improve within six months or he would be without hope. He was treated with electric currents applied under his arms and, as his chest pains increased, with doses of morphine that came closer together as the weeks passed. Late in May he, Julia, the children, and the servants all returned to Woodstock, aware that he most likely was going home for the last time. Billings lived to see most of the work on the Mount Tom road system, the farm manager's house, and all the remodeling of the Congregational church completed. He was unable to attend the dedication ceremonies for the church, though he did go to sit in the pews "to see that they were right" and pronounced the building satisfactory. The town lay beneath the hill, going about its busy routine, knowing that his death was only a matter of weeks, as Billings dreamed through the cicadas of late summer.[9]

Billings gave careful attention to his will, liberally providing for his brothers and sisters, servants, and old friends, and arranging that both the Woodstock and the New York homes should be well maintained for as long as Julia lived. He did not have a great fortune by the standards of the railway kings. When Jay Gould died in 1892, he left an inheritance of $72,000,000: he bequeathed nothing to charity and nothing outside the family. Billings had perhaps $10,000,000 in Northern Pacific and other shares and property, perhaps a like value in the Woodstock and New York homes and their accoutrements. He owned eleven square blocks in Superior, Wisconsin, still held lots in San Francisco, and had other properties in North and South Dakota, Minnesota, Montana, and Missouri. Billings's stocks and bonds were distributed across fifty or more companies. Estimates of his total worth ranged between $10,000,000, which was clearly too low, and $30,000,000, which probably was too high.[10]

One does not, even at the end, abandon the habits of a lifetime. Billings described the funeral service he wanted and wrote to a few key newspapers to indicate who he hoped would be asked to write his obituary. For the New York Times, the obituary that most likely would be read by the largest number of people, he chose Augustin Snow, a veteran correspondent who had been a classmate at Kimball Union. Henry Swan Dana almost certainly wrote the Vermont Standard obituary, which was most important for local sensibilities, Alfred Barstow prepared the principal California notice, while James Davie Butler wrote the main midwestern obituary.[11]

One could prepare for the idea of death—not, of course, for the actuality—in several ways. One could make ready with rage, following Dylan Thomas's advice long before he uttered it. One could prepare with joy, knowing that one was being taken up to the Lord. One could regard death as a duty, the final obligation of life in a time when duty was a guiding concept for many. Billings had a strong sense of duty, though he had never allowed duty to become a burden; he simply assumed that it was what one did. Certainly he accepted the widespread Christian beliefs of his day about laying down one's burdens to take one's place in eternity. But for the most part he consciously prepared for death, accepting it largely as a matter of dutiful expectation. There was nothing dour about this, since duty too could be a joy. There were last things to be done, which could draw together the strands of a life and express its meaning, at least to oneself.

On the first Sunday in September the Reverend Peter McMillan came up from the Congregational church to give Frederick the sacrament, as he could no longer sit up. He regretted that he had been "such a formal Christian," and he spoke of his sins. He was, in his conversation, borne back ceaselessly into the past. On September 10 he was delirious. On Tuesday night, September 30, he and Julia read and prayed together, and shortly after half past nine, a week after his sixty-seventh birthday and six months to the day granted by his doctor, he died in his sleep. For the last weeks he had taken to saying a child's prayer each night: "If I should die before I wake / I pray Thee Lord my soul to take / And this I ask for Jesus' sake." The trees outside were a bright red—as Henry David Thoreau observed, this was the best season for a journey into the Green Mountains.[12]

NOTES

In the interest of economy, many printed primary and most secondary citations have had to be omitted here. Those interested in fuller documentation may find it in the original typescript for this book, which is on deposit in the Mansion Archives in Woodstock.

CHAPTER 1

1. Zadock Thompson, *A Gazetteer of the State of Vermont* (Montpelier: E. P. Walton and the Author, 1824), pp. 232–35, 296–98, 309–11.
2. Harold A. Meeks, *Time and Change in Vermont: A Human Geography* (Chester, Conn.: Globe Pequot Press, 1986), pp. 87–103.
3. Various genealogies, short biographies of family members, and obituaries on Oel, Sophia, or Frederick in the Mansion Archives contain numerous contradictions and misconceptions. Some are scarcely logical internally. One such account says that John Billings had nine children. On such matters I follow throughout the admirable family trees (which show by name eleven children of John Billings, with three unnamed children who died in infancy) in Mansion, box A31: *Billings and Related Families: Genealogical Charts and Notes, 1750–1950, with Photo Portraits*, comp. by Janet R. Houghton (unpublished typescript, rev. 1985). Hereafter references to Mansion box numbers without elaboration are to the Billings Mansion Archives, Woodstock, Vermont. Names of correspondents are spelled in full, with these exceptions:

    FB   = Frederick Billings
    JPB = Julia Parmly Billings, his wife
    OB   = Oel Billings, his father
    SWB = Sophia Wetherbee Billings, his mother

4. On Lafayette's visit see Woodstock *Observer*, June 25, 1825; Jay Read Pember, *A Day with Lafayette in Vermont* (Woodstock: Elm Tree Press, 1912); and Mary Grace Canfield, *Lafayette in Vermont* ([Rochester, N.Y.: Canfield & Tack], 1934).
5. The evidence on the offer of the superintendency is ambiguous. See Mansion, box A1: Edward Billings to OB, March [1832]. On Oel as a J. P. (recorder of mar-

riages) see Royalton Town Clerk's Office, South Royalton, Vital Records, IX, 151.

6. See H. J. Conant, "Imprisonment for Debt in Vermont: A History," *Vermont Quarterly* 19 (April 1951): 67–80.

7. The best account of this episode is by John McDill, "Frederick Billings: Vermont '49er, from the Green Mountains to the Golden Gate," *Vermont Life* 4 (Spring 1950): 20–23.

8. It is as impossible to separate fact from legend in this episode as fact from nostalgic memory half a century after any event. The turnpike ledger books have not survived so that toll payment cannot be verified. Since the family was together, there are no letters to attest to the circumstances. Certainly the Billings family did arrive as described, with wagon, cow, pig, and children, and certainly such a slow progress over the steep and rough road would have taken its toll. Even today, over a lessened gradient, the walk is tiring. In observation of the approximate anniversary of the move, on September 27, 1986 I walked the old turnpike route from Royalton to Woodstock. The road is rough, there are many hills, and the climb up from Fox Stand is wearying still. Today the distance is 14.6 miles, and even without a pig, one would want the fullness of a day for it. For the Billingses the walk may have been longer. The bridge over the river at Fox's, built in 1833, was out in 1835, not to be replaced until 1837, and while the family may have forded the river, since water was low in September, they may also have gone down river to the next bridge, adding a mile to their journey.

9. Thompson, *Gazetteer*, pp. 196–98.

10. Jane Curtis, Peter Jennison, and Frank Lieberman, *Frederick Billings: Vermonter, Pioneer Lawyer, Business Man, Conservationist* (Woodstock, Vt.: The Woodstock Foundation, 1986), pp. 11–12, and Henry Swan Dana, *The History of Woodstock, Vermont* (Boston: Houghton Mifflin, 1889), pp. 475–76, 478, 482–85. See also John Donald Hicks's entry on Billings in Allan Johnson, ed., *Dictionary of American History* (New York: Scribners, 1929), II, 265–66.

11. Mansion, box A1: OB to Edward Billings, note added to SWB to Edward, October 3, 1833.

12. Dana, *Woodstock*, p. 217.

13. See Mary Louise Kelley, *Woodstock's U.S. Senator, Jacob Collamer* (Woodstock: Woodstock Historical Society, 1944).

14. See *Catalogue of the Trustees, Instructors and Students of Kimball Union Academy, at Meriden Village, Plainfield, N.H., for the Year Ending April 1840* (Newport, N.H.: Argus, 1840); Kimball Union Academy, *A Remarkable History: K.U.A. from 1813–1913* (Boston: Fort Mill Press, 1913); Jerold Wikoff, "Alumni Profile: Frederick Billings 1840," *K.U. Alumni* (Spring 1984), pp. 25–26; and Ernest L. Sherman to John H. McDill, November 1, 1948, in Kimball Union Academy archives.

15. Mansion, box A1: OB to FB, September 25, 1839.

16. *Ibid.*: OB to FB, September 25, 28, 1839, April 20, 1840, and Edward to FB, September 28, 1839; box A2: Richard to OB, April 23, 1840.

17. Mansion, box A1: OB to FB, April 2, 1839, January 4, 1840; box A2: SWB to FB, March 27, June 17, September 9, October 13, December 5, 1839, and February 27, 1840.

18. Mansion, box A8: FB to M. Richards Porter, n.d. [1840], and Young to FB, June 15, 1840.

CHAPTER 2

1. Julian Ira Lindsay, *Tradition Looks Forward: The University of Vermont, A History 1791–1904* (Burlington: The University, 1954); Charles Andrew Huntington, *The University of Vermont Fifty Years Ago* (Burlington: Whitney & Stanley, 1892); F. E. Goodrich, *A Sketch of the History of the University of Vermont* (Burlington: The University, 1899); Edward Younger, *John A. Kasson: Politics and Diplomacy from Lincoln to McKinley* (Iowa City: State Historical Society of Iowa, 1955), pp. 17–26.

2. Mansion, box A8: James Barrett to FB, September 13, 1840, FB to Barrett, March 14, 1841, January 15, 1842. In 1882 Barrett went through his files and returned all of Billings's letters to him (Barrett to FB, March 2, 1882).

3. Mansion, box A2: SWB to FB, September 27, November 7, 1840, March 11, April 1, November 13, December 6, 18, 1841, October 1, 1843, January 6, February 27, April 28, 1844; OB to FB, October 8, December 17, 1840, May 10, July 23, October 21, 1841; and Laura to FB, July 30, 1843 (on back of OB to FB of the same date), and October 29, 1843; FB to parents, November 29, 1840, March 7, May 7, September 5, 1841, January 4, November 25, and undated letter "Wednesday evening," all 1842, February 19, April 2, 1843, March 9, June 21, undated Wednesday note (though apparently June 26), 1844; box A8: FB to Barrett, September 13, November 26, 1840, March 14, October 17, 1841, January 15, 1842, Barrett to FB, March 2, 1882; box A8: Augustus Young to FB, May 14, 1844, Wheeler to FB, May 4, 1847, July 29, 1853, Jacob Converse to FB, November 30, 1869, Charles Marsh to FB, December 12, 1870, FB to President N. H. Buckham, May 28, 1885, G. W. Benedict to FB, January 16, 1871; and box A10: grade reports and perfect attendance certificate.

4. Mansion, box A2: SWB to FB, May 29, 31; box A1: SWB to Edward Billings, October 29, 1843, and OB to FB, June 10, 23, 1844; box A31: FB to J. Savage, May 20, 1889.

5. J. Kevin Graffagnino, "Zadock Thompson and The Story of Vermont," *Vermont History* 47 (Fall 1979): 237–57.

CHAPTER 3

1. Mansion, box A1: OB to FB, October 9, 1845. See also Charles Jason Billings to FB, October 28, 1844.

2. Frederick Billings's relationship to Horace Eaton, and quotations not otherwise attributed in the following paragraphs, derive from Mansion, box A2: SWB to FB, October 18, 28, 1846, November 12, 1847, FB to parents, September 17, [1846]; box A8: Eaton to FB, June 3, July 24, September 22, 1846, June 12, September 12, 15, 1848; box A10: certificates of appointment, October 9, 1846, the same, 1847; and Vermont Historical Society, MSC 139, Papers of Judge Philip Fairbanks: Eaton to Fairbanks, September 10, 22, October 4, 1847, November 17, 1848.

3. John C. Huden, *Development of State School Administration in Vermont* (Burlington: Vermont Historical Society, 1953), pp. 71–73, 125–29, 144–50; David S. Ludlum, *Social Ferment in Vermont, 1791–1850* (New York: Columbia University Press, 1939), pp. 229–36; William Doyle, *The Vermont Political Tradition: And Those Who Helped Make It* (Barre: Northlight Studio Press, 1984), pp. 116–31.

The key reports are *Circular of the State Superintendent of Common Schools to the County Superintendents; An Address to the Teachers of Common Schools in the State of Vermont* (both St. Albans: Messenger, 1845); and *First Annual Report of the State Superintendent of Common Schools, Made to the Legislature October, 1846* (Montpelier: Eastman & Danforth, 1846).

4. Mansion, box A8: Eaton to FB, January n.d., and March 28, 1846. Billings's warm friendship with another leading Whig, DeWitt Clinton Clarke, who was secretary of the Vermont Senate from 1840 to 1851, is shown in [Abby Maria] Hemenway, *Clarke Papers: Mrs. Meech and Her Family* (Burlington: The Author, 1878), pp. 212–15.

5. Rising Lake Morrow, "The Liberty Party in Vermont," *New England Quarterly* 2 (April 1929): 234–48; Reinhard O. Johnson, "The Liberty Party in Vermont, 1840–1848: The Forgotten Abolitionists," *Vermont History* 47 (Fall 1979): 258–75; T. D. Seymour Bassett, "Vermont Politics and the Press in the 1840s," *ibid.* (Summer 1979): 196–213.

6. Mansion, box A8: Horace B. Jones to FB, May 30, 1848.

CHAPTER 4

1. James Davie Butler, *Butleriana, genealogica et biographica; or Genealogical Notes Concerning Mary Butler and Her Descendants, as Well as the Bates, Harris, Sigourney, and Other Families, with which They Have Intermarried* (Albany: Joel Munsell's Sons, 1888), p. 88; Mansion, box A8: Memorandum on Agnes Bates, prepared by Janet Houghton.

2. Simpson, *History of Old Pendleton District with a Genealogy of the Leading Families of the District* (Anderson, S.C.: Oulla Printing), pp. 40–41.

3. Mansion, box A1: Laura Billings to her parents, November 17, 1844, undated 1841 letter, and April 15, 1842. Laura's reports on Pendleton, and quotations concerning her attitude toward her marriage throughout this chapter, are taken from her diary as transcribed in 1985 from the Mansion Archives by Nancy Cooper.

4. Mansion, box A1: OB to FB and Charles Billings, January 18, 1844; Laura to FB, April 18, 1844, n.d. 1845, and April 16, 1846, and to her parents, December n.d. 1846, and January n.d., 1847; OB to FB, October [17], 1846; box A2: SWB to FB, January 24, 29, 1845.

5. Frederick J. Teggart, ed., *Around the Horn to the Sandwich Islands and California, 1845–1850: Being a Personal Record Kept by Chester S. Lyman* (New Haven: Yale University Press, 1924), p. 241. Laura's attitude toward Bezer is clearly revealed in her journal for 1840–1848, especially entries for March 26 and December 31, 1848, in Mansion, box A15.

6. Mansion, box A2: FB to parents, December 28 [1848].

7. Mansion, box A8: Eaton to FB, January 11, 1849.

8. Mansion, box A10: C. P. Peck to FB, January 15, 1849, and file of letters of introduction; University of California, Los Angeles, Halleck, Peachy and Billings Records, collection 2030: box 1, folder 3, Billings's letters of appointment; Library of Congress, Hamilton Fish Papers: Letter Book H, p. 584.

9. The description of the isthmian crossing that follows is drawn from a variety of sources. The most important are a detailed journal kept by Laura from February 1 to 20 and a diary from January 22 to March 31. The originals, together with

typescripts of them, are in Mansion, box A15. Unless otherwise noted, direct quotations come from these two sources, and from Frederick's letter of February 2, 1849, to "dear ones," in box A1.

There are dozens of accounts of the Chagres route, but most describe the crossing at a later date and therefore are not fully reliable for the Simmons-Billings experience. However, nine accounts are clearly applicable, and I have drawn freely on them. Albert Williams's notebook, Huntington Library, San Marino, HM 632: pp. 1–19; and the Thomas Varney Papers, California Historical Society Library, San Francisco, MS 2232, diary: I, February-August, 1849 are extremely helpful. Theodore T. Johnson, *Sights in the Gold Region, and Scenes by the Way* (New York: Baker and Scribners, 1849), is a rich account devoting over a hundred pages to the voyages of the *Crescent City* and *Oregon*. In 1850 James Delavan published, as One Who has Been There, *Notes on California and the Placers: How to Get There and What to Do Afterwards* (New York: H. Long). This account is limited to the *Falcon* and has much that is clearly fictitious in it; it presents an unnamed Frederick as a stock Vermont yokel. William S. McCollum, *California as I Saw It: Pencillings by Way of Its Gold and Gold Diggers! and Incidents of Travel by Land and Water* (Los Gatos, Cal.: Talisman Press, 1960), edited by Dale L. Morgan, has much detail but McCollum was obviously confused between Laura and Mrs. John B. Geary. George Holton Beach, "My Reminiscences," *Quarterly of the Society of California Pioneers* 9 (December 1932): 231–48, is more useful ashore than at sea. In 1879 Albert Williams published *A Pioneer Pastorate and Times: Embodying Contemporary Local Transactions and Events* (San Francisco: Wallace & Hasset), which is a much-altered version of his journal. A partial list of passengers, with some errors, is contained in [C. W. Haskins], *A Pioneer: The Argonauts of California Being the Reminiscences of Scenes and Incidents that Occurred in California in Early Mining Days* (New York: Fords, Howard & Hulabert, 1890), p. 433. Frederick is erroneously listed as H. Billings. A copy of the log of *Falcon* is in the William T. Thompson Records, HD 25, in the San Francisco Maritime National Historical Park Historic Documents Collection.

See also the letters of Dr. George F. Turner to his wife Mary Stuart Turner, in the Minnesota Historical Society, St. Paul, box 2, describing the isthmian crossing and life in the first months in San Francisco, with frequent references to Laura and Frederick. On Turner, consult James Eckman, "The First Contribution to a Medical Journal from Minnesota," *Minnesota Medicine* 34 (October 1951): 996–1006, and Helen Stuart Mackay-Smith Marlatt, ed., *Stuart Letters of Robert and Elizabeth Sullivan Stuart and Their Children 1819–1864* (N.p.: privately printed, 1961). I have also used [J. M. Letts], *California Illustrated: Including a Description of the Panama and Nicaragua Routes, etc.* (New York: W. Holdredge, 1852).

10. Both Johnson, *Sights in the Gold Region*, p. 70, and McCollum, *California as I Saw It*, pp. 92–93, comment on the palanquin.

11. On the *Oregon* see Society of California Pioneers Library, San Francisco: Clipping file, S.S. *Oregon*.

12. The mutiny and arrival are most fully described in John Haskell Kemble, "The Genesis of the Pacific Mail Steamship Company," *California Historical Society Quarterly* 13 (December 1934): 398–401.

13. Beach, "Reminiscences," p. 237.

14. Frederick always believed that Laura contracted her fever on the Chagres crossing. However, she may not have had pernicious malaria but yellow fever and, if so,

given the incubation period, Laura most likely contracted it in Panama City or possibly when the *Oregon* put in for fuel at Taboga or San Blas.

15. Turner Papers: Turner to his wife Mary, June 14, 1849.

16. When Theodore T. Johnson published his memoir in 1849, he asserted that Laura was "the first American lady who died in San Francisco" (p. 215), and this claim was repeated by William McCollum, when his reminiscences were first published in Buffalo in 1850. While the claim may have been true, it depends upon a narrow and legalistic definition of "American," since women from the United States had died in Yerba Buena earlier and non-Anglo women had died in San Francisco after California was annexed and before Laura's arrival. Albert Williams tells of Laura's death without making such a claim for her in *A Farewell Discourse, Delivered Sunday, October 8th, 1854* (San Francisco: B. F. Sterrett, 1854), p. 15. Of course Laura surely was *one* of the very first "American" women to die in San Francisco and quite possibly the first Anglo woman to die after California became an American territory.

CHAPTER 5

1. Minnesota Historical Society, Turner Papers, box 2: List of Lots in San Francisco the Deeds for which are in the hands of Fred Billings of San Francisco who has my Power of Attorney (c 1851).

2. Mansion, box A2: FB to parents, April 27; box A1: Bowie, April 29 with note added by Turner, to OB and SWB, and a file of condolence letters, including Mary Bates to same, July 23, all 1849, and Henry Mower, March 3, 1886, sending a poem by Mrs. L. H. Sigourney, "Laura," said to have been occasioned by Laura's death; box A8: FB to Anna Bates Butler, May 1, 1849; box A9: S. Griffiths Morgan to FB, March 1, 1865; box A8: Chapman to FB, February 6, 1874. Several printed accounts state that Laura died four days after arrival; this is not the case.

3. Franklin's letters to his parents, Mansion, box A2: January 28, 1848 (in error for 1849) and August 29, 1849, and journal entries, especially for February 8, March 19, 21, April 5, August 4, 5, 1849. For a printed account of the voyage see S. Mortimer Collins, *Journal of a Voyage around Cape Horn in 1849, on Board the Good Ship "Magnolia," B. Frank Simmons, Master* (San Francisco: privately printed, 1892).

4. For the paragraphs that follow see New York *Herald*, January 23, 1849; California State Library, Sacramento, Biographical Information File: Bezer Simmons; San Francisco *Alta California*, July 2, 1849, September 27, 28, 1850; Mansion, box A2: Franklin Billings to his parents, September 25, 30, October 30, 1849, September 14, 26, 1850, SWB to FB and Franklin, June 20, 1850 and April 14, 1851; box A10: D. P. Eldredge to Bezer Simmons, November 20, 1849, Dan Murphy to FB, September 31, Richard Hammont to FB, September 8, 1850, and flyer and catalogue of real estate to be sold by order of assignees of Bezer Simmons, and Simmons, Hutchinson & Co., May 1852; Bancroft Library, University of California, Berkeley, John L. Warren Papers: box 14, October 29, 1851; Halleck, Peachy and Billings Papers, Huntington Library, HM 43004: deed for Rancho Llano Seco; California Historical Society, John T. Doyle Collection, MS 760: box 3, folder 30, Geary vs. Billings, 1872, in re Bezer Simmons; Ira B. Cross, *Financing an Empire:*

History of Banking in California (Chicago: S. J. Clarke, 1927), I, 63, 92; Joseph A. McGowan, *History of the Sacramento Valley* (New York: Lewis Historical Publications, 1961), pp. 71–72; Frank Soulé, John H. Gihon, and James Nisbet, *The Annals of San Francisco . . .* (New York: Appleton, 1855), pp. 186, 228, 291, 557–58, 690, 723.

5. Beinecke Library, Yale University, WA uncatalogued manuscripts S-209, John W. Geary Papers: FB to Geary, April 16, 1872; Warren Papers, box 14: Billings, Bolton & Halleck, January 8, May 4, 1852; Bancroft Library, Halleck, Peachy, and Billings Papers, folder 169: Opinion of the court in the case *U.S.* v. *Frederick Billings, et al.*, Assignees of Bezar *(sic)* Simmons; Mansion, box A1: OB to FB, June 21, 1850.

6. FB to OB, printed in the Woodstock *Mercury*, June, 1849; Mansion, box A2: FB to parents, December 19 [1858]; and box A9: Peachy to FB, May 18, 1857.

7. Huntington Library, HM 17477, Journal of B. B. Harris, "Crumbs of '49," p. 23; Zoeth Skinner Eldredge, ed., *History of California* (New York: Century History Co., [1915]), IV, 384–85; John Williamson Palmer, "Pioneer Days in San Francisco," *Century Illustrated Monthly Magazine* 43 n.s. 21 (February 1892): 556–58; Roger W. Lotchin, *San Francisco, 1846–1856: From Hamlet to City* (New York: Oxford University Press, 1974).

8. The story appears in several accounts. It is best told in McDill, "Frederick Billings," p. 21.

9. Society of California Pioneers: Caspar Thomas Hopkins, autobiographical typescript, pp. 87–88. See also "The California Recollections of Caspar T. Hopkins," *California Historical Society Quarterly* 25 (June 1946): 97–120.

10. Mansion, box A2: Franklin Billings to his parents, September 25, 1849; California Historical Society, "List of Lawyers Papers"; Huntington Library, HM 35191: FB to John Townsend, August 1, 1849, and HM 464: Riley to FB, August 13, 1849; University of California, Los Angeles, Halleck, Peachy and Billings Records, box 1: Halleck to FB, May 1, 1849; Soulé et al., *Annals*, pp. 727–28.

11. Mansion, box A2: SWB to FB and Franklin, June 20, 1850.

12. See Kenneth M. Johnson, ed., *San Francisco As It Is: Gleanings from the* Picayune (Georgetown, Cal.: Talisman Press, 1964), p. 59 n. 39. For photographs of the point, see pages 117 and 164. Albert Shumate has provided a short history in *A Visit to Rincon Hill & South Park* (San Francisco: E. Clampus Vitus, 1963).

13. Erasmus Darwin Keyes, *Fifty Years' Observation of Men and Events, Civil and Military* (New York: Charles Scibner's, 1884), pp. 224–25.

14. Mansion, box A10: FB to Riley, October 30, 1849.

15. Keyes, *Fifty Years' Observation*, pp. 293–94. Keyes suggests he took fifty-two men with him and that a crowd of seven thousand had formed, but most contemporary accounts suggest the confrontation was somewhat smaller.

CHAPTER 6

1. Secondary accounts suggest that Sutter felt Peachy had cheated him. In 1849 Sutter had two legal agents, Peachy for San Francisco and Henry Schoolcraft for Sacramento. Sutter was locked in a battle with his son, August, as he tried to regain control of property lost because of bad management, and it was Peachy who drew

up the deed of retransfer by which Sutter received the proceeds of the sale of his Sacramento property. Sutter later said that Peachy "made a fortune out of me in a short time," and one of his biographers, James Peter Zollinger, concludes that Peachy made $80,000 off Sutter and calls him a "shyster," (pp. 363–64). Zollinger's account is unclear and undocumented, however, and he uses dates oddly, stating (p. 279) that Peachy was Sutter's agent in July while observing (p. 298) that Sutter had revoked the power of attorney he had granted Peachy in June. See Zollinger, *Sutter: The Man and His Empire* (New York: Oxford University Press, 1939). Certainly Sutter did sell some of his Sacramento property to Peachy, whether with full understanding of what he was doing or not.

2. For Peachy's story see College of William and Mary, Swem Library, Williamsburg, Virginia: William and Mary College Papers, folders 61 and 146b; Thomas Roderick Dew Papers: Dew to B. F. Dew, February 13, 1844, Peachy to T. R. Dew, January 15, 1847; Nathaniel Beverley Tucker Papers: Peachy to Tucker, January 26, 1844, February 5, 1846, December 10, 1848, and Thomas G. Peachy to Tucker, March 13, 1829, May 30, July 7, 1834; Galt Family Papers: Elizabeth J. Galt to John H. Strobia, undated [1848] description of student protest; College Records, Alumni file: Peachy's records and letter of resignation, September 20, 1848. The dispute is related in Anne W. Chapman, "The College of William and Mary 1849–59: The Memoirs of Silas Totton," unpubl. M.A. thesis, College of William and Mary (1978), pp. 57–59. (Peachy's middle name is given as both Cary and Carey in the sources.)

3. Mansion, box A1: OB to FB and Franklin, February 20, 1821 (*sic* for 1851); box A9: Peachy to FB, September 1852; California State Library, Sacramento, SLC-Miscellaneous, box 17: R. P. Thompson, November 21, 1851, John Mason, January 20, 1852, F. G. Peachy, January 5, 1852, and March n.d., 1853, Philip Hamilton, August 5, 1853, Mother, January 5 [1854?], and Warren McLane, February 4, 1856, all to Peachy; San Francisco *Call*, April 18, 1883; Sacramento *Record-Union*, April 18, 1883.

4. See Mansion, box A9: Halleck to FB, April 4, 1855; Bancroft Library, Bancroft Collection 55501, C-D 326: MS. Sketch for the life of Major Gen. Henry Wager Halleck, c. 1880; Stephen E. Ambrose, *Halleck: Lincoln's Chief of Staff* (Baton Rouge: Louisiana State University Press, 1962).

5. *Alta California*, January 1, 1850; California Historical Society, Samuel Hopkins Willey Letters: FB to Willey, July 10, 1883; Milton H. Shutes, "Henry Wager Halleck, Lincoln's Chief-of-Staff," *California Historical Society Quarterly* 16 (September 1937): 199. One account suggests that HPB must have been founded after September 1, 1850, because the firm's name does not appear in the San Francisco city directory published by Charles P. Kimball at that time, but this simply overlooks the fact that Kimball included the firm under "Omitted Names" in an addendum, and, indeed, that HPB is present out of alphabetical order as well. See Kimball, *The San Francisco City Directory* (San Francisco: Journal of Commerce Press, 1850), pp. 13, 138.

6. There is no full biography of Trenor Park though he well merits one. I have drawn on the Park Papers in the Park-McCullough House in North Bennington, Vermont; Virginia Bell, "Trenor Park: A New Englander in California," *California History* 60 (Summer 1981): 158–71; Eliza Hall Park McCullough, *Within One's Memory* (North Bennington, Vt.: Catamount Press, 1923); and Lisa Matthews,

"The Hall & Park Families in Vermont & California," typescript (1970), Park-McCullough House archives, box 04, folder 6. Halleck's initial opposition to admitting Park is attested to in Mansion, box A9: Halleck to FB, September 8, 12, 18, 27, 1852. Peachy's decision not to admit Park to HPB's land cases before the Commission is in *ibid.*, Peachy to FB, September 18, 1852. For Park at work during this time see the Letter Book for Charles H. S. Williams, Oscar Shafter, and Trenor Park, dated October 17, 1855, and largely written by Park, in the Bancroft Library (C8/153c, I).

7. HPB has been given other names in a variety of memoirs and secondary accounts to sufficient extent to lead to confusion. It would be well to clear the matter here. The firm was Halleck, Peachy and Billings except for the period when Trenor W. Park was also a partner (September 1852 to August 1855). Several memoirs refer to the firm in 1850 as Peachy and Billings, as though Halleck joined later in the year, but this is not the case. Peachy and Billings were working together by June 1849, however, so the term is correct for that year.

8. Kenneth M. Johnson, "Two Mines: Two Books," *Journal of the West*, 17 (January 1978): 36.

9. See Susanna Bryant Dakin, *The Lives of William Hartnell* (Stanford: Stanford University Press, 1949); and Joseph A. Thompson, *El Gran Capitan: José de la Guerra* (Los Angeles: Franciscan Fathers of California, 1961).

10. Mansion, box A1: OB to FB, December 1, 1854; Flora Haines Loughead, ed., *Life, Diary and Letters of Oscar Lovell Shafter* (San Francisco: Blair-Murdock, 1915), pp. 68, 118–19, 123–25, 131, 150–51, 157, 162–64, 169, 175–77; Oscar T. Shuck, ed., *History of the Bench and Bar of California . . .* (Los Angeles: Commercial Printing, 1901), pp. 573–75; and Arthur Quinn, *Broken Shore: The Marin Peninsula, A Perspective on History* (Salt Lake City: Peregrine Smith, 1981), pp. 137–43.

11. Shutes, "Halleck," p. 199. The HPB Papers held by the Bancroft Library provide material on 249 cases while those at UCLA contain files on another 70. Most but not all of the remainder are in the HPB Papers at the Huntington Library.

12. For an analysis of the business cycles see Peter R. Decker, *Fortunes and Failures: White-Collar Mobility in Nineteenth-Century San Francisco* (Cambridge, Mass.: Harvard University Press, 1978), pp. 25–44.

13. See Gates, "Adjudication of Spanish-Mexican Land Claims in California," *Huntington Library Quarterly* 21 (May 1958): 213–36; and "The California Land Act of 1851," *California Historical Quarterly* 50 (December 1971): 399–404.

14. See Oscar Hoffman, *Reports of Land Cases Determined in the United States District Court for the Northern District of California* (San Francisco: Sumner Whitney, 1862), I, *seriatim*.

15. For my description of the legal work of HPB here and in subsequent chapters I have also drawn on three collections in the Bancroft Library: Thomas O. Larkin Papers (C-B 44, items 31, 34, 36, and copybook, VIII); Elbert R. Jones Papers (C-B 464, folders 17, 19, 23); and Bale Family Papers (C-B 746). See also the Henry Wager Halleck Collection (MS 917), which contains several HPB items, and the Frederick Billings Collection (MS 158), in the California Historical Society.

16. Keyes, *From West Point to California* (Oakland: Biobooks, 1950), p. 78. See also Hubert Howe Bancroft, *Chronicles of the Builders of the Commonwealth: Historical Character Study* (San Francisco: History Company, 1891), I, 113.

CHAPTER 7

1. The water lot controversy is set out in several books. One of the most succinct is John P. Young, *San Francisco: A History of the Pacific Coast Metropolis* (Chicago: S. J. Clarke), I, 147–55. The fullest legal history is contemporary: John W. Dwinelle, *The Colonial History of San Francisco* (San Francisco: Towne & Bacon, 1863). The location of most of Billings's lots may be derived from Yale University, Map Room: Map of San Francisco, January 12, 1852, as prepared by William M. Eddy, City Surveyor.

2. Lotchin, *San Francisco, 1846–1856*, p. 174.

3. L. E. Fredman, "Broderick: A Reassessment," *Pacific Historical Review* 30 (February 1961): 40 n. 6.

4. I have not seen an original copy of the *Pile*; it is reprinted in *The Magazine of History with Notes and Queries*, extra no. 190, 47 (no. 2), by William Abbatt (New York, 1933): 92–107.

5. On Borax Lake see Henry K. Mauldin, *History of Lake County*, I: *Clear Lake and Mt. Konocti* (San Francisco: East Wind Printers, 1960), pp. 54–55; and J. Arthur Phillips, *Report on the Property of the California Borax Company* (San Francisco: Towne and Bacon, 1866). Phillips was an Englishman who wrote a public testimonial on the borax lake. Peachy's copy, inscribed to Benjamin Silliman, is in the Beinecke Library at Yale University.

6. See Vincent P. Carosso, *The California Wine Industry: A Study of the Formative Years* (Berkeley: University of California Press, 1951). Simi and Guenoc appear in Robert G. Cowan, *Ranchos of California: A List of Spanish Concessions 1775–1822 and Mexican Grants 1822–1846* (Fresno: Academy Library Guild, 1956). On their sequencing see Bancroft Library, HPB, Notebook relating to California land claims.

7. Watson wrote of his work in California in a diary, a typescript of which is held by Mrs. J. Callander Heminway of West Cornwall, Connecticut. I am most grateful to Mrs. Heminway for allowing me to read the diary. There is substantial Watson correspondence in the Mansion that I draw upon here.

8. Mansion, box A9; there are many letters from "Griff,'" used here *seriatim*, and a large file of correspondence with Carrie Morgan.

9. The Patterson correspondence in the Mansion is extensive, animated, and detailed, and I have drawn heavily on it.

10. Mansion, box A10; D. W. C. Thompson to FB, May 12, 1851, James A. Sutter to HPB, July 23, 1851, Nathan Simmons to FB, June 8, 1852, July 11, 1854, John Skinner to Maynard, Jan. 11, 1853, tax assessment November 16, 1853, A. K. Grim to FB, October 1853.

CHAPTER 8

1. On the approach of its centennial, the Montgomery Block was the subject of a richly anecdotal history by Idwal Jones, *Ark of Empire: San Francisco's Montgomery Block* (New York: Doubleday, 1951). I have drawn heavily on it; on the more restrained lecture prepared by James Disque Hall for E Clampus Vitus in January 1940, and printed as "Halleck's Majestic Folly" in the *Pony Express Courier* 6 (March 1940): 3–7, 16; and a series of articles in the San Francisco *Bulletin*,

May 4, 1912, the San Francisco *Chronicle*, March 1, April 7, 9, and 11, 1947, May 20, 1951, September 8, 1955, June 15, 1956, January 25, February 21, March 5, 6, 25, 1959, and May 10, 1960; and San Francisco *Examiner*, December 8, 1958, February 21, March 3, May 7, June 24, 1959. There are numerous photographs of the block, the earliest in G. R. Fardon, *San Francisco Album: Photographs of the Most Beautiful Views and Public Buildings of San Francisco* (San Francisco: Herre & Bauer, [1856]).

2. Samuel Augustus Pleasant, *Fernando Wood of New York* (New York: Columbia University Press, 1948), p. 27.

3. Cummings, who was not yet twenty-seven, had been trained in England. See San Francisco *Chronicle*, August 30, 1904. One account attributes the Montgomery Block to W. D. M. Howard, co-founder of Mellus & Howard (Louis E. Taber, *California Gold Rush Days* [San Francisco: Stanford University Press, 1936], p. 23). See also Richard Dillon, *Fool's Gold: The Decline and Fall of Captain John Sutter of California* (New York: Coward McCann, 1967), pp. 312–13.

CHAPTER 9

1. These reflections are drawn from Barth's synoptic and speculative book *Instant Cities: Urbanization and the Rise of San Francisco and Denver* (New York: Oxford University Press, 1975).

2. Reproduced in Lotchin, *San Francisco, 1846–1856*, portfolio of photographs following p. 190.

3. Turner Papers: Turner to his wife Mary, June 14, August 25, November 10, May 1850.

4. See *Empire . . .* (2 vols., New York: George P. Putnam, 1850), and Frank Marryat, *Mountains and Molehills, or Recollections of a Burnt Journal* (London: Longman, Brown, Green, and Longmans, 1855), especially Chapter X, and the present writer's introduction to the 1962 reprint edition (Philadelphia: Lippincott).

5. Mansion, box A2: FB to parents, July 4, 20, 1856, and box A1: OB to FB and Franklin, May 7, 1851. See Mary Floyd Williams, *History of the San Francisco Committee of Vigilance of 1851: A Study of Social Control on the California Frontier in the Days of the Gold Rush* (Berkeley: University of California Press, 1921), for a detailed account. A rich selection of documents is printed in her *Papers of the San Francisco Committee of Vigilance* (Berkeley: University of California Press, 1919).

6. Keyes, *Fifty Years' Observation*, p. 232.

7. See T. H. Watkins and R. R. Olmsted, *Mirror of the Dream: An Illustrated History of San Francisco* (San Francisco: Scrimshaw Press, 1976), pp. 37, 62–63, and Williams, *San Francisco Committee of Vigilance*, pp. 179–80.

8. Mansion, box A2: FB to parents, May 30, June 30, 1851; SWB to her sons in California, April 4, June 20, 23, 1851, and box A1: OB to same, July 4, 23, August 8, 23, 1851.

9. See Leroy Armstrong and J. O. Denny, *Financial California: An Historical Review of the Beginnings and Progress of Banking in the State* (San Francisco: Coast Banker, 1916), pp. 33–76; Cross, *Financing an Empire*, I, 58, 67, 71, 175, 181–96; Lotchin, pp. 58–60; Dwight L. Clarke, *William Tecumseh Sherman: Gold*

Rush Banker (San Francisco: California Historical Society, 1969), pp. 107–18, 420–24; and Decker, *Fortunes and Failures*, pp. 36–37.

10. The travails of Adams & Company and all those involved in these several cases are set out with high coloring in the daily and weekly press of the time. A running account may, in particular, be pieced together from the San Francisco *Daily Herald* from February 24 through August 20, 1855, and to a lesser extent from Soulé, Gihon, and Nisbet, *Annals*. Also of considerable value are the scrapbooks of W. G. Cohen (AM 74), volumes 4 and 5, in the Huntington Library, pp. 106–59 and 46–109 respectively; the letterbook of Alfred A. Cohen and others in the Huntington (AM 42); James King's affidavit of March 8, 1855, also in the Huntington (AM 106); the Records of Adams and Company for 1849–1854, MS. group 18 at the California Historical Society; and Mansion, box A1: OB to FB, August 16, 1854, March 17, May 3, August 4, September 17, 1855; box A2: SWB to FB, August 28, 1855; and box A3: R. S. Watson to FB, July 30, December 3, 1854, December 18, 1855, and Park to FB, March 29, 1855.

11. Jones, *Ark of Empire*, p. 136.

12. Sherman to Johnson, June 9, 1856, in Herbert G. Florcken, ed., "The Law and Order View of the San Francisco Vigilance Committee of 1856, Taken from the Correspondence of Governor J. Neely Johnson," *California Historical Society Quarterly* 14 (December 1935): 367. Sherman gives his views in *Memoirs of General William T. Sherman* (New York: Appleton, 1875), I, 118–32. His contemporary account appears in Clarke, *Sherman*, pp. 206–18.

13. San Francisco *Daily Town Talk*, August 9, 1856; Columbia University, Allan Nevins Papers: Memorandum on Frederick Billings and Vigilance Committee, August 1856, with San Francisco *Daily Evening Bulletin*, August 12, 1856.

14. Bancroft, *Builders of the Commonwealth*, I, 108–47.

CHAPTER 10

1. Mansion, box A10: George A. Worn to FB, July 6, 1857, and Latham to FB, undated [1856?]; HPB, UCLA, box 7, folders 1–3; *The Pioneer* 2 (November 1854): 310–12; *ibid.* 3 (January 1855): 45; Bancroft Library, J. L. L. Warren Papers, C-B 418, box 1: FB to F. W. Macondray and others, October 2, 1854, and box 14: October 28, 1851; W. A. Scott: *Oration, Opening of a Seminary for Young Ladies in Oakland* . . . (San Francisco: n.p., 1856); *LeCount & Strong's City Directory for the Year 1854* (San Francisco: LeCount & Strong, 1854); William Taylor, *California Life Illustrated* (New York: Carlton & Porter, 1860), p. 71. John Williamson Palmer says that Billings was "a conspicuous pioneer in all good works": "Pioneer Days in San Francisco," *Century Illustrated Monthly Magazine*, n.s. 21 (February 1892): 541–60.

2. Beach, "My Reminiscences," 236; James Woods, *Recollections of Pioneer Work in California* (San Francisco: Winterburn, 1878); Albert Williams, *A Pioneer Pastorate and Times: Embodying Contemporary Local Transactions and Events* (San Francisco: Bacon, 1882), with Billings's statement, p. [257]. Hewes's statement is in the Bancroft Library, C-D 929, folder 3: March 6, 1891.

3. See Huntington Library, HM 256, Albert Williams Ms: "1st sermon to the new Society Meeting with a view to the Organization of the First Presbyterian Church

of San Francisco"; Huntington Library, DA 2 (216): penciled memorandum [by William Heath Davis] on Williams's school; Huntington Library, HM 236, Diary of James Woods: January 13, 17, 1850; Mansion, box A10: L. C. Bayles to FB, June 3, 1863; box 4: W. C. Anderson to FB, July 22, December 11, 1855, W. K. Waller to FB, September 25, 1856; and S. A. Woodbridge Jr. to FB, December 17, 1860; *The Watchman*, I (April 1, 1850); and William Warren Ferrier, "The Origin and Growth of the Protestant Church on the Pacific Coast," in Charles Sumner Nash and John Wright Buckham, eds., *Religious Progress on the Pacific Slope* (Boston: Pilgrim Press, 1917), p. 58. Ferrier, in *Pioneer Church Beginnings and Educational Movements in California* (Berkeley: n.p., 1927), p. 42, offers evidence that the first Methodist church may have preceded the Presbyterian by about a week.

4. Billings's hostile remark is in Mansion, box A2: FB to parents, April 20, [1856]. On the growth of the relationship see box A8: Bushnell to FB, August 7, 1861, April 17, 1863, and October 11, 1869.

5. Willey, *Thirty Years in California: A Contribution to the History of the State from 1849 to 1879* (San Francisco: A. L. Bancroft, 1879), p. 59; Nicholas C. Polos, "John Swett: The Rincon Period, 1853–1862," *The Pacific Historian* 19 (Summer 1975): 133–49; Billings, *Address Delivered at the Dedication of the School House in the Fifth District of San Francisco, September 23, 1854* (San Francisco: n.p., 1854).

6. Consult William Warren Ferrier, *Origin and Development of the University of California* (Berkeley: Sather Gate Book Shop, 1930), pp. 89–185, which contains frequent references to Billings's role.

   On Billings's work with the board, consult Bancroft Library, Archives of the University of California: Records of the Board of Trustees, October 17, 1855– August 31, 1869; Bancroft Library, E. B. Walsworth Correspondence: Bushnell to Walsworth, October 24, 1856, Durand to Walsworth, October 25, 1859, and FB to Willey, August 14, 1860; Bancroft Library, College of California, Reports made to the Board of Trustees: Report of the Building Committee, February 6, 1863; Mansion, box A10: Walsworth to FB, August 25, 1860, and notices of election to board of College of Oakland, November 19, 1854, and University of California, June 14, 1859; and Willey, *Thirty Years*, p. 59.

7. Mary Bushnell Cheney, comp., *Life and Letters of Horace Bushnell* (New York: Harper & Brothers, 1880), p. 388: Bushnell to Willey, July 10, 1856; Frederick Billings, *Movement for a University in California: A Statement to the Public, by the Trustees of the College of California, and an Appeal, by Dr. Bushnell* (San Francisco: Pacific Publishing, 1857).

8. *Alta California*, October 19, 1859. This episode is related in Ferrier, *Origin and Development*, pp. 196–204.

9. See William Warren Ferrier, *The Story of the Naming of Berkeley* (Berkeley: n.p., 1929); Lillian Davies, "Founders Rock," in Phil McArdle, ed., *Exactly Opposite the Golden Gate: Essays on Berkeley's History, 1845–1945* (Berkeley: Berkeley Historical Society, 1983), pp. 106–8; and Leon L. Foufbourow, "Concerning the Origins of the University of California," *Pacific Historian* 12 (Spring 1968): 18–24. In a typescript document in the Berkeley Public Library, "The Naming of Berkeley" (1927), Ferrier omits Founders Rock and says that Billings first quoted Berkeley at lunch.

10. Ferrier, *Naming of Berkeley*, pp. 105–106; Yale University Archives, Manuscript Group 550: Weir Papers, series IV, box 11, folder 13; and Add. MSS. 1977, box 1, folder 3: Gilman to Weir, April 6, May 28, 1873.

CHAPTER 11

1. On the broad problem of land claims consult Paul W. Gates, "California's Embattled Settlers," *California Historical Quarterly* 61 (June 1962): 99–130; Gates, ed., *Four Persistent Issues: Essays on California's Land Ownership Concentration, Water Deficits, Sub-State Regionalism, and Congressional Leadership* (Berkeley: Institute of Governmental Studies, 1978), pp. 3–30; and Gates, "The Suscol Principle, Preemption and California Latifundia," *Pacific Historical Review* 39 (November 1970): 453–71. These, taken with Gates, "Land Act of 1851," cited previously, make a compelling argument in favor of the federal government's position. Also of interest are William Henry Ellison, *A Self-Governing Dominion: California, 1849–1860* (Berkeley: University of California Press, 1950), pp. 102–36; and Leonard Pitt, *The Decline of the Californios: A Social History of the Spanish-Speaking Californians, 1846–1890* (Berkeley: University of California Press, 1966).
2. *Ibid.*, p. 91. There is substantial correspondence between Halleck and de la Guerra in the archives room of the University of Southern California, Los Angeles, and other de la Guerra papers in the Mission Archives in Santa Barbara and in the Bancroft Library. The Vallejo connection is shown in the Mansion archives, and in the Mariano Guadalupe Vallejo Documentos para la Historia de California, 1780–1875, in the Bancroft Library (C-B, 1–36), especially volumes 13 and 14.
3. The literature on the New Almaden Mine is voluminous. For the general story I have relied most directly on Kenneth M. Johnson, *The New Almaden Quicksilver Mine, with an Account of the Land Claims Involving the Mine and its Role in California History* (Georgetown, Cal.: Talisman Press, 1963). I have also drawn on the documentation prepared by William C. Everhart and Charles W. Snell of the National Park Service in preparation for designating the community a National Historic Landmark, and on the various HPB papers, particularly those at the Huntington Library, which contain over a hundred relevant letters and many more supporting documents; some 350 pages of letters in the Mansion, box A10: California Affairs, *seriatim*; and a recently acquired collection of New Almaden documents held by the Beinecke Library at Yale University, which contains copies of letters from William and Eustace Barron, Reverdy Johnson, John A. Rockwell, and Jeremiah Black. There are also a number of printed records and contemporary pamphlets espousing one side or the other.
4. George Tays, "Captain Andrés Castillero, Diplomat," *California Historical Society Quarterly* 14 (September 1935): 230–68.
5. Henry Winfred Splitter, "Quicksilver at Almaden," *Pacific Historical Review* 26 (February 1957): 33–49.
6. Paul W. Gates, "The Land Business of Thomas O. Larkin," *California Historical Society Quarterly* 54 (Winter 1975): 323–44.
7. *United States* versus *Castillero*, March 10, 1863, 67 U.S. (2 Black) 17, 102.
8. Mansion, box A9: Halleck to FB, September 8, 12, 18, 27, 1852, April 19, May 18, June 3, 1854; box 4: Peachy to FB, September 8, 1852; UCLA, box 3 HPB: folder 2, May 2, 1854. On Walker consult James P. Shenton, *Robert John Walker:*

A Politician from Jackson to Lincoln (New York: Columbia University Press, 1961). Walker's claim on New Almaden is first mentioned in Barron, Forbes & Company to Messrs. Goodhue, June 22, 1850, and again in Eustace Barron to Goodhue, March 20, 1851 (copies in Beinecke Library, New Almaden Papers).

9. Phillis F. Butler, "New Almaden's Casa Grande," *California Historical Quarterly* 54 (Winter 1975): 315–33; Mansion, box A10: FB to Theodore Sedgwick, February 15, 1853.

10. On Hoffman see Kermit L. Hall, "Mere Party and the Magic Mirror: California's First Lower Federal Judicial Appointments," *Hastings Law Journal* 32 (March 1981): 819–37; and Christian G. Fritz, "Judicial Style in California's Federal Admiralty Court: Ogden Hoffman and the First Ten Years, 1851–1861," *Southern California Quarterly* 64 (Fall 1982): 179–203. In 1854 Billings had tried to get Hoffman's salary raised (Trenor Park Papers, Park-McCullough House, box 15: FB to J. S. Morrill, May 3, 1854).

11. UCLA, HPB, folder 7: November 4, 1853; William F. Heintz, *San Francisco's Mayors, 1850–1880* (Woodside, CA.: Gilbert Richards, 1975), pp. 20–25, 31–32; Alston G. Field, "Attorney-General Black and the California Land Claims," *The Pacific Historical Review* 9 (iii/1935): 235–45; William Norwood Brigance, *Jeremiah Sullivan Black: A Defender of the Constitution and the Ten Commandments* (Philadelphia: University of Pennsylvania Press, 1934), pp. 46–55 and 131–44; Jeremiah Black Papers, Library of Congress: April 16, 19, August 1, 1858.

12. Jan Bazant, "Joseph Yves Limantour (1812–1885) y su aventura California-II," *Historia Mexicana* 29 (March 1980): 353–74; Benjamin P. Thomas and Harold M. Hyman, *Stanton: The Life and Times of Lincoln's Secretary of War* (New York: W.W. Norton, 1953), pp. 59–75. Both accept Stanton's view that the New Almaden claim was fraudulent. Kenneth Johnson, *New Almaden*, after examination of the original documents, concludes that the claim was valid. I take Johnson's view, though there is no doubt that, on a narrow technicality rather than a charge of fraud, the government had a good case.

13. *In the Claim of Andres Castillero in the Mine of New Almaden: Deposition of James Alexander Forbes, a Witness Called by the United States, on the 14th of July, 1858* (San Francisco: Commercial Steam Book and Job Printing, 1858), pp. 4, 39–45; *Examination of James Alexander Forbes, a Witness for the United States, on His Recall for Further Cross-Examination by Claimant, on the 30th June, 1858, in the Claim of Andres Castillero, for the Mine of New Almaden* (San Francisco: Whitton, Towne, 1858); and Beinecke Library, New Almaden Papers: John Rockwell, Judah Benjamin, J. J. Crittenden, and Reverdy Johnson to Secretary of State Lewis Cass, November 19, 1858.

14. New Almaden Papers: Black to Cass, April 23, 1859.

15. Billings's mission to Mexico is presented in *Letters from Mexico, 1859*, edited by his daughter Mary M. Billings French (Woodstock, Vt.: Elm Tree Press, 1936). In addition, I have drawn upon Frederick Billings's diary, in the Mansion, box A11, and Robert S. Watson to FB, February 14, 1859, in box A9, together with the John Arnold Rockwell Collection, Huntington Library, box 34: FB to Rockwell, February 18, March 17, April 18, 24, May 19, June 4, 1859.

16. The extensive notes for Billings's research and interviews are in the Mansion, box A10: New Almaden file.

17. New Almaden Papers: Rockwell and Johnson to Cass, March 18, April 19, and Black to Cass, March 28, April 23, all 1858; Rockwell and Johnson, list of wit-

nesses, April 19, John Black to Eustace Barron, June 21, Rockwell and Johnson to Cass, October 4, November 29; Cass to Rockwell and Johnson, October 26, December 7, Buchanan to do., December 17, all 1859; *Diario Oficial* 2 (December 6, 1859): 1–2.

18. UCLA, HPB, box 4: FB to Yale, February 19, March 19, 23, 30, April 3, 4, June 4, 1860; Mansion, box A9: Peachy to FB, March 21, 1860, Mary Watson to FB, December 5, 1859; box A10: Barron to FB, April 19, May 3, 14, June 3, July 1, 14, Milton S. Latham to FB, March 10, April 20, 27, June 5, July 1, all 1860; Milton Slocum Latham Collection, California Historical Society Library, MS 1245, box 3: George Wallace to Latham, March 16, 22, 1860; Rockwell Collection, box 35: FB to Rockwell, March 19, 23, April 13 [1860]; box 36: June 2, 4, 29, July 23, 25, September 11, 29, 1860.

19. Benjamin destroyed his papers. On Benjamin see Robert Douthat Meade, *Judah P. Benjamin, Confederate Statesman* (New York: Oxford University Press, 1943); and Eli N. Evans, *Judah P. Benjamin: The Jewish Confederate* (New York: Free Press, 1988). John W. Wills, in an analysis with which I disagree, examines "Benjamin's Ethical Strategy in the New Almaden Case," *Quarterly Journal of Speech* 50 (October 1964): 259–65. See also Edgar M. Kahn, "Judah Philip Benjamin in California," *California Historical Society Quarterly* 47 (June 1968): 157–73.

20. For the proceedings see *In the United States District Court, Northern District of California: The United States vs. Andres Castillero . . . New Almaden . . .* (San Francisco: Commercial Steam Book and Job Printing, 1860); *. . . on Cross Appeal: Claim for the Mine and Lands of New Almaden, Opening Argument, by Archibald C. Peachy, for the Claimant, Begun on the 8th October, 1860* (San Francisco: Commercial Steam Book and Job Printing, 1860); *Proofs, Indexes & Authorities in Behalf of Claimants* (San Francisco: Commercial Steam Book and Job Printing, 1861); and Rockwell Collection, box 36: Eustace Barron (the younger) to Rockwell, July 25, 1860, William Barron to Rockwell, September 11, 29, October 10, December 20, 1860. Copies of many of the depositions are in the Mansion, box A10: New Almaden file.

21. *Ibid.*: William Barron to FB, January 25, February 10, 1861.

22. Mansion, box A8: William Barron to FB, July 1, 1862.

23. See Milton H. Shutes, "Abraham Lincoln and the New Almaden Mine," *California Historical Quarterly* 55 (March 1936): 3–20; Leonard Ascher, "Lincoln's Administration and the New Almaden Scandal," *Pacific Historical Review* 5 (March 1936): 38–51; and Samuel C. Wiel, *Lincoln's Crisis in the Far West* (San Francisco: privately printed, 1949).

24. *War of the Rebellion: A Compilation of the Official Records of the Union and Confederate Armies* (Washington: Government Printing Office, 1897), series I, L, part 2, 515–18; Howard K. Beale, *The Diary of Edward Bates, 1859–1866* (Washington: Government Printing Office, 1933), p. 354.

25. Billings was so disappointed, indeed, that in later years he was unaccustomedly petty in a matter of little moment for him, except emotionally. Edmund Randolph had died in September 1861. Billings disliked Randolph, who had embarrassed him by making it appear, in an unrelated case in which Billings was summoned as a witness, that Billings was something of a snob and a dandy. To discredit Billings's remarks about a witness, Randolph argued that the lawyer was too elevated to know the man well enough to pass judgment on his truthfulness, comparing Billings to a fish from the banks of Newfoundland that had once fed at the bottom of

the ocean and now, raised high to the surface, had swollen up so that it could not remember what the dark depths had been like. Billings had thought this exceptionally personal and was deeply wounded by it. Nor did he feel Randolph had behaved well in his interrogation of the "distinguished gentlemen" who had come back from Mexico City with him. When His Excellency Joaquín Maria de Castillo y Lanzas, a very prominant Mexican who had been ambassador to several nations, including the United Kingdom, had been called to the stand, Randolph had tried to prove him an imposter, browbeating him mercilessly, asking him for detailed information about how he acquired his English, where he went to school, what offices he had held, testing him on the names of all the presidents of Mexico, putting no fewer than 657 questions to him. When Castillo, asked whether he knew anyone in San Francisco, named an individual, that person too was put on the witness stand and, having identified Castillo, was then forced to prove that he too was not an impostor. Billings believed in thoroughness, and he understood that the government was out to prove conspiracy, but this was a man he had accompanied across Mexico and he was very angry.

So Billings may indeed have been small-minded in the matter of Edmund Randolph, who had been paid only $5,000 for prosecuting the government's case in 1859–1860. His widow, joined by fifty lawyers in San Francisco, petitioned Congress for an allowance of $75,000, a sum more in keeping with the fees other attorneys had received. Congress allowed only $12,000, with Senator George Edmunds of Vermont leading the opposition to the larger sum. Years later, in 1888, Oscar T. Shuck, a lawyer and historian of his profession who had a clear dislike for Billings, wrote that rumor held that Edmunds's opposition was inspired by Billings. Shuck concluded his observation that, "Whether Mr. Billings believed that Mr. Randolf [sic] had not earned his fee, or rather his widow's demand, or whether he permitted a personal grievance to influence his action, is not for me to say." One may only conclude with a Scot's verdict: "Not proven." See Shuck, *Bench and Bar in California, History, Anecdotes, Reminiscences* (San Francisco: Occident Printing, 1889), pp. 264–67. Shuck adds circumstantial detail which is very doubtful, though the case itself is real enough.

26. Yale, *Legal Titles to Mining Claims and Water Rights, in California* . . . (San Francisco: A. Roman).

Chapter 12

1. California Historical Society: Title to Las Mariposas, in MS 1289/1, Rufus A. Lockwood Papers. There is a fine description of the Frémont estate in *Alta California*, August 2–17, 1860.
2. M. Colette Standart, "The Sonora Migration to California, 1848–1856: A Study in Prejudice," *Southern California Quarterly* 58 (Fall 1976): 333–57.
3. Allan Nevins, *Frémont, The West's Greatest Adventurer* . . . (New York: Harper & Brothers, 1928), II, 444–45.
4. The best history of Las Mariposas remains unpublished: Charles Gregory Crampton, "The Opening of the Mariposa Mining Region, 1849–1859, with Particular Reference to the Mexican Land Grant of John Charles Frémont," Ph.D. diss., University of California, Berkeley, 1941. See also Paul W. Gates, "The Frémont-Jones Scramble for California Land Claims," *Southern California Quarterly* 56

(Spring 1975): 13–44; and Raymond F. Wood, *California's Agua Fria: The Early History of Mariposa County* (Fresno: Academy Library Guild, 1954), pp. 18–25.

5. Newell D. Chamberlain, *The Call of Gold: True Tales on the Gold Road to Yosemite* (rev. ed., Fresno: Valley Publishers, 1972), pp. 63–85.

6. *Sketch of the Life of Stephen J. Field, of the U.S. Supreme Court* ([New York]: N.p., 1880), pp. 5–6.

7. Trenor Park Papers: F. F. Fargo to Park, July 31, G. W. Tyler to Park, July 17, August 1, Cyrus Palmer to Park, September 20, all 1862, J. W. Wilson to Park, November 8, 1865, and file of eleven letters of John Conness to Park. Box 13 also contains seven undated FB letters to Park and seventeen business folders on Las Mariposas. In box 15 there are eighty-eight letters, FB to Park, spanning 1859 to 1863. I wish to thank those responsible for the Park-McCullough archive for granting me access to this material. On Billings's purchase see Mansion, box A10: McAllister to Billings, December 27, 1860.

8. Mansion, box A2: FB to parents, July 4, 20, 1856; box A10: Frémont to FB, May 2, and Agreement between Frémont and FB, November 26, 1860. For typical Clark letters see Clark to FB, August 11, 21, 29, 1860. See also Day to FB, October 9, 1860.

9. [Frémont and Billings], *The Mariposa Estate* (London: n.p., 1861). Copies are in the California Historical Society, the Beinecke Library at Yale University, and the Mansion Archives.

CHAPTER 13

1. Bancroft Library, C-D 929: Dictation of Alfred Barstow on Frederick Billings, March 6, 1891.

2. Mansion, box A2: FB to parents, September 26, [1851].

3. On 1852 see Winfield J. Davis, *History of Political Conventions in California, 1849–1892* (Sacramento: California State Library, 1893), pp. 21–23, 37; and Oscar T. Shuck, ed., *History of the Bench and Bar of California . . . Comprehending the Judicial History of the State* (Los Angeles: Commercial Printing, 1901), pp. 478–49. For 1855 see Mansion, box A1: OB to FB, February 16, March 10, 1855. The following summary, as it relates to Frederick Billings, is in part from the *Alta California*, files of which are in the Bancroft Library and the Yale University Library. See also H. Brett Melendy and Benjamin F. Gilbert, *The Governors of California: Peter H. Burnett to Edmund G. Brown* (Georgetown, Cal.: Talisman Press, 1965), pp. 25–165.

4. Mansion, box A2: FB to parents, July 4, 1856.

5. Bancroft Library, C-Y 185: Trenor Park, form letter, September 2, 1856; and C-B 105, folder 94: Republican State Central Committee, November 6, 1856; *ibid.*, Manuel de Jesús Castro Documentos para la historia de California, C-B 52, II: Park to Castro, October 16, 1856; California Historical Society, Barnaby W. Hathaway Collection, MS 959: May 30, 1860.

6. Bancroft Library, C-D 929: Extract of letter to JPB from John T. Doyle, 1890.

7. Ward M. McAfee, "California's House Divided," *Civil War History* 33 (June 1987): 115–30. A more fanciful account appears in Herbert L. Phillips, *Big Wayward*

Girl: An Informal Political History of California (Garden City, N.Y.: Doubleday, 1968), pp. 16–21.

8. On the issue of slavery in California politics see Gerald Stanley, "Slavery and the Origins of the Republican Party in California," *Southern California Quarterly* 60 (Spring 1978): 1–16, and "Racism and the Early Republican Party: The 1856 Presidential Election in California," *Pacific Historical Review* 48 (May 1979): 171–87.

9. Edgar Eugene Robinson, ed., "The Day Journal of Milton S. Latham," *Quarterly of the California Historical Society* 11 (March 1932): 3–28; William Fletcher Thompson Jr., "The Political Career of Milton Slocum Latham of California," unpubl. M.A. thesis, Stanford University, 1952; and Mansion, box A8: Downey to Billings, April 27, December 27, 1860; box A2: Oliver Billings to FB, December 5, 1860.

10. There is no recent study of California in the Civil War. Older works include John J. Earle, "The Sentiment of the People of California with Respect to the Civil War," *American Historical Association Annual Report* 1 (1907): 123–36; Imogene Spaulding, "The Attitude of California to the Civil War," Historical Society of Southern California, *Annual Publications* 9 (1912–1913): 104–31; Joseph Ellison, *California and the Nation, 1850–1869: A Study of the Relations of a Frontier Community with the Federal Government* (Berkeley: University of California Press, 1927); Etta Olive Powell, "Southern Influences in California Politics before 1864," unpubl. M.A. thesis, University of California, Berkeley, 1919; and William Penn Moody, "The Civil War and Reconstruction in California Politics," unpubl. Ph.D. diss., University of California, Los Angeles, 1950.

11. Yale University, Beinecke Library, Thomas Starr King Papers: King to Russell Alger, May 9, 1861.

12. On Starr King see George L. Andreini, "An Historical Evaluation of Thomas Starr King's Public Address with Special Reference to the Retention of California as a Union State," unpubl. Ph.D. diss. (University of Southern California, 1951); William Day Simonds, *Starr King in California* (San Francisco: Paul Elder, 1917); Charles W. Wendte, *Thomas Starr King, Patriot and Preacher* (Boston: Beacon Press, 1921); and Arnold Crompton, *Apostle of Liberty: Starr King in California* (Boston: Beacon Press, 1950). Howard Allen Bridgman equates Billings with Starr King in *New England in the Life of the World: A Record of Adventure and Achievement* (Boston: Pilgrim Press, 1920), p. 192.

13. *The White Hills: Their Legends, Landscape, and Poetry* (Boston: Crosby Nichols, 1860) [actually printed in 1859].

14. Thomas Starr King Collection, California Historical Society, MS 1192: King to N. A. Haven [Ball], n.d.; Mansion box A31: programs and tickets folder, October 30, 1862.

15. Warren Papers: FB to F. W. Macondray, October 2, 1854; Paul W. Gates, ed., *California Ranchos and Farms, 1846–1862: Including the Letters of John Quincy Adams Warren in 1861, Being Largely Devoted to Livestock, Wheat Farming, Fruit Raising, and the Wine Industry* (Madison: State Historical Society of Wisconsin, 1967), pp. x–xii; Mansion, box A23: Speeches made in California to promote the Union cause in the Civil War.

16. The sale was announced in the *Alta California* April 11, 1861. A copy of the sale catalogue is in the Bancroft Library: *Catalogue of a Valuable Law Library, containing Three Thousand Volumes being the Entire Library of Messrs. Halleck,*

Peachy, and Billings (San Francisco: n.p., 1861). The final sale was on April 22 (*Alta California*, April 22, 1861). Halleck delayed his formal resignation. It is in California State Library, SLC—Miscellaneous Manuscripts, box 2: Halleck to Downey, September 18, 1861.

17. Mansion, box A8: King to Bellows, October 22, 1863; box A4: JPB to FB, March 6, 1864; California Historical Society, San Francisco MS 1192: King to Haven, Oct. 4, 1862; Massachusetts Historical Society, Henry W. Bellows Papers: December 3, 1862, June 13, August 8, 1864. California's contribution is attested to in Charles J. Stillé, *History of the United States Sanitary Commission: Being the General Report of Its Work during the War of the Rebellion* (Philadelphia: J.B. Lippincott, 1866), pp. 197–243; William Y. Thompson, "The U.S.Sanitary Commission," *Civil War History* 2 (June 1956): 41–64; and Russell M. Posner, "Thomas Starr King and the Mercy Million," *California Historical Society Quarterly* 43 (December 1964): 291–307.

18. Bancroft, "Dictation of Alfred Barstow," *The Pioneer* 14 (September 15, 1899): 105.

19. *Alta California*, September 15, 1862.

CHAPTER 14

1. Park Papers: FB to Park, January 14; Mansion, box A2: FB to Goldsmith Bailey, December 10, 1860. Though Billings's letter to Park contains a plea that Park tear it up rather than leaving it lying about "as you usually do," Park obviously kept it. On Wright see Mary Lee Spence, "George W. Wright: Politician, Lobbyist, Entrepreneur," *Pacific Historical Review* 58 (August, 1989): 345–59.

2. Mansion, box A10: Frémont to FB, February 19, 1861; Park Papers: FB to Park, February 4, 6, 11, 14, 22, 27, [1861].

3. Mansion, box A8: Benjamin to FB, December 18, 1860.

4. See Robert A. Burchell, "The Loss of a Reputation, or, The Image of California in Britain before 1875," *California Historical Quarterly* 53 (Summer 1974): 115–30.

5. Park Papers: FB to Park, March 18, 21, 28, April 12, 20, 27, May 11, 24, June 15, July 31, August 13, 1861; Mansion, box A2: FB to parents, March 22, April 11, 24, May 8, 18, June 5, 1861; box A9: Elizabeth Patterson to FB, June 12, 1861; Charles Francis Adams Papers, microfilm edition: May 15, June 4, 7, 1861; Sarah Agnes Wallace and Frances Elma Gillespie, eds., *The Journal of Benjamin Moran, 1857–1865* (Chicago: University of Chicago Press, 1948), I, 792.

6. Mansion, box A8: Tennett to FB, June 12, 1861; box A2: FB to parents, July 4, 10, 13, 14, 1861; Richard Monckton Milnes, First Baron Houghton Papers, Trinity College, Cambridge University: letters, March–June, 1861; William L. Clements Library, University of Michigan, Henry Stevens Papers: Frémont to FB, August 17, 1861; Park Papers: Selover to Park, April 18, 1861; James Pope-Hennessy, *Monckton Milnes: The Flight of Youth, 1851–1885* (London: Constable, 1951), p. 45.

7. Henry Shelton Sanford Memorial Library and Museum, Sanford, Florida, Henry S. Sanford Papers: FB to Sanford, July 7 [1861].

8. Mansion, box A2: SWB to "dear sons," April 10, 23, June 7, July 12, 25, 1861, and FB to parents, July 4, 10, 14, 1861; Park Papers: FB to Park, June 26, 1861.

9. [Daniel Lord], *The Effects of Secession upon the Commercial Relations between*

the North and South, and Upon Each Section (London: Henry Stevens, 1861). I have not seen the first printing. The Yale University Library has a copy of the second printing, issued later in the year.

10. Adams, *The Union and the Southern Rebellion: Farewell Address of Mr. Adams to His Constituents Upon His Acceptance of the Mission to England, and Speech of Mr. Everett at Roxbury, in Behalf of the Families of the Volunteers* (London: Henry Stevens, 1861). I wish to thank Dr. Anton Sande, Director of the Western Reserve Historical Society, Cleveland, for providing me with a Xerox copy of this pamphlet.

11. University of Michigan, Clements Library, Henry Stevens Papers: Frank Stevens to Frémont, August 31, 1861, and to FB, August 31, September 21, 25, October 12, 16, 19, 1861, and FB to Frederic Claudet, August 22, 1861; Buckinghamshire County Archives, Aylesbury, Fremantle Papers: box 66, bundle 2; Mansion, box A2: FB to OB, September 19, [1861]; box A10: Sanford to FB, May 5, July 30, [1861], and certificate as bearer of dispatches, July 29, 1861; Sanford Papers: FB to Sanford, October 13, [1861]; Park Papers: FB to Park, May 31, July 31, [1861], Laura Park to Trenor Park, November 27, December 5, 1861; New York Public Library, John Bigelow Papers, box 1: Bigelow to George Perkins Marsh, September 10, 1861, June 4, 1863.

12. Both Frémonts wrote frequently to Billings while he was in Europe and after his return. The originals of these letters are in the Mansion, box A10, and there is a portfolio of typescripts, which omits some letters and portions of letters. I draw on Frémont to FB, July 1, 13, 17, 1861. See also Wyman W. Parker, *Henry Stevens of Vermont: American Rare Book Dealer in London, 1845–1886* (Amsterdam: N. Israel, 1963), pp. 239–58.

13. R. Gordon Wasson, *The Hall Carbine Affair: A Study in Contemporary Folklore* (New York: Pandick Press, 1948).

14. Mansion, box A6: Sanford to FB, August 16, Wilson to FB, August 22, 1861.

15. Mansion, box A10: Frémont to FB, September 11, October 1, 10, November 10, 20, 1861.

16. Mansion, box A2: FB to parents, November 30 [1861], January 21, 1862; box A10: Eustace Barron to FB, November 28, 1861, Jessie Frémont to FB, November 18, [1861]; Park Papers: FB to Park, August 13, December 2, 9, 15, 20, 1861. See Allan Nevins, *Frémont: Pathmarker of the West* (New York: Appleton-Century, 1939), p. 491.

17. Quoted in Ephraim Douglass Adams, *Great Britain and the American Civil War* (London: Longmans, Green, 1925), p. 217.

18. Park Papers: FB to Park, May 18, July 31, 1861, June 25, [1862].

19. Mansion, box A11: FB diary, February 29 (in error for March 1), March 2, 3, 1862.

CHAPTER 15

1. Mansion, box A2: FB to parents, October 4 [1857].

2. Mansion, box A10: certificate of safe passage, November 22, 1861.

3. Mansion, box A2: SWB to sons, April 4, 1851, April 18, 1853, April 2, 1857, Oliver Billings to FB, January 5, February 23, 1861.

4. Judith Walzer Leavitt, *Brought to Bed: Childbearing in America, 1750 to 1950* (New York: Oxford University Press, 1986).

5. Mansion, box A2: Sophia Bailey to FB, May 17, 1859.

6. Mansion, box A8: James Davie Butler to FB, January 16, 1852, Anna Bates Butler to FB, October 17, 1853; box A4: Louisa Cotton Parmly to JPB, February 22, 1869, JPB to FB, June 10, 1881; box A8: Memorandum on Agnes Bates, prepared by Janet Houghton.

7. Mansion, box A2: Oliver Billings to FB, September 16, 1853, Richard Billings to FB, February 13, 1855, March 16, 1858, FB to parents, July 4, 1856; box A8: Mary P. Caswell to JPB, June 6, 1859, April 8, 1862; box A9: Mary Watson to FB, July 2, 1854, Eber Whitmore to FB, August 4, n.d. [1852?], December 5, 1853, May 20, 1854, June 20, August 26 [1855], February 20, May 5, June 5, August 20, December 22, 1856, May 20, 1857, October 25 [1859], Whitmore to SWB, October 28, 1853; box A4: Francis Ehrick, June 24, n.d., and JPB to FB [July, 1865].

8. Lawrence Parmly Brown, "The Greatest Dental Family," *Dental Cosmos* 66 (1923): 251–60, 363–73, 428–91. The Mansion archives, box A31, contains a Parmly genealogical record.

9. Mansion, box A15: Anna Maria Valk Smith Parmly commonplace book, with copied letters and memorandum on Eleazar Parmly by Hannah Burridge, dated October 1881; box A31: typescript "Memoranda of the Billings and Parmly Families for my dear daughter Laura Billings Lee," prepared by JPB, August 20, 1893.

10. Mansion, box A8: Abraham Coles to JPB, April 30, May 1, July 18, December 26, 1861, Caswell to JPB, April 8, 1862; box A9: Mary E. W. Sherwood to JPB [August 20, 1883], box A4: JPB to FB, October 16, 19, 1876.

11. Mansion, box A4: JPB to FB, March 11, 12, 19, 21, 24, 25, 1862, and telegram to D. B. Northrup, April 1, 1862; box A10: Jesse Frémont to FB, May 7 [1862]; box A4: JPB to FB, November 9, 1880; box A8: Nellie Ford to JPB, March 8, 1912, with diary, March 31, April 9, 1862; box A11: FB diary, March 3, 1862; Park Papers: FB to Park, March 20, 29, 1862; box A12: "Memorandum of Events in the Life of J.B.," with a typescript transcription.

12. Mansion, box A28: Eleazar Parmly will.

13. Mansion, box A15: JPB to FB, January 4, 1870, April 5, 1875, May 30, 1880; box A8: Moody to FB, September 28, 1886.

14. *Ibid.*: diary of European trip, Louisa Cotton Parmly, June 4, October 5, 20, 22, November 8, 1862; box A4: JPB to FB, February 24, 1870.

15. Mansion, box A4: JPB to FB, April 13, 1869, February 4, 24, 26, 1873, December 8, 1876; box A9: Sylvia Watson to FB, March 31, and Robert Watson to FB, also March 31, 1862.

16. Mansion, box A4: JPB to FB, April 8, 1871.

17. Mansion, box A31: draft of obituary of Charles Loomis Dana, apparently by FB. The obituary appeared in the Woodstock *Vermont Standard*, July 11, 1884.

18. Mansion, box A6: Parmly to and from his parents, *seriatim*, and to Laura Billings Lee, November 30, 1887; box A2: Bailey to FB, December 26, 1880, and Ethan Hitchcock to FB and JPB, 1883–1884 throughout; box A9: Elizabeth Patterson to JPB, September 23, n.d.; box A6: Parmly to FB, December 15, 1872, August 22, September 16, 1885; box A1: Sarah Towne Billings to JPB, June 26, 1881; box A4: JPB to FB, June 9, 1881; box A10: John MacMullen to JPB, April 26, 28, May 8, July 14, 22, 29, August 12, 19, 1883; box A8: Julius H. Seelye to FB, September

29, 1888; Billings *Weekly Gazette*, May 10, 1888; Parmly Billings Memorial Library, Billings, Montana: typescript memoir of I.D. O'Donnell.

19. Mansion, box A10: John S. White to JPB, October 14, 1886; box A31: undated document [October 17, 1872]; box A6: Ehrick to FB, June 20, 1888; box A31: Announcements and invitations file.

20. Mansion, box A12: "Memorandum of Events in the Life of J.B.," entry for December 1909.

Chapter 16

1. Mansion, box A4: JPB to Eleazar Parmly, May 18, 27, August 17, and JPB to Louisa Cotton Parmly, May 27, June 22, 27, July 27, August 3, 28, September 20, all 1862, January 22, May 29, June 21, 1863; box A10: Selover to Frémont, August 10, 1862, Frémont to FB, June 15, July 19, August n.d., September 9, 10, 20, 21, 1862; Park Papers: Park to FB, March 1, 1862, F. F. Fargo to Park, September 6, 1862, FB to Park, July 6, October 18, 23 (twice), 25, 29.

2. *Ibid.*: FB to Park, October 31 (twice), November 1, 4, 8, 12, December 9, 23, 1862.

3. *Ibid.*: November 14, December 9, 1862; Mansion, box A10: FB to Doyle, November 29, 1862.

4. Bancroft Library, John T. Doyle Papers: Summary of the condition of the Mariposas Estate as to title encumbrances & debts; Francis J. Weber, "John Thomas Doyle: Pious Fund Historiographer," *Southern California Quarterly* 49 (September 1967): 297–99.

5. Doyle Papers: FB to Frémont, November 7, 10, 12, 13, November 25, December 9, 13, 14, 22, 1862, Doyle to FB, December 13, 15, 20, 1862, Frémont to FB, December 20, Doyle to Selover, December 28, all 1862.

6. *Ibid.*: FB to Doyle, December 29, 31, Doyle to FB, December 27, 30, 31, Ketchum to FB, December 29, Brumagim to Frémont, December 29, J. H. Brumagim to Mark Brumagim, December 29, 20, and reply, December 31, all 1862.

7. *Ibid.*: Doyle to FB, January 2, 6, 7, 9 (three times), 10, FB to Ketchum, January 3, Mark Brumagim to J. H. Brumagim, January 2, FB to Doyle, January 6, 8, 9, 13, all 1863.

8. *Ibid.*: FB to Doyle, January 12, 14, 16, 20, February 28, Doyle to FB, January 14, 19, 20, February 18, March 4, 5, all 1863; Mansion, box A3: Ehrick Parmly to JPB, March 14, 1863.

9. Mariposa, Office of County Auditor and Recorder: Deed from Frémont to Ketchum, January 13, and indenture, January 12, 1863.

10. Park Papers: FB to Park, January 12, 1863; Mansion, box A10: FB to Doyle, January 17, 1863.

11. *Ibid.*: FB to Doyle, November 28, 1862.

12. *Ibid.*: FB to Doyle, January 20, March 2, 16, 30, April 8, May 16, 21, 1863; and box A4: JPB to FB, September 8, 1863. Julia's diary for 1863, in box A12, reveals the hectic social pace.

13. Park Papers: FB to Park, June 12, 1863, and copy, Frémont to FB, January 19, 1864.

14. Jane Turner Censer, ed., *The Papers of Frederick Law Olmsted*, IV, *Defending the Union: The Civil War and the U.S. Sanitary Commission, 1861–1863* (Baltimore:

The Johns Hopkins University Press, 1986), 34–61; Laura Wood Roper, *FLO: A Biography of Frederick Law Olmsted* (Baltimore: The Johns Hopkins University Press, 1973), pp. 231–37.

15. Mariposa Estate, *The Manager's General Report, January 1st* (New York: Wm. C. Bryant, 1864), p. 13; [Benjamin M. Stillwell] Chairman of the Late Committee of Investigation, *The Mariposa Estate: Its Past, Present and Future: Comprising the Official Report of J. Ross Browne (U.S. Commissioner) Upon Its Mineral Resources, Transmitted to Congress on the 5th of March, 1868, by the Hon. Hugh McCullough, Secretary of the Treasury of the United States, and Other Documents* . . . (New York: Russells' American Steam Printing House, 1868), pp. 34–41.

16. David McCullough, *The Path between the Seas: The Creation of the Panama Canal, 1870–1914* (New York: Simon and Schuster, 1977), pp. 67, 135–36.

17. Younger, *Kasson*, pp. 315–19; Mansion, box A26, correspondence covering 1882 to 1890 from Daniel Ammen, Hiram Hitchcock, S. E. Kilner, S. L. Phelps, and others; University of California, Los Angeles, Daniel Ammen Papers, Collection 811, box 3: FB to Ammen, October 12, 1887, and box 2, folders 1 through 21, on the Nicaraguan canal route.

18. New York *Tribune*, December 21, 1882, January 5, 1883; Park Papers: box 14: Park obituaries; H. A., "Who Remembers Trenor Park?," *Wall Street Journal*, February 28, 1927; Cecil G. Tilton, *William Chapman Ralston, Courageous Builder* (Boston: Christopher, 1935), p. 252.

CHAPTER 17

1. Mansion, box A1: Charles Jason Billings to OB and SWB, October 22, 1843, September 23, 1846, January 21, 1847, to FB, June 18, 1856.

2. See Mansion, box A1: SB to sons in California, May 2, 1851; box A2: Elizabeth Billings to FB, September 30, 1853, February 14, May 29, 1861, April 5, 1870, and Elizabeth Sprague Billings Allen file, *seriatim*, George Washington Allen to FB, May 3, October 14, 1855, February 10, May 13, August 8, October 21, 1856, December 6, 1858, June 14, June 29, September 8, 1859, May 18, 1860, December 14, 1861, March 23, 1863, March 1, 1864, and to SB, November 18, 1857 and December 6, 1858.

3. Mansion, box A2: Sophia Farwell Billings Bailey to FB, November 1, 1845, December 8, 1851, July 1 [1852], May 17, 1859, July 11, [1861], April 24, July 23 [1862], and to JPB, February 20, 1865; Goldsmith Bailey to FB, June 19, September 3, 1860, February 5, 1861; Elizabeth Sprague Billings Allen to FB, February 14, May 29, 1861; Goldsmith Bailey to FB, December 18, 1860; Memorial, *Mrs. Sophia F. Billings Wallace* (Fitchburg: privately printed, 1895), pp. 4, 7, 8, 18; William A. Emerson, *Fitchburg, Past and Present* (Fitchburg: Blanchard & Brown, 1887), pp. 284–85.

4. Mansion, box A2: Richard Oel Billings to his parents, June 9, 1850, July 27, 1851, January 9, 1853, to FB, June 8, 1851, January 26, 1852, to Elizabeth, December 1 [1851], Franklin to his parents, November 8, 1853, July 15, 1854, and to Goldsmith Bailey, January 16, 1855; box A1: OB to FB and Franklin, July 5, 7, 23, 1851, January 17, February 3, April 2, May 5, June 22, July 1, September 13, 16, 1852, OB to FB, December 31, 1853, January 16, February 15, March 1, May 17,

June 2, 16, October 2, 16, November 16, 1854; box A10: Catlin to FB, July 17, 1854.

5. Mansion, box A2: Nancy Swift Hitch Billings to FB file, *seriatim* from 1859; box A1: OB to FB, March 15, April 1, 1854; box A2: Franklin to FB, October 14, 1853, August 2, 1859, February 25, 1861, March 28, 1862; and box A10: file on the purchase of the Mower house.

6. Mansion, box A2: Oliver Phelps Chandler Billings to FB file, *seriatim*, but especially November 21, 1851, February 11, 1853, April 18, September 16, 1854, April 2, June 30, 1857, December 2, 1858, January 6, December 2, 1859, May 3, August 20, November 21, 1860, and Goldsmith Bailey to FB, June 18, 1860; box A1: OB to sons in California, April 2, August 16, 1852, and University of Vermont Commencement program August 6, 1856.

7. Mansion, box A8: Barstow to Billings, December 6, 1863, Doyle to Billings, November 4, 1863, Conness to Billings, December 27, 1863; box A9: Fanny Williams to JPB, November 15, 1864; Chase Papers, reel 21: Conness to Chase, August 19, October 12, 20, 24, 1863.

8. Edward Lurie, *Louis Agassiz: A Life in Science* (Chicago: University of Chicago Press, 1960), pp. 346–48; Louise Hall Tharp, *Adventurous Alliance: The Story of the Agassiz Family of Boston* (Boston: Little, Brown, 1959), pp. 159–68.

9. The voyage is described in Mansion, box A4: FB journal of trip to California, filed in FB to JPB. On the milch cow see Albert D. Richardson, *Beyond the Mississippi: From the Great River to the Great Ocean . . . 1857–1867* (Hartford: American Publishing Co., 1867), p. 387, and on the barber's chair, Olmsted to his wife, October 31, 1863, quoted in Victoria Post Ranney, ed., *The Papers of Frederick Law Olmsted*, V, *The California Frontier, 1863–1865* (Baltimore: The Johns Hopkins University Press, 1990), 126.

10. Elizabeth Hardwick, ed., *The Selected Letters of William James* (Boston: David R. Godine, 1980), p. 19: James to his parents, April 21, 1865; Henry James, ed., *The Letters of William James* (Boston: Atlantic Monthly Press, 1920), I, 57: James to his parents, [March 30], 1865.

11. M.A. DeWolfe Howe, *Memoirs of the Life and Services of the Rt. Rev. Alonzo Potter, D.D., Ll.D. Bishop of the Protestant Episcopal Church in the Diocese of Pennsylvania* (Philadelphia: J.B. Lippincott, 1871), pp. 352–70.

12. Mansion, box A9: Morgan to FB, March 1, 1865; box A4: FB to JPB, July 2, 4, 1865.

13. Mansion, box A11: FB diary for 1865, entries for September 4–October 9.

14. Mansion, box A4: JPB to FB, August 15, September 13, December 31, 1865, January 26, 30, 1866, FB to Louisa Cotton Parmly, August 13, November 5, 1865.

15. Mansion, box A23: original copy of California Senate Joint Resolution Number Eight.

16. For the diagnosis see Mansion, box A10: Brown-Séquard to FB, April 5, 1867. The thought of epilepsy was not publicly acceptable at the time, and Billings always described his condition as "a congestive attack of the brain." See Barstow Papers: FB to Barstow, February 17, 25, April 17, 30, May 16, 30, July 10, 1867; Yale University, Samuel Bowles Papers: FB to Bowles, June 6, 1871, February 4, March 5, July 4, 15, 1874, and three undated letters; Mansion, A8: Bowles to JPB, January 26, February 2, 5, 1874. Billings discussed his health at length with both Barstow and Bowles. On Billings's physician see André Role, *Le vie étrange d'un grand savant: Le professeur Brown-Séquard, 1817–1894* (Paris: Plon, 1977).

17. Barstow Papers: FB to Barstow, September 11, December 16, 1866, February 17, 1867; Catherine Coffin Phillips, *Cornelius Cole, California Pioneer and United States Senator: A Study in Personality and Achievements Bearing upon the Growth of a Commonwealth* (San Francisco: John Henry Nash, 1929), p. 143.

18. Davis, *Political Conventions*, p. 240. Billings's views on reconstruction are illustrated in his draft notes for speeches given in 1865 (Mansion, box A23).

19. *Ibid.*: Barstow obituary, with annotation and commentary, March 6, 1891; California State Library, SLC—Miscellaneous, box 2: FB to Ball, December 24, 1856; *Constitution, By-Laws and Lists of Members of the Society of California Pioneers, Since Its Organization* (San Francisco: The Society, 1874), p. 49.

20. Mansion, box A4: JPB to FB, November 15, 1865; box A12: "Memorandum of Events in the Life of J.B.," November 1863.

21. Badé Institute and Library, Pacific School of Religion, Berkeley, Samuel Hopkins Willey Papers: FB to Willey, March 17, 1874.

22. Van Dyke was minister of New York's Brick Presbyterian Church, which Frederick and Julia attended when at their Madison Avenue house. The line is from van Dyke's poem "America for Me."

CHAPTER 18

1. Carl I. Wheat, "A Sketch of the Life of Theodore D. Judah," *California Historical Society Quarterly* 4 (September 1925): 220–24; Gilbert H. Kneiss, *Bonanza Railroads* (4th ed., Stanford: Stanford University Press, 1954), pp. 3–48; SVRR, *Articles of Association, Report of Board of Directors, and Report of Chief Engineers,* in the Bancroft Library's Pamphlets on California Railroads, volume VII; Mansion, box A1: OB to FB, February 1, 1855.

2. *Alta California*, August 9, November 15, 1866; Decker, *Fortunes and Failures*, pp. 156–58.

3. The reference is, of course, to the argument made by Leo Marx in *The Machine in the Garden: Technology and the Pastoral Ideal in America* (New York: Oxford University Press, 1964).

4. These changes are summarized in Lewis D. Stilwell, "Migration from Vermont (1776–1860)," *Proceedings of the Vermont Historical Society,* n.s., V (ii/1937), 214–29; and Meeks, *Time and Change in Vermont,* pp. 106–39.

5. On railroads in Vermont, see T. D. Seymour Bassett, "500 Miles of Trouble and Excitement: Vermont Railroads, 1848–1861," *Vermont History* 48 (Summer 1981): 133–54.

6. Nevins, *Frémont*, pp. 586–91; Cardinal Goodwin, *John Charles Frémont: An Examination of His Career* (Stanford: Stanford University Press, 1930), pp. 240–56; Alan W. Farley, "Samuel Hallett and the Union Pacific Railway Company in Kansas," *The Kansas Historical Quarterly* 25 (Spring 1959): 1–16.

7. On the Atlantic & Pacific under Frémont, see H. Craig Miner, *The St. Louis-San Francisco Transcontinental Railroad: The Thirty-fifth Parallel Project, 1853–1890* (Lawrence: University Press of Kansas, 1972), pp. 43–59. Billings came back on to the board and was a member of the executive committee in 1870. He resigned a second time in 1872.

8. Barstow Papers: FB to Barstow, December 16, 1866, October 28, 29, 1867, February 26, April 28, June 3, July 14, October 8, 23, 1868, July 7, August 14, 1869.

9. William Howard Tucker, *History of Hartford, Vermont . . .* (Burlington: Free Press Association, 1889), pp. 161–65; Edgar T. Mead, Jr., *Over the Hills to Woodstock: The Saga of the Woodstock Railroad* (Brattleboro, Vt.: Stephen Greene Press, 1967), pp. 5–20; Mead, "Roots of the Woodstock Railway," The Railway and Locomotive Historical Society *Bulletin* 120 (April 1969): 48–59.

10. Mansion, box A10: inventory of property, Grannis to FB, June 15, 1869, D. B. Northup to FB, March 4 and undated, 1869.

CHAPTER 19

1. There are two histories of the Northern Pacific Railroad. The first, by Eugene V. Smalley, *History of the Northern Pacific Railroad* (New York: G.P. Putnam's, 1883), appeared in conjunction with the completion of the line and is uncritical on events after 1873. On Smalley as publicist see Mike Jordan's unpublished paper, "Eugene Smalley: Boomer and Believer in the Great Northwest," read at the Centennial West Symposium in Billings, Montana, June 24, 1989. The second, *The History of the Northern Pacific Railroad* (Fairfield, Wash.: Ye Galleon Press, 1980), by Louis Tuck Renz, is a labor of love that draws heavily on the *Annual Reports*. Renz also wrote *The Construction of the Northern Pacific Railroad Main Line during the Years 1870 to 1888* (Walla Walla, Wash.: The Author, 1973), and compiled a useful statistical compendium, *Northern Pacific Data Tables* (Walla Walla, Wash.: The Author, 1978).

2. William S. Greever, "A Comparison of Railroad Land-Grant Policies," *Agricultural History* 25 (April 1951): 83–90.

3. Lloyd Mercer, "Building Ahead of Demand: Some Evidence for the Land Grant Railroads," *Journal of Economic History* 34 (June 1974): 492–500.

4. Smalley, *Northern Pacific*, pp. 141–47. On Canfield, consult *Life of Thomas Hawley Canfield: His Early Efforts to Open a Route for the Transportation of the Products of the West to New England . . .* (Burlington: Donohue & Henneberry, 1889).

5. Eugene V. Smalley, comp., *Northern Pacific Railroad Book of Reference, for the Use of the Directors and Officers of the Company* (New York: E. Wells Sackett & Rankin, 1883), p. 60.

6. There are two essential studies of Cooke: Henrietta M. Larson, *Jay Cooke, Private Banker* (Cambridge, Mass.: Harvard University Press, 1936); and Ellis Paxson Oberholtzer, *Jay Cooke, Financier of the Civil War* (Philadelphia: George W. Jacobs, 1907). On Minnesota, see John L. Harnsberger, *Jay Cooke and Minnesota: The Formative Years of the Northern Pacific Railroad, 1868–1873* (New York: Arno Press, 1981).

7. Canfield, *Life*, pp. 33–34; Mansion, box A5: FB to JPB, August 6, 1871.

8. Mansion, box A10: W. B. Ogden to FB, March 18, 1872.

9. Billings was also buying land in Superior. See Mansion, boxes A24 and A26: deeds and other records.

10. Historical Society of Pennsylvania, Philadelphia, Jay Cooke Papers: Cooke to FB, May 17, 1871.

11. *Ibid.*: Billings to Cooke, February 22, 1871, March 19, 1872, Cooke to Billings, March 4, 23, [1872].

12. Mansion, box A4: JPB to FB, November 27, 1871, November 23, 1872, April 24,

1875; box A12: JPB diary, January 22, 25, 1872, "Memorandum of Events in the Life of J.B.," entries for March and April 1870, November 1871, December 1874.

13. Cooke Papers: Cooke to Billings, February 19, March 1, April 27 [1872]; Billings to Cooke, April 29, 1872; Oberholtzer, *Cooke*, II, 325–26.

14. *Ibid.*, 340–41.

15. For a succinct summary, see Vincent P. Carosso, *Investment Banking in America: A History* (Cambridge, Mass.: Harvard University Press, 1970), pp. 23–25. On the role of Cooke & Company in the Panic of 1873, see Mark W. Summers, *Railroads, Reconstruction, and the Gospel of Prosperity* (Princeton: Princeton University Press, 1984), pp. 268–98.

16. See James B. Hedges, "The Colonization Work of the Northern Pacific Railroad," *Mississippi Valley Historical Review* 13 (December 1926): 311–42; two unpublished M.A. theses from the University of Minnesota, Siegfried Mickleson, "Promotional Activities of the Northern Pacific Railroad's Land and Immigration Departments, 1870 to 1902: A Case Study of Commercial Propaganda in the Nineteenth Century" (1940), and Harold Fern Peterson, "Railroads and the Settlement of Minnesota, 1862–1880" (1927); Lars Lgungmark, *For Sale—Minnesota: Organized Promotion of Scandinavian Immigration, 1866–1873*, Studia Historica Gothoburgensia, XIII (Stockholm: Akademiförlaget, 1971); and William M. Bomash, *Guide to a Microfilm Edition of the Northern Pacific Land Department Records* (St. Paul: Minnesota Historical Society, 1983), pp. 6–13.

17. Benjamin Wade's few papers are in the Library of Congress. His work for the Northern Pacific is discussed briefly in H. L. Trefousse, *Benjamin Franklin Wade: Radical Republican from Ohio* (New York: Twayne, 1963), pp. 315–16.

18. Cooke Papers: FB to Cooke, February 22, 1871.

19. *Ibid.*: FB to Cooke, February 23, March 13; *Letter of John S. Loomis to Frederick Billings, Chairman of Land Committee, February 20, 1871, Recommending a Plan for the Organization and Operation of Land Department, Including Plans for Promoting Emigration and Land Settlement* (New York: n.p., 1871).

20. *Land Department of the Northern Pacific Railroad Company, Bureau of Immigration for Soldiers and Sailors, George B. Hibbard, Superintendent of Immigration* (New York: Northern Pacific, 1873); Minnesota Historical Society Research Center, St. Paul, Northern Pacific Papers: George Sheppard to FB, March 16, 18, April 1, 8, 17, 30, May 4, 7, 13, 17, 18, 23, 24, June 1, 4, 14, 15, 21, July 10, September 13, October 10, 12, 1872; Nettleton to FB, March 27, October 7, 27, 1871, July 10, 16, 19, 26, 1872; FB to Sheppard, February 24, 1872; FB to Hibbard, April 1, 1872; Hibbard to FB, July 11, 1872; Wright to FB, June 19, 1872.

21. Mansion: Browne to FB, March 12, 1871, a letter tipped inside a copy of Browne's *Resources of the Pacific Slope: A Statistical and Descriptive Summary* . . . (New York: Appleton, 1869); Cooke Papers: Cooke to FB, October 2 [1871], April 7 [1872]; FB to Cooke, March 11, 13, April 2, 13, May 17, 1872; Oberholtzer, II, 311.

22. Cooke Papers: FB to Cooke, March 14, 1871; Mansion, box A8: Bowles to FB, May 13, 1874.

23. Cooke Papers: Cooke to FB, March 1, 4 [1872], FB to Cooke, February 23, 1872; University of Oregon Library, Eugene, John C. Ainsworth Papers: May 14, November 21, 1873, May 31, 1878; Yale University, Beinecke Library, Elwood Evans Scrapbooks, "Pioneer Days," I, 8–9; Mansion, box A8: Colfax to FB, May 10, 11, 16, 1872, January 18, 1881.

24. Cooke Papers: Cooke to Roberts, March 1, 1872.
25. Quoted in Larson, *Cooke*, pp. 363–64.
26. John L. Harnsberger, "Land Speculation, Promotion and Failure: The Northern Pacific Railroad, 1870–1873," *Journal of the West* 9 (January 1970): 33–45.
27. For the most part, the papers of the Lake Superior and Puget Sound Company, held by the Minnesota Historical Society, proved unrewarding for the purposes of this study, but they await closer research. There are sixty-six volumes and eleven folders, and I examined only the Land Records and the early volumes of the general records.
28. On Wright see Thomas Porter Harney, *Charles Barstow Wright, 1822–1898: A Builder of the Northern Pacific Railroad and of the City of Tacoma, Washington* (N.p., 1956).
29. *History of the Pacific Northwest: Oregon and Washington* . . . (Portland, Ore.: North Pacific History Company, 1889), II, 48.
30. Bancroft Library, E. W. Nolan Northern Pacific Collection, XIV: FB to Sprague, December 13, 15, 16, 17, 1872, October 7, 1875, George B. Hibbard to FB, February 18, 23, March 9, 22, [May 20?], July 20, 1875, FB to William K. Mendenhall, January 25, 1875; Ainsworth Papers: FB to Ainsworth, March 21, 1874, April 27, 1875.
31. *Ibid.*: FB to Ainsworth, May 14, June 10, 1873; Mansion, box A8: Olmsted to FB, September 18, 1873; Olmsted Papers: FB to Olmsted, September 19, Wright to Olmsted, September 27, and Olmsted to FB, September 25, all 1873. Olmsted's original drawing for Tacoma is held by the Washington State Historical Society in that city.
32. Cooke Papers: FB to Cooke, January 16, 1873.
33. Northern Pacific Papers: Josiah P. Tustin to FB, August 23, 29, September 6, 8, 11, 16, 27, November 4, 9; Nettleton to FB, March 13, April 27, July 3, August 23, 28; Sheppard to FB, September 18, October 10; L. H. Tenney to FB, January 20, November 14, December 25; H. L. Turner to FB, March 9, 16; FB to Tenney, February 10, all 1872; John L. Harnsberger and Robert P. Wilkins, "New Yeovil, Minnesota: A Northern Pacific Colony in 1873," *Arizona and the West* 12 (Spring 1970), 5–22.
34. William Watts Folwell, *A History of Minnesota* (reprint ed., St. Paul: Minnesota Historical Society, 1969), III, 362–88.

CHAPTER 20

1. Smalley, *Northern Pacific*, pp. 204–05; Alvin F. Harlow, *Steelways of New England* (New York: Creative Age Press, 1946), pp. 424–25; Agnes C. Laut, *The Romance of the Rails* (New York: Robert M. McBride, 1929), II, 453–57.
2. The figure of $5.40 an acre is adduced by Ljungmark, *For Sale*, p. 206 and n. 55.
3. On Power see his "Bits of History Connected with the Early Days of the Northern Pacific Railway and the Organization of Its Land Department," *Collections of the State Historical Society of North Dakota* 3 (1910): 337–49; and Stanley N. Murray, "James B. Power: The Second President of North Dakota Agricultural College," *North Dakota Quarterly* 42 (Autumn 1974): 5–7.
4. Barnes County Historical Society, Valley City, N.D.: Barnes County Land Records, transfers of property November 16, 1878, filings of property March 13, 1879.

5. North Dakota Institute for Regional Studies, North Dakota State University, Fargo, James B. Power Papers: Power to Nettleton, October 30, to L. J. Cravath, November 27, and to George D. Hubbard, December 14, 1873, to George A. Johnson, December 3, 1875, and to FB, April 10, 25, June 19, 20 [1874]. At one point Billings apparently contemplated buying the Dalrymple Farm for himself: see FB to Power, March 29, 1877, March 14, 1878. See also Harold E. Briggs, "The Great Dakota Boom, 1879 to 1886," *North Dakota Historical Quarterly* 4 (January 1930): 78–108, and "Early Bonanza Farming in the Red River Valley of the North," *Agricultural History* 6 (January 1932): 26–37.

6. Ross Ralph Cotroneo, *The History of the Northern Pacific Land Grant, 1900–1952* (New York: Arno Press, 1979), pp. 44–46.

7. The words are Smalley's, *Northern Pacific*, p. 208.

8. Mansion, box A11: FB diary, July 31–August 13, 1875; Smalley, *op cit.*, pp. 209–10.

9. Portland *Oregonian*, September 9, 1883.

10. Mansion, box A11: FB diary, September 26, 1877.

11. Bancroft Library, C-D 929, folder 4, Letters in re Northern Pacific Railroad: Stark to FB, n.d., and FB to Stark, December 11, 1875.

CHAPTER 21

1. Mansion, box A4: JPB to FB, November 18, continued on a letter dated November 15, 1865.

2. Ainsworth Papers: FB to Ainsworth, November 11, 1878.

3. Stuart Daggett, *Railroad Reorganization* (Boston: Houghton, Mifflin, 1908), pp. 270–71.

4. Ainsworth Papers: FB to Ainsworth, July 22, December 3, 1878, March 29, 1879; Northern Pacific Papers, Secretary, Corporate Documents: Final notice to holders of bonds, by FB, December 18, 1878.

5. See Albro Martin, "James J. Hill," in Robert L. Frey, ed., *Encyclopedia of American Business History and Biography: Railroads in the Nineteenth Century* (New York: Facts on File, 1988), pp. 7–20.

6. James J. Hill Reference Library, St. Paul, James J. Hill Papers: Hill to Stephen, October 18, 1878, Hill to Charles Elliot Furness, October 31, 1878, Stephen to FB, December 9, 1878, Hill to Smith, July 26, 1879. See Martin, *James J. Hill and the Opening of the Northwest* (New York: Oxford University Press, 1976).

7. Ainsworth Papers: FB to Ainsworth, July 22, 25, August 17, October 29, November 9, 19, 25, 1878.

8. *Ibid.*: FB to Ainsworth, October 2, 12, 29, 1878. Italics in original.

9. *Ibid.*: FB to Ainsworth, November 9, 11, 1878, January 6, 1879.

10. *Ibid.*: March 25, April 10, 1879; Mansion, box A11: FB diary, April 17, 1879.

11. Ainsworth Papers: FB to Ainsworth, May 24, 1879; Mansion, box A11: FB diary, May 6, 24, 1879; box A4: JPB to FB, May 24, 26, 31, 1879.

12. Mitchell, *The Northern Pacific and Portland, Salt Lake and South Pass Railroads* (Washington: n.p., 1877), and *The Columbia River: Its Freedom Must be Established* (Washington: n.p., 1878); Harold Hathaway Dunham, *Government Handout: A Study in the Administration of the Public Lands, 1875–1891* (New York: Columbia University, 1941), pp. 90–92.

13. Northern Pacific Papers: FB to Henry E. Johnston, December 19, 1879.

14. Billings, *Official Statement to the New York Stock Exchange, of the Conditions of the Northern Pacific Railroad Company, July 1st 1879,* printed copy in Beinecke Library, Yale University. On the Utah Northern see Merrill Beal, "The Story of the Utah Northern Railroad," *Idaho Yesterdays* 1 (Spring 1957): 3–10, and (Summer 1957): 16–23.

15. Smalley, *Northern Pacific,* p. 232.

16. George Bliss to Pasco du Pré Grenfell, August 10, 27, 1880, as cited in Dolores Greenberg, *Financiers and Railroads, 1869–1889: A Study of Morton, Bliss & Company* (Newark, Delaware: University of Delaware Press, 1980), p. 251, n. 12. I have searched the Bliss Papers, which are held by the New-York Historical Society, but they provide no information.

17. Mansion, box A11: FB diary, September 13, October 19, 1880. Smalley, *Northern Pacific* p. 232, says that Winslow, Lanier asked Billings to call in, and that only after weeks of deliberation did Drexel, Morgan join the emerging syndicate. Vincent P. Carosso, in *The Morgans: Private International Bankers, 1854–1913* (Cambridge, Mass.: Harvard University Press, 1987), p. 249, says that Billings approached Drexel, Morgan in October. Of course, one should not make too much of these personal relationships, since in business they often did not count for much.

18. Greenberg, *Financiers and Railroads,* p. 139; Thomas C. Cochran, *Railroad Leaders, 1845–1890: The Business Mind in Action* (Cambridge, Mass.: Harvard University Press, 1953), pp. 48–51, 70, 102.

19. Northern Pacific Papers: Billings to Cheney, November 18, 20, 1880; quoted, with a slight alteration, in Cochran, *Railroad Leaders,* pp. 51 and 258.

20. Lewis Corey, *The House of Morgan: A Social Biography of the Masters of Money* (New York: G. Howard Watt, 1930), p. 114.

21. Indiana University, Lilly Library, Hugh McCulloch Papers: FB to McCulloch, November 20, 1881; Renz, *Northern Pacific,* p. 74.

22. Mansion, box A11: FB diary, July 27, August 23, September 13, 16, 20, 23, 25, 29, October 31, November 5, 9, 10, 13, 17–20, 1880.

23. *Ibid.:* November 19, 21–27, 1880.

24. Minnesota Historical Society, Alexander Ramsey Papers, microfilm roll 25: FB to Ramsey, November 27, and reply, 1880.

25. Mansion, box A11: FB diary, November 24, December 1, 2, 4, 9, 16, 20, 30, 1880; Rutherford B. Hayes Presidential Center, Fremont, Ohio: Hayes to FB, November 29, 1880.

26. *Compendium of History and Biography of North Dakota . . .* (Chicago: Geo. A. Ogle, 1900), pp. 138–40.

27. Helena *Daily Herald,* October 16, 1879, in the Mansion, box A19: NP Scrapbooks, I.

28. Olympia *Transcript,* September 27, 1879. Frederick (or more likely Julia for him) kept scrapbooks of news clippings concerning the construction of the Northern Pacific, and that story can be told in close detail from these scrapbooks. There are seven volumes in the Mansion, box A19.

29. Northern Pacific Railroad, *Annual Report and Proceedings of the Stockholders at Their Annual Meeting, September 29th, 1880* (New York: Evening Post Presses, 1880), pp. 10–11.

30. Northern Pacific Papers: FB to James F. Wilson, April 18, 1880. The bare outline

of the trip is given in Mansion, box A11: FB diary, May 30–June 15, 1879. See also Ainsworth Papers: FB to Ainsworth, June 18, October 9, 1879. William Bloss, an old California friend, encountered Billings on the journey and rode part way with him: see Chester McArthur Destler, ed., "Diary of a Journey into the Valleys of the Red River of the North and the Upper Missouri, 1879," *Mississippi Valley Historical Review* 33 (December 1946): 432.

31. Ainsworth Papers: FB to Ainsworth, October 17, 1879.

32. Ceremonial spikes always appear as silver or gold in the contemporary literature. The term is symbolic: usually the spikes were of iron, gilded for the occasion. The silver spike in the library of the Mansion is a reproduction of the one used later, in 1883.

33. Mansion, box A11: FB diary, September 3, 1880, December 3, 8, 1880; Power Papers: Power to FB, September 6, 1879, November 4, December 13, 1880, February 15 [1881]; Ainsworth Papers: FB to Ainsworth, October 9, 1879.

34. This is the opinion of Stanley N. Murray in "Railroads and the Agricultural Development of the Red River Valley of the North, 1870–1890," *Agricultural History* 31 (October 1957): 63.

35. Ainsworth Papers: FB to Ainsworth, June 17, October 9, November 7, 1879.

36. Wisconsin State Historical Society, Joseph Thompson Dodge Papers: FB to Dodge, January 30, 1880.

37. The Staff of the Montana Historical Society, *Not in Precious Metals Alone: A Manuscript History of Montana* (Helena: Montana Historical Society Press, 1976), p. 114.

38. Robert M. McLane, *Reminiscences, 1827–1897* (Wilmington, Del: Scholarly Resources, 1972), p. 121; Northern Pacific Papers: FB to Joseph Potts, January 15, 1880.

39. Billings, *The Northern Pacific Railroad: Its History and Equitable Rights. Address of Frederick Billings, President, to the Committee on Pacific Railroads of the House of Representatives, April 15th, 1880.* There is a copy in the Beinecke Library, Yale University.

40. *Legal Rights of the Company, and the Equitable Rights of Its Stockholders—Opinion of Hon. Jeremiah S. Black* (York, Pa.: N.p., 1880). There is a copy in the Beinecke Library.

41. Mansion, box A11: FB diary, March 30, April 1, 3, 7, 9–17, 19–23, May 2, 6, 8–10, 13, 18, 22–24, 28, 29, 1880; Northern Pacific Papers: FB to Black, April 23, 1880.

42. Mansion, box A23: Carl Schurz to FB, November 4, Ramsey to FB, November 24, 1880; box A8: Bowles to FB, November 24, 1880; box A19: Boston *Globe*, August 15, 1884.

43. *Annual Report of the Commissioner of Indian Affairs to the Secretary of the Interior for the Year 1886* (Washington: Government Printing Office, 1886), pp. 317–19.

44. Yale University, Beinecke Library, WA MS.: "Incidents in the Life of Rev. C. A. Huntington Written at the Request of His Children, 1898," typescript, pp. 83–111.

45. On the Utah & Northern in Montana see Merrill D. Beal, *Intermountain Railroads, Standard and Narrow Gauge* (Caldwell, Idaho: Caxton, 1962), pp. 73–109. On Potts consult Clark C. Spence, *Territorial Politics and Government in Montana, 1864–89* (Urbana: University of Illinois Press, 1975), pp. 74–149. Nettleton

and Billings had thought of Potts as a "semi-representative" (see Northern Pacific Papers: Nettleton to Potts, August 24, 1872).

46. Loring Benson Priest, *Uncle Sam's Stepchildren: The Reformation of United States Indian Policy, 1865–1887* (New Brunswick, N.J.: Rutgers University Press, 1947), p. 230.

47. H. Craig Miner, *The Corporation and the Indian: Tribal Sovereignty and Industrial Civilization in Indian Territory, 1865–1907* (Columbia: University of Missouri Press, 1976); and a perceptive essay by Geoffrey Blodgett, "A New Look at the American Gilded Age," *Historical Reflections* 1 (Winter 1974): 231–46.

48. Library of Congress, Carl Schurz Papers: FB to Schurz, September 28, December 1, 1880; Dodge Papers: FB to Dodge, October 22, 1880, Anderson to Dodge, January 8, March 17, April 12, 30, May 19, June 26, July 13, August 10, 1881.

49. *Ibid.*: FB to Anderson, May 7, 12, Williams to Kirkwood, May 24, and A. Bell, Acting Secretary of the Interior, to FB, May 27, all 1881. On Schurz and Indian affairs see Hans L. Trefousse, *Carl Schurz: A Biography* (Knoxville: University of Tennessee Press, 1982), pp. 242–47. On Kirkwood consult Dan Elbert Clark, *Samuel Jordan Kirkwood* (Iowa City: State Historical Society of Iowa, 1917), pp. 364–66.

50. Northern Pacific Papers: FB to John Douglas, August 2, FB to Doane, November 24, FB to Ainsworth, November 24, FB to Johnston, December 19, all 1879, FB to Power, February 9, October 27, November 19, December 13, 1880, FB to Sprague, June 28, 1880, FB to Villard, September 16, 1880, February 21, 1881, FB to P. N. Laird, January 10, 1880. These letters are all quoted in Cochran, *Railroad Leaders*, pp. 250–60.

51. Yale University Library, Herman Haupt Papers, R.G. 269, box 5: FB to Haupt, March 29, 1881, and FB's public announcement of appointment, April 1, 1881, with Haupt's circular, October 27, 1881.

CHAPTER 22

1. This is evident from the *Annual Reports*. Typical was a series of meetings on the day before Christmas, the day after Christmas 1888, and early in January 1889, when a special committee on which Billings served, to deal with problems relating to the East Side Line in St. Paul, met at inconvenient times yet he was nonetheless present at all meetings (Northern Pacific Papers: Special Committee Records, pp. 159–73).

2. Exceptions include Enoch A. Bryan, *Orient Meets Occident* (Pullman, Wash.: Students' Book Corporation, 1936); Clinton A. Snowden, *History of Washington: The Rise and Progress of an American State* (4 vols., New York: Century History Company, 1909); and Cy Warman, *The Story of the Railroad* (New York: D. Appleton, 1898). See also Chapter 20, n. 1.

3. *Memoirs of Henry Villard, Journalist and Financier, 1835–1900* (Boston: Houghton, Mifflin, 1904), II, 309, n. 1. Though the title page does not say so, the *Memoirs* were edited by Villard's daughter, Fanny Garrison Villard. Villard's granddaughter, Katherine Neilley Villard, in "Villard: The Years of Fortune," unpubl. Ph.D. diss., University of Arkansas, 1988, p. iii, calls the dictated chapter in Smalley's book "an amusingly camouflaged account" and notes that in his *Memoirs* Villard did not get even the date right for the Northern Pacific's 1883 celebration.

4. On close examination even the sketch on Villard in the *DAB* 19: 273–75, by James Blaine Hedges, is a summary of the Villard-Smalley memoir. Hedges' *Henry Villard and the Railways of the Northwest* (New Haven: Yale University Press, 1930) is a sound history but it is not a biography. See also Villard's *The Early History of Transportation in Oregon*, edited by his son Oswald Garrison Villard (Eugene: University of Oregon Press, 1944), which Villard wrote in 1899, and which was first printed in serial form in the Portland *Oregonian* in 1926, and Benjamin MacLean Whitesmith, *Henry Villard and the Development of Oregon*, University of Oregon Thesis Series No. 14 ([Eugene]: Works Progress Administration, 1940). Dietrich G. Buss, *Henry Villard: A Study of Transatlantic Investments and Interests, 1870–1895* (New York: Arno Press, 1978), is strong on financial matters. Nonetheless, Villard still awaits his biographer.

5. Harvard Business School, Baker Library, Henry Villard Papers, box 26: Kindred to Villard, April 23, 1881, with copies of Billings to Power, January 18, March 18, August 28, September 7, 11, 23, October 12, 28, 1878. Billings's brothers apparently purchased land which they had deeded to him, thus obscuring precisely how much property he had bought. Billings asked Power to find him choice lands, and then purchased three or more sections but, having lost his land book, could not recall what he had paid per acre. Clearly Billings felt worried about using Power as his personal agent in the matter. None of these acts were illegal, and Billings can scarcely have intended to hide for long his purchase of land through his brothers, so that the precise use Villard intended to make of these letters is unclear. Billings's letter to Homer Sargent about instructing Power to dismiss Kindred is in Cochran, p. 259: December 13, 1880.

6. John T. Ganoe, "The History of the Oregon and California Railroad," *Oregon Historical Society Quarterly* 25 (September 1924): 236–83, and (December 1924): 330–52.

7. Villard, *Memoirs*, II, 283; Dorothy R. Adler, *British Investment in American Railways, 1834–1898* (Charlottesville: University Press of Virginia, 1970), p. 191, n. 7.

8. On the OSN see P. W. Gillette, "A Brief History of the Oregon Steam Navigation Company," *Oregon Historical Quarterly* 5 (June 1904): 120–32. The fullest exploration of the relationship between the Villard companies and the Northern Pacific is Peter J. Lewty, *To the Columbia Gateway: The Oregon Railway and the Northern Pacific, 1879–1884* (Pullman, Wash.: Washington State University Press, 1987).

9. Hedges, *Henry Villard*, pp. 21–29.

10. Northern Pacific Papers: FB to Sprague, June 17, 1879, FB to Cheney, February 24, 1880, both quoted in Cochran, pp. 251, 255.

11. Ainsworth Papers: FB to Ainsworth, November 18, 24, 1879, February 12, March 29, April 7, 1880. On the route see Eugene V. Smalley, *Completion of the Cascade Division Northern Pacific R.R. and a Description of Tacoma, the Western Terminus* (Saint Paul: Pioneer Press, 1887).

12. Villard Papers, box 26: FB to Villard, December 26, 1879, January 28, February 9, March 5, 14, 19, 1880; Mansion, box A11: FB diary, January 30, February 7, 19, 23, 24, 28, March 1, 15, 30, 31, 1880.

13. Northern Pacific Papers: FB to Sprague, February 19, March 19, 1880, and FB to Villard, February 23, 1880, both quoted in Cochran, pp. 255–56.

14. University of Washington Library, Seattle, Oregon Improvement Company Pa-

pers: FB to Villard, October 8, 1880; Mansion, box A11: FB diary, April 5, 24, June 25, July 15–17, August 6, September 17, October 18, 29, 1880; Hedges, *Villard*, pp. 71–73. For the traffic agreement see Villard Papers, box 58: October 20, 1880. On gauge see Ainsworth Papers: FB to Ainsworth, December 18, 1879.

15. Hedges, *Villard*, pp. 76–79. The blind pool document of February 11, 1881, is in the Villard Papers, box 75. Villard put in $900,000.

16. Villard, *Memoirs;* Hedges, *DAB*, XIX, 274; Cochran, pp. 49–51; Matthew Josephson, *The Robber Barons: The Great American Capitalists, 1861–1901* (New York: Harcourt, Brace, 1934), p. 242.

17. Cochran, pp. 480–81: Villard to C. E. Bretherton, December 14, 1880, January 24, February 19, 1881, Villard to William Endicott, February 4, 1881.

18. Mansion, box A11: FB diary, January 15, 1881, and February to May, *seriatim;* box A4: JPB to FB, May 7, 14, 18, June 10, 1881.

19. Cochran, p. 259: FB to Potts, February 8, 1881.

20. *Ibid.,* p. 260: FB to Sprague, February 21, 1881.

21. Villard Papers, box 26: FB to Villard, March 14, 1881.

22. Mansion, box A11: FB diary, March 13, 14, 1881.

23. *Ibid.:* March 17, 1881.

24. See Buss, *Henry Villard*, p. 126.

25. Hedges, *Villard*, p. 79; Villard Papers, box 33: Billings's affidavit.

26. For an example see New York *Tribune*, May 14, 1881.

27. Mansion, box A11: FB diary, March 21–May 9, 1881; Villard papers: FB to Villard, May 18, 1881; St. Paul *Pioneer-Press*, April 3, 1881.

28. Mansion, box A11: FB diary, April 19, 20, 1881.

29. Various figures have been given for Billings's sale, but the documents support $4,500,000. See Villard Papers: Drexel, Morgan to Villard, May 21, 28, 1881, Tracy, Olmstead & Tracy to Villard, May 25, 1881, FB to Villard, June 10 [1881]; Mansion, box A10: Contract between FB and Villard, May 21, 1881. Though the agreement called for full payment from Villard by September 1, Billings gave him some leeway, and he received the final million in October. (See Mansion, Miscellaneous Papers: Drexel, Morgan to FB, October 1, 1881.)

30. Ainsworth described this maneuver in his journal, written for his children. It is printed as "Steamboating on the Columbia River: The Pioneer Journal of Captain J. C. Ainsworth, Covering the Development of the Oregon Steam Navigation Company," edited by Henry H. and Lucetta A. Clifford, in the Los Angeles Corral *Westerners Brand Book* 9 (1967): 117–68. See especially p. 156.

31. Mansion, box A11: FB diary, May 10–June 12, 1881; Villard Papers: FB to Villard, June 10 [1881].

32. Dodge Papers: Dodge to Anderson, April 5, Anderson to Dodge, March 17, June 24, 26, 27, 1881.

33. Mansion, box A11: FB diary, June 13–September 27, 1881; box A8: Eliza Hunt to JPB, May 2, 1881; Villard Papers: FB to Villard, July 8, December 21, 1881.

CHAPTER 23

1. There are many eyewitness accounts of the excursion, which was to be expected given the number of writers who went on it. The description here is drawn from Thomas J. McCormack, ed., *Memoirs of Gustave Koerner, 1809–1896: Life-sketches*

Written at the Suggestion of His Children (Cedar Rapids: Torch Press, 1909), II, 671–722; H. A. Rattermann, *Gustav Körner, Deutsch-amerikanischer Jurist, Staatsmann, Diplomat und Geschichtschreiber* (Cincinnati: Verfassers, 1902), pp. 373–74; Gustav Schwab, *Die Festreise zur Eröffnung der Nord-Pacific Eisenbahn im September 1883* (New York: n.p., 1884), and Nicholas Mohr, *Ein Streifzug durch den Nordwestern Amerikas: Festfahrt zur Northern Pacific-Bahn im Herbste 1883* (Berlin: R. Oppenheim, 1884); Harney, *Wright*, pp. 37–38; Edmund Ions, *James Bryce and American Democracy, 1870–1922* (London: Macmillan, 1968), pp. 110–13; George A. Bruffey, *Eighty-one Years in the West* (Butte: Butte Miner Company, 1925), pp. 92–94; Joseph H. Hanson, *Grand Opening of the Northern Pacific Railway . . .* (St. Paul: Brown & Treacy, 1883); New York Public Library Annex, U.S. Railroads Pamphlet Collection, 22: *Northern Pacific Railroad, Opening Excursions: Programme for Guests from England, and for Other Guests Traveling in Private Cars* (N.p.: n.p., 1883); and *Northern Pacific R.R. Opening Excursion, List of Guests, Portland, Oregon, September 1883* (Portland: Himes, [1883]; *Frank Leslie's Illustrated Newspaper*, September 22, 1883; *Harper's Weekly* 27 (September 22, 1883): 594–95; and Mansion, box A11: FB diary, August 30–September 9, 1883.

2. New-York Historical Society, Henry Villard Manuscripts: Bryce to Villard, August 7, 1883; Oxford University, Bodleian Library, James Bryce Papers: Villard to Bryce, July 25, 1883.

3. Library of Congress, Chester Arthur Papers, microfilm: reel 2, Roscoe Conkling to Arthur, July 23, Williams to Arthur, July 26, Arthur to [Villard], July 28, 1883.

4. George N. Hillman, *Driving the Golden Spike: Story of a Great Achievement* (St. Paul, 1883), records Billings's speech.

5. Carroll Van West, "Coulson and the Clark's Fork Bottom: The Economic Structure of a Pre-Railroad Community, 1874–1881," *Montana, The Magazine of Western History* 35 (Autumn 1985): 42–55; Waldo O. Kliewer, "The Foundation of Billings, Montana," *Pacific Northwest Quarterly* 31 (July 1940): 255–83; Albro Martin, *If Not Billings, What?: An Essay on the Naming of Montana's Biggest City* (N.p.: Burlington Northern, 1982). In 1982 the city unveiled a statue of Frederick Billings, the work of sculptor Mike Capser and the gift of the Billings Exchange Club. I wish to thank Mr. Capser for discussing the public reception of his work with me.

6. See John C. Hudson, *Plains Country Towns* (Minneapolis: University of Minnesota Press, 1985), pp. 70–71.

7. Billings *Herald*, October 15, 16, 1883, April 26, September 7, 1884; Billings *Gazette*, June 30, 1927; Billings *Post*, November 22, 1883; Mansion, box A23: Shuart to FB, December 4, 1882. Shuart's relationship to Frederick Billings is revealed throughout a thick file of his letters in the Mansion Archives, and is well recounted by Carroll Van West in chapters six through eight of his forthcoming book on Clark's Fork Bottom. I wish to thank him for sharing his manuscript with me.

8. On Hauser see William L. Lang and Rex C. Myers, *Montana: Our Land & People* (Boulder, Colo.: Pruett, 1979), pp. 45–48; and John William Hakola, "Samuel T. Hauser and the Economic Development of Montana," unpubl, Ph.D. diss., Indiana University, 1961.

9. Montana Historical Society, Helena, Samuel T. Hauser Papers: Billings to Hauser, August 30, 1880, May 11, October 22, 1886, July 22, 1887, June 21, September

28, 1888, July 17, 1889, Kilner to Hauser, September 4, 1886, Bailey to Hauser, October 8, 1894; Dodge Papers: FB to Dodge, October 29, 1880; Mansion, box A23: Shuart to FB, December 3, 1884, with affadavits; Villard Papers, Baker Library: Billings to Villard, June 13, 1883; Great Falls *Tribune*, April 24, 1886, August 13, 1887; Billings *Herald*, September 7, 1884, March 31, April 22, 24, 28, May 23, June 14, 20, October 31, 1886, February 14, 28, April 15, 25, September 18, 22, 29, 1887; Western Heritage Center, Billings: Assessment Book, 1887.

10. Montana State Historical Society, Helena, Photo Archive, holds ten views that Haynes made of Billings, Montana on the day of Frederick's talk. Haynes faithfully recorded all the Villard Arches, as they were called, but it was too dark for a photograph of the driving of the last spike, which had to be reenacted.

11. Mansion, box A23: Kilner to Franklin N. Billings, September 9, 1883, and "Address by Ex-President Frederick Billings on the Opening of the Northern Pacific Railroad"; Villard, *Memoirs*, II, 309–311; Katherine Villard Seckinger, ed., "The Great Railroad Celebration, 1883: A Narrative by Francis Jackson Garrison," *Montana, The Magazine of Western History* 33 (Summer 1983): 12–23; George A. Bruffey, "The Last Spike," in Joseph Kinsey Howard, ed., *Montana Margins: A State Anthology* (New Haven: Yale University Press, 1946), pp. 459–62.

12. Renz, *Construction of the Northern Pacific*, pp. 37, 42; Bette E. Meyer and Barbara Kubik, *Ainsworth: A Railroad Town* (Fairfield, Wash.: Ye Galleon Press, 1983), pp. 20–23. I wish to thank the Franklin County Historical Society for supplying an original photograph of the *Billings*.

13. Lewty, *Columbia Gateway*, pp. 105–11.

14. Mansion, box A11: FB diary, November 30, December 1, 12, 15, 1883.

15. For an analysis of Villard's decline see Grodinsky, *Railway Strategy*, pp. 202–208.

16. Mansion, box A11: FB diary, January, 1884, *seriatim*.

17. Hauser Papers: FB to Hauser, January 3, 1885, January 25, June 21, 1888, Bailey to Hauser, March 22, 1888, Kilner to Hauser, July 18, 1889; Mansion, box A10: J. Fletcher Toomer to S. Chapman, June 16, 1899.

18. There is no history of most of these companies. An exception is *A Century of Progress: History of The Delaware and Hudson Company, 1823–1923* (Albany: J.B. Lyon, 1925). Frederick Billings is listed as a "manager" from April 28, 1886, until his death in 1890; his brother Oliver succeeded him in 1892 (p. 724). Billings was also a director of the Nicaragua Canal Company, the Overland Stage Company, Manhattan Life Insurance, the National Bank, the Manhattan Savings Institute, and the Presbyterian Hospital of New York City.

19. Minnesota Historical Society, St. Paul: the Papers of the American Steel Barge Company, microfilm, 1885–1896, include letters to and from Billings about routine matters. The original documents are in the Baker Library at Harvard University.

20. Oregon Historical Society, Miscellaneous Manuscripts: Billings to Matthew P. Deady, January 14, 1884.

21. James A. Ward, *Railroads and the Character of America, 1820–1887* (Knoxville: University of Tennessee Press, 1986), p. 163.

Chapter 24

1. Doyle, *Vermont Political Tradition*, pp. 451–50.

2. Lyman Jay Gould and Samuel B. Hand, "A View from the Mountain: Perspectives

of Vermont's Political Geography," in H. Nicholas Muller II and Samuel B. Hand, eds., *In a State of Nature: Readings in Vermont History* (Montpelier: Vermont Historical Society, 1982), pp. 186–90; Charles T. Morrissey, *Vermont: A Bicentennial History* (New York: W. W. Norton, 1981), p. 42.

3. Mansion, box A10: Petitioners to Grant, January 13, Huntington to FB, January 13, 21, Dennison to Huntington, January 20, Flanders to FB, February 24, Flanders to Grant, February 24, all 1869; box A8: Field to FB, February 14, 1869.

4. The conventions of 1872 are discussed in newspaper clippings taken from a wide range of Vermont papers in Mansion, box A20: Scrapbook on Vermont Politics.

5. Mansion, box A23: Bellows Falls speech; Woodstock *Standard*, May 2, Rutland *Daily Herald*, May 2, 3, Burlington *Free Press*, May 3, 1872; Harvard University Library: broadside, Vermont State Republican Convention, May 1 [1872]; University of Vermont, Burlington, Benedict Family Papers, II: FB to Benedict, April 18, 1871.

6. Mansion, box A23: speech, "Mr. President and Gentlemen of the Convention."

7. These events were widely reported throughout the Vermont press. The fullest account is in the Burlington *Free Press*, June 27, 1872. On the Stewart-Billings-Benedict triangle see Benedict Family Papers, II: FB to Benedict, March 29, June 24, 1872, Stewart to Benedict, February 22, May 8, 10, 27, June 20, 21, 1872.

8. *Free Press*, December 5, 1879.

9. Yale University Library, William Maxwell Evarts papers, series II, box 8: FB to Evarts, January 23, April 5, 1880; Mansion A4: JPB to FB, June 5, 1880; box A11: FB diary, February 23, 25, March 4, 5, May 30–June 5, 1880; box A23: speech, "Chicago Convention, 1880."

10. Mansion, box A11: FB diary, June 6–9, 1880; *Proceedings of the Republican National Convention Held at Chicago, Illinois, Wednesday, Thursday, Friday, Saturday, Monday and Tuesday, June 2nd, 3d, 4th, 5th, 7th and 8th, 1880* (Chicago: Jno. B. Jeffery, 1881), pp. 190–91, 297, Appendix p. 29; Henry J. Clancy, *The Presidential Election of 1880* (Chicago: Loyola University Press, 1958), pp. 82–121; Walter Hill Crockett, *Vermont: The Green Mountain State* (New York: Century History Company, 1921), IV, 110–11.

11. Library of Congress, James A. Garfield Papers: reel 79, FB to Garfield, December 1, 1880; Bancroft Library, C-D 929, folder 3, document 5: "Impromptu Speech delivered in the Town Hall in Woodstock, Vt., on the occasion of a Republican Rally, September 5th, '80 by Mr. Frederick Billings"; Mansion, box A11: FB diary, August 3–5, September 6, 7, November 2, 3 1880; Curtis, Jennison, Lieberman, *Billings*, p. 85.

12. Mansion, box A4: JPB to FB, June 4, 7, 1884; box A8: Evarts to FB, August 1, 1884; box A1: C. T. Billings to JPB, November 1, 1884; Crockett, *Vermont*, IV, 135–42.

13. Mansion, box A8: Grant to FB, May 31, 1885, Harrison to FB, October 25, 1886.

CHAPTER 25

1. Marsh, *Man and Nature; or, Physical Geography as Modified by Human Action* (New York: Charles Scribner, 1864), pp. 2–3, in a marked copy in Billings's private library. The 1874 edition had a revised title: *The Earth as Modified by Human Action*.

2. Hope Nash, *Royalton, Vermont* (Royalton: Royalton Historical Society, 1975), p. 106.

3. Mansion, box A2: SWB to her sons in California, n.d. 1851 and July 2, also 1851. On tourism in the White Mountains see Diane M. Kostecke, ed, *Franconia Notch* (Concord, N.H.: Society for the Protection of New Hampshire Forests, 1975), pp. 4–20. The contemporary appreciation of landscape is discussed in Roderick Nash, *Wilderness and the American Mind* (New Haven: Yale University Press, 1967).

4. California Historical Society, Thomas T. Seward Papers: no. 332.

5. Mansion, box A2: FB to parents, June 18, 1856.

6. Cheney, comp., *Horace Bushnell*, pp. 377–439. See King, *Mountaineering in the Sierra Nevada* (New York: James R. Osgood, 1871); Whitney, *The Yosemite Book: A Description of the Yosemite Valley and the Adjacent Region of the Sierra Nevada, and of the Big Trees of California* (New York: Julius Bien, 1868); and Edmund A. Schofield, "John Muir's Yankee Friends and Mentors: The New England Connection," *The Pacific Historian* 29 (Fall 1985): 65–89.

7. The cave is described in the Sacramento *Bee*, June 5, 1966, and the journey to it is discussed in *Hutchings' California Magazine* 5 (December 1860): 219–27.

8. Mansion, box A9: Fanny Williams to JPB, September 2, 1866.

9. The early history of Yosemite is told in Margaret Sanborn, *Yosemite: Its Discovery, Its Wonders and Its People* (New York: Random House, 1981). On the Calaveras groves see Rodney Sydes Ellsworth, "Discovery of the Big Trees of California, 1833–1852," unpubl. M.A. thesis in Forestry, University of California, Berkeley, 1933. Billings's visit is mentioned in Mansion, box A2: FB to parents, May 4 [1856]. See also the excellent analysis of Alfred Runte, *Yosemite: The Embattled Wilderness* (Lincoln: University of Nebraska Press, 1990).

10. King's book, *A Vacation Among the Sierras, Yosemite in 1860* (San Francisco: The Book Club of California, 1962), edited by John A. Hussey, is based on his articles in the Boston *Transcript* for December 1, 15, 31, 1860, and January 2, 19, 26, February 2, 9, 1861. His sermon on Yosemite is recorded in Boston Public Library, Thomas Starr King Papers, MS 280: "A Visit to the Yo-Semite: Its Religious Lessons." For a sensitive use of King see Walter L. Creese, *The Crowning of the American landscape: Eight Great Spaces and Their Buildings* (Princeton: Princeton University Press, 1985), of which I have paraphrased pp. 108–09.

11. On Watkins see Peter E. Palmquist, *Carleton E. Watkins: Photographer of the American West* (Albuquerque: University of New Mexico Press, 1983). A memoir by Charles B. Turrill, "An Early California Photographer: C. E. Watkins," *News Notes of California Libraries* 13 (January 1918): 29–37, remains useful. *California History* has devoted a full issue to Watkins: 57 (Fall 1978). Specific to Yosemite are Thomas Weston Fels, *Carleton Watkins, Photographer* (Williamstown, Mass.: Sterling and Francine Clark Art Institute, 1983), and George Dimock, *Exploiting the View: Photographs of Yosemite & Mariposa by Carleton Watkins* (North Bennington, Vt.: Park-McCullough House [1983]).

12. I have followed Palmquist, *Watkins*. A different view on some of the facts appears in Mary Pauline Grenbeaux, "Before Yosemite Art Gallery: Watkins' Early Career," *California History* 57 (Fall 1978): 220–29, and "The Early Yosemite Photographs of Carleton E. Watkins," unpubl. M.A. thesis, University of California, Davis, 1977.

13. Peter E. Palmquist, "Carleton E. Watkins's Oldest Surviving Landscape Photograph," *History of Photography* 5 (July 1981): 223–24.

14. Dimock, *Exploiting the View*, pp. 13, 16, states that Park commissioned Watkins's Mariposa series and that the Yosemite photographs of 1861 were made with Park's help and in his presence. Fels, *Watkins*, p. 20, more generally credits both Park and Billings with having commissioned the Mariposa series.

15. Sets of the Mariposa series are to be found in the Park-McCullough House in North Bennington, Vermont; the Bancroft Library at the University of California, Berkeley; and the California Historical Society Library in San Francisco. The Billings Mansion and the Yale University Library, among others, have sets of Watkins's stereographs of Mariposa. The photograph of the Guadalupe silver mine is in the National Archives in Washington.

16. Sets of New Almaden photographs of 1863 are in the Billings Mansion, the Bancroft Library, and The Huntington Library.

17. See Nanette Sexton, "Carleton E. Watkins: Pioneer California Photographer (1829–1916): A Study in the Evolution of Photographic Style During the First Decade of Wet Plate Photography," unpubl. Ph.D. diss., Harvard University, 1982.

18. Paul Hickman, "Art, Information, and Evidence: Early landscape Photographs of the Yosemite Region," *Exposure* 22 (Spring 1984): 26–29.

19. I thank Mrs. Jack Ellis Haynes of Bozeman, Montana, for giving me permission to examine the F. Jay Haynes Papers (officially the Yellowstone National Park and Manuscripts Collection) at Montana State University. There are two published volumes of Haynes's photographs: Montana Historical Society, *F. Jay Haynes, Photographer* (Helena: Montana Historical Society Press, 1981), and Edward W. Nolan, *Northern Pacific Views: The Railroad Photography by F. Jay Haynes, 1876–1905* (Helena: Montana Historical Society Press, 1983). The Haynes Photograph Collection, held by the Montana Historical Society in Helena, consists of 23,500 photographs. See also Charles S. Fee, "Pioneering the Northwest and the Yellowstone National Park," *Haynes' Bulletin* (May 1923), pp. [2–4].

20. Strahorn, *The Resources of Montana Territory and Attractions of Yellowstone National Park . . .* (Helena: The Legislature, 1879), and *Montana and Yellowstone National Park* (Kansas City: Ramsey, Millet & Hudson, 1881).

21. Winser, *The Yellowstone National Park: A Manual for Tourists* (New York: Putnam's, 1883), and *The Great Northwest: A Guide-Book and Itinerary for the Use of Tourists and Travellers over the Lines of the Northern Pacific Railroad . . .* (New York: Putnam's, 1883).

22. Alfred Runte, *Trains of Discovery: Western Railroads and the National Parks* (Flagstaff, Ariz.: Northland Press, 1984). A less favorable view of the railroad's intentions for the park emerges from Aubrey L. Haines, *The Yellowstone Story: A History of Our First National Park* (2 vols., Boulder: Yellowstone Library and Museum Association in cooperation with Colorado Associated University Press, 1977). On Kelley's advocacy, see Oberholtzer, II, 236. Richard A. Bartlett, in *Nature's Yellowstone* (Albuquerque: University of New Mexico Press, 1974), concludes that "the inspiration for the creation of the Yellowstone National park [should be credited] to officials of the Northern Pacific Railroad" (p. 208). See also the convincing analysis by John F. Reiger, *American Sportsmen and the Origins of Conservation* (rev. ed., Norman: University of Oklahoma Press, 1986), which emphasizes the alliance between hunters, anglers, and the national parks movement.

23. Cooke Papers: FB to Cooke, March 22, 1872.

24. On Muir see, in particular, Stephen Fox, *The American Conservation Movement: John Muir and His Legacy* (Boston: Little, Brown, 1981), and Frederick Turner,

Rediscovering America: John Muir in His Time and Ours (San Francisco: Sierra Club Books, 1985).

25. Mansion, box A8: Barnum to JPB, December 26, 1883, January 11, February 15, 1884; *The Nation,* September 6, 1883. The chief exponent of Glacier National Park was George Bird Grinnell, senior editor and publisher of *Forest and Stream.*

26. On an earlier time, see Richard J. Orsi, " 'Wilderness Saint' and ''Robber Baron': The Anomalous Partnership of John Muir and the Southern Pacific Company for Preservation of Yosemite National Park," *The Pacific Historian* 29 (Summer 1985): 136–56.

27. Alfred Runte, *National Parks: The American Experience* (2nd ed., Lincoln: University of Nebraska Press, 1987), pp. 48–105; Bill McMillon, *The Old Lodges & Hotels of Our National Parks* (South Bend: Icarus Press, 1983), pp. 17–55, 87–107; Keith R. Widder, "Mackinac National Park, 1875–1895," *Reports in Mackinac History and Archaeology,* no. 4 (Williamstown, Mich.: Mackinac Island State Park Commission, 1975). The Arkansas Hot Springs had been designated a national park in 1832, but the title was an empty one quite unrelated to the concept as applied to Yellowstone.

28. Renz, *Northern Pacific,* pp. 112–13.

29. Yale University, Beinecke Library, Carroll T. Hobert Papers: Oakes to Hobart, September 15, 1882.

30. National Archives, Records of the Office of the Secretary of the Interior, Patents and Miscellaneous Division, Letters Received: Pat Conger (park superintendant) to Teller, November 6, 1882, and Hatch to David W. Wear (park superintendent), October 31, 1885; Hobart Papers: William Starbuck to Villard, June 17, 1881, Hobart to Directors of the NPRR, n.d., and to Harris, November 3, 1885, and 1883 file *seriatim,* especially Hobart to Allie (Alice, his wife), August 18; St. Paul *Pioneer-Press,* October 1, 1882 (clipping in Hobart Papers).

31. Northern Pacific Papers: Harris to Billings, August 3, 1886 (and quoted in Cochran, p. 361); 49th Congress, 1st Session, Senate Executive Document 51 (1885), pp. 28–29; 52nd Congress, 1st Session, "Inquiry into the Management and Control of the Yellowstone National Park" (1892), pp. 241–46; Richard A. Bartlett, *Yellowstone: A Wilderness Besieged* (Tucson: University of Arizona Press, 1985), pp. 155–61, 170–79; Barbara H. Dittl and Joanne Mallmann, "Plain to Fancy: The Lake Hotel, 1889–1929," *Montana, The Magazine of Western History* 34 (Spring 1984): 32–45.

32. McElrath, *The Yellowstone Valley: What It Is, Where It Is, and How to Get to It* . . . (St. Paul: Pioneer Press, 1880), p. 94.

33. Haines, *Yellowstone,* II, 30–48.

34. Sherman and Philip Henry Sheridan, *Travel Accounts of General William T. Sherman to Spokan [sic] Falls, Washington Territory, in the Summers of 1877 and 1883* (Fairfield, Wash.: Ye Galleon Press, 1984), p. 160.

35. Nash, *Wilderness and the American Mind,* pp. 108–21.

36. Quoted in Gustavus Myers, *History of the Great American Fortunes* (New York: Modern Library, 1936), first published in 1907: Myers states that Pettigrew wrote these words to him (pp. 233–34, n. 23).

37. This point is most forcefully argued by Alfred Runte in his controversial book, *National Parks: The American Experience,* cited in Note 29. While I disagree with some parts of Runte's "worthless lands thesis"—that national parks were carved out of lands viewed as economically worthless—I have had my chance to complain

in the *Journal of Forest History* 27 (July 1983): 142–43, and there is much merit in his broad argument. John A. Jakle, in *The Tourist: Travel in Twentieth-Century North America* (Lincoln: University of Nebraska Press, 1985), also stresses the importance of tourism to the national park movement. In 1899 Mount Rainier became the nucleus for a national park when the Northern Pacific agreed to give up its grants within the proposed boundaries in exchange for productive land elsewhere.

38. Beinecke Library, Haupt Papers, box 12: typescript summary, "1881–1883: The Northern Pacific Railroad."

CHAPTER 26

1. Michael Williams, "Clearing the United States Forests: Pivotal Years, 1810–1860," *Journal of Historical Geography* 7 (1/1982): 12–28; David C. Smith, "The Logging Frontier," *Journal of Forest History* 18 (October 1974): 96–106; James Elliott Defebaugh, *History of the Lumber Industry in America* (Chicago: The American Lumberman, 1906), II, 148–77.

2. Fred C. Kohlmeyer, "Northern Pine Lumberman: A Study of Origins and Migrations," *Journal of Economic History* 16 (December 1956): 530.

3. Estimates vary substantially. I use the figure in Richard C. Davis, ed., *Encyclopedia of American Forest and Conservation History* (New York: Macmillan, 1983), II, 670.

4. Harold A. Meeks, *Vermont's Land and Resources* (Shelburne, Vt.: New England Press, 1986), pp. 239–41; and Howard S. Russell, *A Long, Deep Furrow: Three Centuries of Farming in New England* (Hanover, N.H.: University Press of New England, 1976).

5. This is the judgment of Charles W. Johnson in *The Nature of Vermont: Introduction and Guide to a New England Environment* (Hanover, N.H.: University Press of New England, 1980), p. 44.

6. Donald J. Pisani, "Forests and Conservation, 1865–1890," *Journal of American History* 72 (September 1985): 352–54.

7. Harold K. Steen, *The U.S. Forest Service: A History* (Seattle: University of Washington Press, 1976), pp. 9–20; Edna L. Jacobsen, "Franklin B. Hough: A Pioneer in Scientific Forestry in America," *New York History* 15 (July 1934): 311–25.

8. Thomas R. Cox, Robert S. Maxwell, Phillip Drennon Thomas, and Joseph J. Malone, *This Well-Wooded Land: Americans and Their Forests from Colonial Times to the Present* (Lincoln: University of Nebraska Press, 1985), pp. 112–88; Jeannie S. Peyton, "Forestry Movement of the Seventies, in the Interior Department, Under Schurz," *Journal of Forestry* 18 (April 1920): 391–405.

9. Mansion, box A26: Vermont Forestry Commission, Sargent to FB, February 4, 1884; S. B. Sutton, *Charles Sprague Sargent and the Arnold Arboretum* (Cambridge, Mass.: Harvard University Press, 1970), pp. 93–97.

10. Andre Rebek, "The Selling of Vermont: From Agriculture to Tourism, 1860–1910," *Vermont History* 44 (Winter 1976): 14–27; Louise B. Roomet, "Vermont as a Resort Area in the Nineteenth Century," *ibid.*: 1–13.

11. There is no biography of Phelps; the nearest approach to an essay is Francis Parsons, *Six Men of Yale* (New Haven: Yale University Press, 1939), pp. 85–101. The only works on Proctor are unpublished: Roger G. Cooley, "Redfield Proctor:

A Study in Leadership, The Vermont Period," unpubl. Ph.D. diss., University of Rochester, 1955; Chester W. Bowie, "Redfield Proctor: A Biography," unpubl. Ph.D. diss., 2 vols., University of Wisconsin, 1980; and Ruth Lois Tweedy, "The Life of Redfield Proctor," unpubl. M.A. thesis, University of Illinois, 1942.

12. A copy is in the Mansion, box A26.

13. The report appears in the *Journal* of the House of Representatives of the State of Vermont, Biennial Session, 1884 (Montpelier: Watchman and Journal Press, 1885), pp. 162–69. There is a copy of the separate pamphlet in the Vermont State Law Library in Montpelier.

14. Northeast Pulpwood Research Center, *Forest Practice Survey Report* (Gorham, N.H.: The Center, 1952), p. 82.

15. Mansion, box A12: JPB diary, March 4, 1869; box A11: FB diary, October 10, 1884; Woodstock *Vermont Standard*, March 4, 1869, July 20, 1882.

16. Mansion, box A29: FB to Justin F. MacKenzie, March 29, Henry S. Dana to FB, April 9, 1869, and FB to Henry Boynton, January 18, 1871.

17. This is not a history of either the mansion or of the farm, though both are fully worthy of a monograph. On the mansion see Janet Houghton McIntyre, "The Billings Mansion: A History of Its Design and Furnishing, 1869–1900: A Summary Report Derived from the Records of the Collections Cataloging Project, 1975–1977," unpubl. typescript in the Mansion Archives. On the farm there is a lively pamphlet, *The Billings Farm*, by John H. McDill, prepared for the opening of a new dairy and cow barn in 1948, and revised in 1971. *Billings Farm & Museum News*, from its first number in 1987, has contained useful short articles on the construction of the farm buildings. *The Vermont Farm Year in 1890*, by Scott E. Hastings, Jr. and Geraldine S. Ames ([Woodstock]: Billings Farm & Museum, 1983), describes farming in Vermont in the year of Billings's death.

18. Mansion, box A25: FB's will, June 30, and codicil, August 29, with Kilner's memorandum on conservation, September 26, 1890; David Lowenthal, *The Vermont Heritage of George Perkins Marsh* (Woodstock: Woodstock Historical Society, 1960), p. 14.

19. Mansion, box A29: Agricultural Society to FB, August 3, 1870; box A4: JPB to FB, November 21, 1870; Curtis, Jennison, and Lieberman, pp. 103–107.

20. *Jersey Herd Register*, 1871: I, Bulls 536, Cows and heifers, 438, 445; Woodstock *Vermont Standard*, June 12, 1888; Hubert Howe Bancroft, *The Book of the Fair: An Historical and Descriptive Presentation of the World's Science, Art, and Industry, as Viewed through the Columbian Exposition at Chicago in 1893* (4 vols., Chicago: Bancroft, 1895).

21. P. B. Day, "The Billings Farm, Woodstock, Vt.," *The Jersey Bulletin* 3 (August 26, 1885): 828–29.

22. Mansion, box A25: FB's supplementary instructions on the administration of his estate, 1890.

23. These views are summarized by Hal Seth Barron in "The Impact of Rural Depopulation on the Local Economy: Chelsea, Vermont, 1840–1900," *Agricultural History*, 54 (April 1980): 318–35. See also his book, *Those Who Stayed Behind: Rural Society in Nineteenth-Century New England* (Cambridge: Cambridge University Press, 1984).

24. Mansion, box A21: "The Billings Family, Woodstock and New York," a scrapbook compiled by Julia Lee McDill.

25. On Emerson see Cynthia Zaitzevsky, *The Architecture of William Ralph Emerson, 1833–1917* (Cambridge, Mass.: Fogg Art Museum, 1969).

26. Mansion, box A31: Bierstadt to FB, May 7, 1886, April 28, 1870, FB to Church, February 10, 1881, Thomas P. Rossiter, October 12, 1865, Harry Chase to FB, March 12, 1875, files on Kensett, Ehrick K. Rossiter, Thomas Waterman Wood; box A4: JPB to FB, January 25, February 13, May 11, July 16, 1865, April 25, 1871.

27. University of Vermont, George Perkins Marsh Papers: FB to Caroline Crane Marsh, January 1, 1883; Mansion, box A8: Marsh to FB, December 26, 1846, Caroline Crane Marsh to JPB, December 20, 1883; box A11: JPB diary, August 26, 1886.

28. Mansion, box A10: FB to Buckham, May 28, 1885; Laurel Ginter, "Building Billings," *Vermont* 1 (Winter 1984): 2–7; Jane and Will Curtis and Frank Lieberman, *The World of George Perkins Marsh, America's First Conservationist and Environmentalist* (Woodstock: Countryman Press, 1982), p. 114.

29. *Harper's Weekly*, 34 (October 11, 1890): 1764; Seneca Haselton, "The Hon. Frederick Billings, LL.D.," *The University Magazine* 5 (November 1891): 1077–79.

CHAPTER 27

1. Mansion, box A23: speech at the centennial of the church in Royalton, 1877; box A29: Windsor County Fair Grounds, file on miscellaneous land transactions, 1884–1905, and on closure and sale, 1933; and file on abandonment of the line of the Woodstock Railway Company; Dennis M. Zembala, *Elm Street Bridge* (Woodstock: Woodstock Historic District Commission, 1977); Bancroft Library, C-D 929: Presentation Speech. . . , October 31, 1880.

2. Mansion, box A31: "Excerpts from Diaries of Julia Parmly Billings and Frederick Billings," pp. 36–37: December 24–26, 1889. Diary citations in this chapter are to this compilation of excerpts.

3. Mansion, box A10: Buckham to FB, January 29, 1872, J. H. Sidge to FB, September 2, 1888; box A4: JPB to FB, May 14, 1881; box A8: Moody to FB, September 28, 1886, October 11, 1889, Rachel C. Burton to FB, April 24, 1890; Stephen B. L. Penrose, *Whitman: An Unfinished Story* (Walla Walla, Wash.: Whitman Publishing, 1935), p. 109; Josiah Bushnell Grinnell, *Men and Events of Forty Years: Autobiographical Reminiscences of an Active Career from 1850 to 1890* (Boston: Lothrop, 1891), p. 364.

4. Theresa D. Weiss, *The Vermont Alumnus* 19 (June 1940): 203; Betty Bandel, "The World Is Becoming a Single Room: Sho Nemoto's Years in Vermont and Japan," *Vermont History* 39 (Winter 1971): 72–76.

5. Woodstock *Vermont Standard*, January 10, 1889; Burlington *Free Press*, June 28, 1889; Mansion, box A9: Nemoto to FB, March 30, June 11, 1889, August 8, 1890, and to JPB, January 10, 1891, August 20, 1913; box A8: Butler to FB, July 16, 1889, and to JPB, August 29, September 23, 1890; box A15: Elizabeth Billings's travel diary, May 15–17, 1898; Preface to Nemoto, trans., *The Nation: The Foundation of Civil Order and Political Life in the United States*, by Elisha Mulford (Tokyo, 1915); and Nemoto, *In Memoriam . . . Frederick Billings* (Tokyo, 1930), with description of laws initiated by Nemoto.

6. Mansion, box A8: Dana to FB and JPB, with Williams to FB, May 24, 1888, and Dana's text on the Norman Williams Public Library, submitted for Billings's ap-

proval; box A23: speech at the presentation of a portrait of Norman Williams to the Windsor County Court, and speech at the dedication of the Billings Chapel; Billings Farm Archives: receipts for repair of school house, November 13, and for architect for parsonage, December 24, 1882; Woodstock *Vermont Standard*, April 1, November 4, 1880, June 22, November 30, 1882, October 11, 1885, July 20, August 4, October 27, 1887, October 3, November 28, 1889.

7. Dana, *History of Woodstock, Vermont* (Boston: Houghton, Mifflin, 1889). The book was reprinted with dates—*1761–1886*—added to the title in 1980 (Taftsville: Countryman Press for The Woodstock Foundation), and with introduction and epilogue (the latter to bring the history to 1890, the year of its patron's death) by Peter S. Jennison. Jennison provides much interesting detail on the gestation of the project and a short biography of the Dana family. Subsequently Jennison wrote a companion volume, *The History of Woodstock, Vermont, 1890–1983* (Woodstock: Countryman Press for the Woodstock Foundation, 1985), which is a model of its kind.

8. Mansion, box A8: Butler to Dana, October 5, 1903, Dana to FB, August 29, 1881, June 10, 1886, April 5, 1888.

9. Mansion, box A31: "Excerpts from Diaries," pp. 39, 41–45.

10. Mansion, boxes A25 and A27: wills of Frederick and Julia Billings, List of the estate of Frederick Billings, and Supplementary instructions on the administration of his estate.

11. *New York Times*, October 14, 1890; Woodstock *Vermont Standard*, October 2, 1890; Bancroft Library, C-D 929, folder 1: Butler's obituary of FB, Barstow's dictation of March 6, 1891, and Memorial Sketch for Society of California Pioneers.

12. Mansion, box A31: Memoranda of Frederick's last days, by JPB; "Excerpts from Diaries," pp. 55–59. There are many descriptions of the funeral, at which President Buckham gave the eulogy. A full account appears in *The University Cynic* 8 (October 11, 1890): 75–81. See Buckham, "Frederick Billings," *Sunset Magazine*, (March 1906): 487–91.

A NOTE ON SOURCES

There are two paramount manuscript collections for this biography. The most important is the large collection of family papers now kept in the former Billings home in Woodstock, Vermont (generally known as the George Perkins Marsh House, or simply as the Mansion). The material is well organized, and is in process of further reorganization by its able archivist, Janet Houghton. Ms. Houghton has prepared *A Guide to the Billings Mansion Archives and Photograph Collection* (unpublished typescript, revised and enlarged, 1988), which readers may consult for greater detail. The archive is not open to the public, though Ms. Houghton will search out answers to legitimate requests. Throughout I refer to the collection as simply the Mansion with names of writers and recipients given in full, with these exceptions: FB for Frederick Billings, OB for his father Oel, SWB for his mother Sophia, and JPB for his wife Julia Parmly Billings.

The Billings family correspondence is rich in social and family history, making possible an unusually personal biography more in keeping with the expectations of late twentieth-century scholarship than most more business oriented and well-pruned family archives allow. There are gaps, of course, especially in the early California years, since Billings was burned out twice in 1849–1851. However, these years are well represented in the letters he sent home and to others. Unhappily, the early diaries have not survived, so that the collection contains only one for c. 1840, those for 1859 and 1865, and a run from 1874 through 1889. Julia Billings's diaries help fill some gaps, however. Drafts of Frederick's public addresses, his school composition books, and a wide range of business records are also present. In the Mansion is most of Billings's library as he left it containing various signed presentation copies. Finally, in the library of the Billings Farm and Museum, also in Woodstock, there are financial records, stud books, runs of contemporary periodicals, and other materials relating to the farm's growth and administration. These are under the protection of Esther Swift. I am most grateful to her and to Ms. Houghton for their unfailing help in providing me with unfettered access to the materials in their care.

The other major collection consists of the official records of the Northern Pacific Railway Company, at the Research Center of the Minnesota Historical Society in St. Paul. In coherence and range, this railway archive is second to none. For methodological reasons I preferred when possible to use the original records at the Research Center, but most may be purchased in a comprehensive microfilm edition. The records of the Land Department, given Billings's chairmanship of that enterprise, and the Annual Reports of the company were especially valuable. Also helpful was a microfilm edition of Great Northern Railway Company advertisements, articles, and other publicity from 1884. There is an excellent guide to much of this material by William M. Bomash.

Research on Billings's crucial years in California is difficult and must be pieced together from dozens of scattered and often quite small collections. The state library lost many records in the great earthquake and fire of April 1906. The surveyor general's office was in the path of the worst of the destruction and the general archives on the open shelves were entirely destroyed except for a few items that had been taken home for study by staff the night before. Four valuable collections of business records for Halleck, Peachy, and Billings have survived, however, though in separate locations: at the Bancroft Library of the University of California, Berkeley; at the California Historical Society in San Francisco; in the library of the University of California, Los Angeles; and at The Huntington Library in San Marino. Taken together these collections provide a good though not full picture of HPB, and must be supplemented with many other collections in these institutions and elsewhere.

The Vermont years present a mixed bag. The Vermont Historical Society in Montpelier houses several collections, none of them central to the Billings story, which helped to supply corroborative detail. Unhappily, no personal papers relating to Governor Horace Eaton, under whom Billings served, appear to have survived, nor have the original records of the Vermont Forestry Commission. For this study, the most important collections in Vermont are at the University of Vermont in Burlington, in particular the papers of George Perkins Marsh, and the Trenor W. Park Papers at the Park-McCullough House in North Bennington. I am most grateful to Marguerite d'Aprile-Smith, the curator, for her guidance while I was there.

Various materials in private hands were made available to me, especially by Billings family members. Julia McDill of Woodstock scoured her home for relevant records and transferred them to the Mansion Archives; the late Eleanor French also searched her Woodstock and New York homes and allowed me to review a substantial collection of her husband's own research notes and business records relating to Frederick Billings's properties in Wisconsin. Peter Jennison of Taftsville, Vermont, author of the exemplary second volume of the history of Woodstock, shared with me his research notes.

A good bit of my research was done by the time-honored principle of boot-legging: that is, as I was working on two other books during this time, I was able to consult various collections as a by-product of other inquiries. All manuscripts or depositories were researched personally save for those marked *; for these, archivists and librarians supplied photocopies of the pertinent records. An asterisk by an institution means that I obtained materials solely by correspondence, while one by a collection means that I visited the depository but did not personally search the cited collection. I wish to thank those who preside over the institutions that helped me in this way, for their assistance was truly indispensable. The list of manuscripts is set out geographically within a time sequence: that is, with Vermont records followed by California, etc.

The economics of publishing prevents supplying a bibliography of printed primary or secondary materials consulted. The most important are to be found in the Notes, and a full bibliography has been supplied to the Billings Mansion Archives, where it may be consulted by interested researchers. However, readers should know without recourse to notes of my debt to the following: Jane Curtis, Peter Jennison, and Frank Lieberman, *Frederick Billings: Vermonter, Pioneer Lawyer, Business Man, Conservationist* (Woodstock, Vt.: The Woodstock Foundation, 1986); Henry Swan Dana, *His-*

tory of Woodstock, Vermont (Boston: Houghton, Mifflin, 1889); Louis Tuck Renz, *The History of the Northern Pacific Railroad* (Fairfield, Wash.: Ye Galleon Press, 1980); and Eugene V. Smalley, *History of the Northern Pacific Railroad* (New York: G. P. Putnam's, 1883). Despite these last two books, there is still a need for a full scholarly history of the railroad.

MANUSCRIPT COLLECTIONS

Vermont

The Mansion Archives, Woodstock, Vermont
  Billings Family Correspondence
  Parmly Family Correspondence
  Correspondence to Frederick and Julia Billings from mutual and individual friends
  General Business Correspondence
  Diaries of Frederick and Julia Billings
  Diaries of Mary Montagu Billings French
  Miscellaneous Diaries
  Miscellaneous Notebooks, Guest Books, and Genealogies
  Account Books
  School Exercises, Grade Reports, and Notes
  Scrapbooks
  Biographies, Obituaries, and Speeches of Frederick Billings
  Property Deeds
  Inventories
  Frederick Billings Estate Papers
  Billings Estate Corporation and Related Papers
  Wills and Financial Papers
  Legal Documents—Local
  Legal Documents—Corporate and General
  Family Histories and Miscellaneous Papers
  Billings, Missouri, and Montana, Centennial files
  Engraved and Lithographed Portraits of Frederick Billings
  Photograph Collection
  Carleton E. Watkins Photographs
The Billings Farm & Museum, Woodstock, Vermont
  Estate and Stud Records
Norman Williams Library, Woodstock, Vermont
  Clipping files
Woodstock Historical Society, Woodstock, Vermont
  Orlando Dana Miller Scrapbooks
  Elm Street Bridge File
  George Perkins Marsh File
  D.A.R. and Historical Society Photograph Collection
Woodstock Town Hall, Woodstock, Vermont
  Town Records
Royalton Historical Society, Royalton, Vermont
  Calvin Skinner Memorabilia

Royalton Town Clerk's Office, South Royalton, Vermont
  Index of Birth and Death Records
  Index of Marriage Records
  Town Records
  Family Records
  Vital Records
Green Mountain Perkins Academy and Historical Association, South Woodstock,
    Vermont
  Academy Endowment Fund Records
Middlebury College Library, Middlebury, Vermont
  Alumni Notices
Proctor Free Library, Proctor, Vermont
  Redfield Proctor Papers
Park-McCullough House, North Bennington, Vermont
  Trenor W. Park Papers
  Lisa Matthews, MS. "The Hall & Park Families in Vermont & California," 1970
St. Albans Historical Society, St. Albans, Vermont
  J. Gregory Smith Papers
Shelburne Farms, Shelburne, Vermont
  Horace Eaton Letters
  Webb Family Correspondence
University of Vermont Library, Burlington
  Benedict Family Papers
  Matthew H. Buckham Papers
  Thomas H. Canfield Papers
  Champlain Transportation Company Collection
  Lucius E. Chittenden Papers
  Jacob Collamer Letters
  Samuel C. Crofts Papers
  Horace Eaton File
  George Edmunds Papers
  Elm Tree Press Papers
  George Perkins Marsh Papers
  James Marsh Papers
  Justin Smith Morrill Papers (microfilm)
  Sho Nemoto File
  Redfield Proctor Papers
  J. Gregory Smith Papers
  Stevens Family Papers
  Zadock Thompson Collection
  University of Vermont Archives
  Peter T. Washburn Papers
  Windsor County Court Records
Vermont Historical Society, Montpelier
  James Barrett File
  Thomas H. Canfield Papers
  Clement Family Papers
  Horace Eaton File
  Philip Fairbanks Papers

Resolution on Mexican War
Justin Smith Morrill Letters
Trenor Park File
Edward J. Phelps Papers
Redfield Proctor Papers
Calvin Skinner Jr. Memoirs
J. Gregory Smith Papers
Charles W. Willard Papers

Elsewhere in New England

Connecticut
Connecticut State Library, Hartford
  Connecticut Railroad Papers
Beinecke Rare Book and Manuscript Library, Yale University, New Haven
  Joshua E. Clayton Correspondence
  Elwood Evans Clippings on "Pioneer Days"
  John W. Geary Papers
  Theodore H. Hittell Scrapbooks
  C. T. Hobart Papers
  Charles A. Huntington Manuscript Autobiography
  New Almaden Quicksilver Mine Papers
  Reports of Railroad Surveys in California
  Thomas Starr King Papers
  Carleton E. Watkins Stereographs
Sterling Memorial Library, Yale University, New Haven
  Simeon Baldwin Papers
  William Henry Brewer Papers
  Bowers Family Papers
  Samuel Bowles Papers
  Chauncey M. Depew Papers
  William Maxwell Evarts Papers
  George Bird Grinnell Papers
  Hermann Haupt Papers
  Kent Healy Papers
  Chester Smith Lyman Papers
  Map Collection
  Pacific Railway Surveys (microfilm)
  William Lyon Phelps Papers
  Silliman Family Papers
  John Ferguson Weir Papers
New Haven Colony Historical Society, New Haven
  Chester Smith Lyman Papers
  Thomas Prichard Rossiter Papers

Massachusetts
Baker Library, Harvard Business School, Harvard University
  Jay Cooke Papers
  Forbes Family Business Records

New York and New Haven Railroad Papers
New York, New Haven, and Hartford Railroad Papers
Northern Pacific Railroad Collection
Henry Villard Papers
Woodstock Railroad Company Records
Houghton Library, Harvard University, Cambridge
  E. L. Godkin Papers
  William James Papers and Brazilian Diary
  Houghton Mifflin Company Records
  C. S. Sargent Papers
  Henry Villard Papers
  Carleton E. Watkins Photographs
Massachusetts Historical Society, Boston
  Charles Francis Adams Papers (microfilm)
  Henry W. Bellows Papers
  John O. Sargent Papers
Boston Public Library, Boston
  John C. Frémont Papers
  Richard Monckton Milnes Papers
  Thomas Starr King Papers
  Carleton E. Watkins Photographs
Amherst College Library, Amherst
  * Alumni Records: Parmly Billings
Phillips Academy Library, Andover
  * Alumni Records: Oliver Phelps Chandler Billings
Fitchburg Historical Society, Fitchburg
  * Charles Jason Billings portrait
  * Obituary files: Bailey, Billings, and Wallace
Forbes Library, Northampton
  Billings Genealogical and Biographical Materials

New Hampshire
Dartmouth College Library, Hanover
  Connecticut and Passumpsic Rivers Railroad Company Papers
  Northern Railroad Company Papers
Kimball Union Academy, Meriden
  * Alumni Records: Frederick Billings
New Hampshire Historical Society, Concord
  Blair Family Papers
University of New Hampshire, Durham
  Papers of the Society for the Protection of New Hampshire Forests

California

Bancroft Library, University of California, Berkeley
  Bale Family Papers
  H. H. Bancroft Collection
  Frederick Billings Biographical Sketches
  David C. Broderick Papers

Peter Hardeman Burnett Papers
Manuel de Jesús Castro Documentos para la historia de California
William Tell Coleman Manuscripts
College of California, Records of the Board of Trustees
College of California, Pamphlet Collection
John Conness Papers
John Thomas Doyle Papers
Joseph Libby Folsom Papers
John C. Frémont Papers
James R. Garniss, MS. "Early Days of San Francisco"
William M. Gwin Papers
Halleck, Peachy and Billings Business Records
David N. Hawley, Joshua S. Henshaw, John T. Little: Statements aboard *Oregon*
Elbert P. Jones Papers
Herbert Coffin Jones, MS. "History of Acquisition of the Big Basin as State Park,"
  1955
William Carey Jones Papers
John Neely Johnson Papers
Robert Underwood Johnson Papers
Edith G. Kettlewell, MS. "Yosemite: The Discovery of Yosemite Valley and the
  Creation and Realignment of the National Park," 1928
Thomas O. Larkin Papers (microfilm)
Anna Pierce Judah Statement
Ruth Lewis, MS. "Stories and Legends of Lake County," 1935
Pamphlets on Marin County
Angus McIsaac Diary
Henry Morris Naglee Papers
Charles Christian Nahl File and paintings
Nolan Northern Pacific Collection
Rodney P. Odall Letters
Overland Monthly Papers
Trenor W. Park Letters
Picture File
Republican State Central Committee Papers
W. F. Reed Journal
San Francisco Collection
William Anderson Scott Papers
Charles Shafter, H. S. Williams, and Trenor Park Letter Book
Richard Goss Stanwood Journals
I. W. Taber Collection of Photographs
George Tays, MS. on Montgomery Block
Van Valkenburgh Collection of Portraits
General Vallejo Document Collection on Early California
Dictation on Vigilance Committee of 1856: C. J. Dempster, James Dows, George W.
  Frank
E. B. Walsworth Correspondence
J. L. L. Warren Papers
Carleton E. Watkins Letters and Photographs
Samuel Hopkins Willey Papers (microfilm)

Henry F. Williams Recollections
Mary Floyd Williams Papers
Gregory Yale Papers
University of California, Los Angeles
  Daniel Ammen Papers
  Halleck, Peachy and Billings Business Records
  Carleton E. Watkins Photographs
California Historical Society Library, San Francisco
  Adams and Company Bank Records
  Alfred Barstow Collection
  Frederick Billings Collection
  Samuel Marsden Brookes Collection
  John Thomas Doyle Collection
  John Whipple Dwinelle Papers
  William M. Gwin Papers
  Henry Wager Halleck Collection
  Barnaby W. Hathaway Collection
  Henry Peter Haun Papers
  John Coffey Hays Collection
  Thomas Starr King Collection
  Milton Slocum Latham Collection
  List of Lawyers Papers
  William Alexander Leidesdorff Collection
  Rufus Allen Lockwood Papers
  Frederick William Macondray Jr. Papers
  New Almaden Quicksilver Mining Company Papers
  Francis W. Page Collection
  Photographic Collection (donated by Elizabeth Billings)
  San Francisco Business Records
  San Francisco Early Documents, Scrapbook
  William Tecumseh Sherman Papers
  Thornton-Creswell Family Papers
  Thomas Varney Papers
  Carleton E. Watkins Photographs
  John B. Weller Papers
  Samuel Hopkins Willey Letters
  Young Men's Christian Association, San Francisco, Records

The above collections are described in Diana Lachatanere, *Preliminary Listing of the San Francisco Manuscript Collections in the Library of the California Historical Society* (San Francisco: California Historical Society, 1980).

California State Archives, Sacramento
  Proclamations
California State Library, Sacramento
  Biographical Information File: Peachy, Simmons
  Records of the Families of California Pioneers
  Clark, David & Company Account Book
  John Daggett Scrapbooks

John Franklin Houghton Records
Harbour White & Company Account Books
H. M. McCartney Railroad Papers
Milton Slocum Latham Papers
Miscellaneous Papers: Henry W. Halleck to John Downey/Archibald Peachy Letters
Northern Pacific Railroad Correspondence
State Library Collection Manuscripts
Carleton E. Watkins Stereographic Views
John B. Weller Papers
Badé Institute Museum and Library, Berkeley
Samuel Hopkins Willey Correspondence and Papers
Berkeley Public Library
W. W. Ferrier, MS. "The Naming of Berkeley," 1927
Hearst Mining Building, University of California, Berkeley
Carleton E. Watkins Mother Lode Photographs
Crocker Art Gallery, Sacramento
Carleton E. Watkins Photographs
DeYoung Memorial Museum, San Francisco
Engineer's Log, S. S. *Oregon*
Society of California Pioneers, San Francisco
Evgenio Bianchi, Jr., MS. "Names of Men Who Occupied Montgomery Block"
Society of California Pioneers Membership Records
Society of California Pioneers Mortuary Records
Caspar Thomas Hopkins Autobiography
Montgomery Block Deeds, Certificates, Specifications
S. S. *Oregon* Clipping File
Edmund Randolph Documents
James A. Rockwell Papers
Jacob Rink Snyder Collection
Carleton E. Watkins Photographs
San Francisco Maritime National Historical Park
New Almaden Mine Manifests
* William T. Thompson Records
* San Francisco Theological Seminary Library, San Anselmo
William Anderson Scott Papers
Timothy Dwight Hunt Papers
San Francisco Museum of Modern Art
Carleton E. Watkins Photographs
California State Railroad Museum, Sacramento
Northern Pacific Railroad Photographs
Folsom Historical Society, Folsom
* Alabaster Cave File
Fort Point and Army Museum Association, San Francisco
Bennet Riley and John C. Frémont Memorabilia
Fresno City & County Historical Society
* Ben Walker Collection
The Henry Huntington Library, San Marino
John D. Adams, MS. "Recollections of Early Theatricals in San Francisco"

Samuel L. M. Barlow Collection
Frederick Billings, Chairman, Board of Inspectors and Judges
Frederick Billings, Appointment as Attorney General
California Scrapbooks
Thomas G. Cary, Sketches of California
Alfred A. Cohen Letterbook
W. G. Cohen Scrapbooks
William Heath Davis Memorandum
John F. Duling Collection
Joseph Libby Folsom Accounts
John C. Frémont Papers
William M. Gwin Papers
Halleck, Peachy and Billings Legal Records
B. B. Harris Journal
Clarence King Papers
James King of William Affidavits
La Motte Family Papers
John A. McDougall Papers
John Muir Papers
New Almaden Mine Collection
Trenor W. Park to Hage and Wilson
John Arnold Rockwell Collection
Abel Stearns Collection
Vigilance Committee Papers
Carleton E. Watkins Photographs
Henry S. Will Collection on California Titles
Albert Williams Manuscripts and Notebook
Benjamin Davis Wilson Collection
James Woods Diary

Most of these collections are described in *Guide to American Historical Manuscripts in the Huntington Library* (San Marino: Huntington Library, 1979).

Mariposa County Courthouse, Mariposa
  Deeds
  Legal Records
Mariposa County History Center, Mariposa
  Horace Snow Letters
New Almaden Museum, New Almaden
  Clipping files and artifacts
The Oakland Museum, Oakland
  Carleton E. Watkins Photographs
Oakland Public Library, Oakland
  Don Morton, MS. "The Montgomery Block"
  Mountain View Cemetery File
  Oakland Views File
*County of Placer Historical Museum, Auburn
  Historic Towns Envelope
*San Jose Historical Museum, San Jose

Lawrence Bulmore Collection: New Almaden Chronology
Henry Naglee and Frederick Billings Legal Documents
Santa Barbara Historical Museum, Santa Barbara
Pablo de la Guerra Papers
Santa Barbara Mission Archives, Santa Barbara
De la Guerra Collection
Santa Cruz Historical Society, Santa Cruz
California State Parks File: Big Basin Redwoods
Simi Public Library, Simi
Local History Collection
Stanford University Library, Division of Special Collections, Stanford
Timothy Hopkins Transportation Collection
William Carey Jones Papers
Milton S. Latham Papers
George E. Tillotson Collection
Carleton E. Watkins Photographs of the Pacific Coast
Holt-Atherton Research Center, University of the Pacific, Stockton
John Muir Papers
University of Southern California, Los Angeles
De la Guerra Collection
The Southwest Museum, Los Angeles
John C. Frémont Letters
Vallejo Naval and Historical Museum, Vallejo
*Mare Island History File
Wells Fargo Museum, San Francisco
Reports and Book inscribed by Frederick Billings
Yosemite National Park Research Library, Yosemite Village
Carleton E. Watkins Letters, Interview, and Photographs

Elsewhere in the West

Arizona
John C. Frémont House, Tucson
Memorabilia

Colorado
Denver Public Library
*Charles Dewar Papers
*Grenville M. Dodge Papers

Idaho
Idaho Historical Society, Boise
Northern Pacific File

Montana
Montana State Historical Society, Helena
Clarence W. Growth, MS. "Montana Banking History, 1864–1954," 1955
F. Jay Haynes Photographs and File
Samuel K. Hauser Papers

T. M. Hoglen Diary
Nathaniel P. Langford Papers
Yellowstone National Park Records (microfilm)
Montana State University Library, Bozeman
F. Jay Haynes Papers
William Milnor Roberts Papers
Parmly Billings Public Library, Billings
Biographical Files: Frederick and Parmly Billings
Biographies taken from the Billings *Gazette*
Centennial Celebration Files
James R. Goss Reminiscences
I. D. O'Donnell Memoir
Typescript volumes on Early Billings History and Billings Pioneers
Western Heritage Center, Billings
Assessors Lists, Yellowstone County
Real Property Assessment Books, Yellowstone County
Memorabilia on Parmly Billings Library
Yellowstone Valley Historical Society, Billings
Memorabilia on settlement and naming of Billings
Park County Historical Society, Livingston
Northern Pacific Railroad and Gardiner Branch Line Memorabilia
*Richey Historical Museum, Richey
Billings, Montana and Billings Family File
Museum of the Yellowstone, West Yellowstone
National Park and Railroad Memorabilia
F. Jay Haynes Photographs

North Dakota
State Historical Society of North Dakota, Bismarck
*David R. Taylor Papers
*North Dakota Institute for Regional Studies, Fargo
James B. Power Papers
*Barnes County Historical Society, Valley City
Barnes County Land Records, 1878

Oklahoma
*University of Oklahoma Library, Norman
Earl L. Bell Collection
Isaac N. Seligman Letter Books
Gilcrease Museum, Tulsa
*Ethan Allan Hitchcock Diaries

Oregon
Oregon Historical Society, Portland
Frederick Billings to Matthew P. Deady
Frank B. Gill Papers
University of Oregon Library, Eugene
*John C. Ainsworth Papers
*Hazard Stevens Papers

Utah
Utah Historical Society, Salt Lake City
  Utah & Northern Photograph File

Washington
University of Washington Library, Seattle
  * Daniel H. Gilman Papers
  * Harry C. Heermans Papers
  * Oregon Improvement Company Papers
East Benton County Historical Museum, Kennewick
  Displays and files on Ainsworth and Wallula
Eastern Washington State University, Cheney
  Press clippings of C. S. Kingston
Franklin County Historical Museum, Pasco
  Displays and files on the *Frederick Billings*
* Jefferson County Historical Society, Port Townsend
  Northern Pacific Railroad Files and Photographs
Washington State Historical Society, Tacoma
  * Frederick Law Olmsted's plan for Tacoma
Whitman College Library, Walla Walla
  File on Billings Hall

Wyoming
* Transportation History Foundation, University of Wyoming, Laramie
  Howard Greene Collection of Railroadiana
Yellowstone National Park Museum and Archives, Mammoth Hot Springs, Wyoming
  "History of Improvements in Yellowstone National Park" by Newell F. Joyner
  Records of explorers and travellers
  Scrapbooks

Washington, D.C.

Library of Congress
  Chester Arthur Papers
  Nathaniel Banks Papers
  August Belmont Papers
  James G. Blaine Papers
  John Bigelow Papers
  Jeremiah S. Black Papers
  Benjamin Barstow Papers
  Salmon P. Chase Papers
  Chauncey M. Depew Papers
  Hamilton Fish Papers
  John C. Frémont Papers
  James A. Garfield Papers (microfilm)
  Horace Greeley Papers
  Henry Wager Halleck Papers
  Benjamin Harrison Papers
  Ethan Allen Hitchcock Papers

Abraham Lincoln Papers
Robert Todd Lincoln Collection
Justin Smith Morrill Papers
Frederick Law Olmsted Papers (microfilm)
Carl Schurz Papers
William H. Seward Papers
William T. Sherman Papers
Edwin M. Stanton Papers
Benjamin Wade Papers
Elihu Washburn Papers
Carleton E. Watkins Stereoptican Slides
The National Archives
Secretary of the Interior, Indian Affairs Bureau, Crow Reservation Records
Secretary of the Interior, Lands and Railroads Division, Railroad Packages
Department of State, Foreign Post Records, Mexico
Department of State, Consular Dispatches, Mexico City
Records of the United States-Mexican Claims Commission
Bureau of Railway Economics, Pamphlet Collection
Records of the Commissioner of Railroads: Northern Pacific Reports
Columbia Historical Society
Brentwood File
Patterson Family File
George Washington University Library
Chauncey M. Depew Papers
National Society, Daughters of the American Revolution Library
*History of Dorchester Pope Family
Association of American Railroads Library
Pamphlet Collection

New York

New-York Historical Society, New York City
August Belmont Papers
Gustavus Vasa Fox Papers
Rufus H. Gilbert Papers
Litchfield Family Papers
Morton, Bliss Papers: George Bliss Letterbooks
Henry Villard Manuscripts
Atlases and Maps of the City of New York
New York Public Library
William F. Allen Papers
August Belmont Sr. Papers
John Bigelow Papers
William Conant Church Papers
Darius Nash Couch Papers
Billings and Hambro Day Book
Horace Greeley Papers
Northern Pacific Railroad Opening Excursion, 1883
U.S. Railroads Pamphlet Collection

U.S. Sanitary Commission, California Branch, Archives
Calvert Vaux Papers
Carleton E. Watkins Yosemite Views
Gideon Wells Papers
Fernando Wood Papers
Museum of the City of New York, New York City
Map Collection
The Pierpont Morgan Library, New York City
John C. Frémont Letters
Butler Library, Columbia University, New York City
Allan Nevins Collection
Charlemagne Tower Letters
William J. Wilgus Papers
James M. Olin Library, Cornell University, Ithaca
*Douglass Boardman Papers
*Davenport Family Papers
*A. L. Langellier Papers
*Justin S. Morrill Papers
George Eastman House, Rochester
Carleton E. Watkins Yosemite Photographs
*New York State Historical Association, Cooperstown
Sophia Witherbe [sic] Pictorial Embroidery
Olana State Historic Site, Hudson
*Frederic Church Papers
*New York State Library, Albany
Malcolm and Thomas H. Canfield Papers
Franklin B. Hough Papers
Rush Rhees Library, University of Rochester
William Henry Seward papers
George Arents Research Library, Syracuse University, Syracuse
*Eugene A. Kozlay Papers
*Justin Smith Morrill Letters

Elsewhere in the United States

Florida
*Henry Shelton Sanford Memorial Library and Museum, Sanford
Henry Shelton Sanford Papers

Illinois
Chicago Historical Society, Chicago
*William R. Ogden Papers
The Newberry Library, Chicago
The Chicago, Burlington & Quincy Archives
Illinois Central Library
Carleton E. Watkins Photographs
Northwestern University Library, Evanston
Transportation Library Collection: Northern Pacific Railroad

Indiana
Indiana University, Lilly Library, Bloomington
   *Hugh McCulloch Papers

Iowa
*Iowa State Historical Department, Des Moines
   Grenville M. Dodge Collection
   John Kasson Papers
*Council Bluffs Free Public Library, Council Bluffs
   Grenville M. Dodge Papers

Michigan
William L. Clements Library, University of Michigan, Ann Arbor
   *Baldwin Family Correspondence
   *Lewis Cass Papers
   *Henry Stevens Papers
*Clarke Historical Library, Central Michigan College, Mount Pleasant
   Lewis Cass Papers

Minnesota
Minnesota Historical Society, St. Paul
   American Steel Barge Company Papers (microfilm)
   George A. Brackett Papers
   Nathan Butler Papers
   Clark Family Papers
   Philip R. Clarke Papers
   Solomon G. Comstock Papers
   Lake Superior and Puget Sound Company Papers
   William P. Murray Papers
   Northern Pacific Railway Company Records
   Henry M. Rice Papers
   H. H. Sibley Papers
   Franklin Steele Papers
   James Wickes Taylor Papers
   George F. and Mary Stuart Turner Papers
   William Windom Papers
   Edwin W. Winter Papers
James J. Hill Reference Library, St. Paul
   James J. Hill Letter Books and Letterpress Books
*Anoka County Historical Society, Anoka
   Thurston Manuscript File
*Clay County Historical Society, Moorhead
   Northern Pacific Railroad Photographs
*Southwest Minnesota History Center, Marshall
   Frederick Billings letters
   Northern Pacific pamphlet
*Verndale Historical Society, Verndale
   Billings Family File

Missouri
* Missouri Historical Society, St. Louis
William Tecumseh Sherman Papers
* St. Louis Mercantile Library, St. Louis
Hermann Haupt Chapman Manuscripts

New Jersey
Princeton University Library, Princeton
William L. Dayton Papers
Henry van Dyke Family Papers
* Shrewsbury Historical Society, Shrewsbury
Old Rumson Burying Ground File
Parmly Family File

North Carolina
Forest History Society, Durham
American Forestry Association Papers
Duke University Library
Richard Monckton Milnes Papers

Ohio
Ohio Historical Society, Columbus
Jay Cooke Papers
Carleton E. Watkins Yosemite Album
Lake County Historical Society Research Library, Painesville
* Obituaries, genealogy, and family reunion files
* Parmly Family Genealogy, prepared by Carl Thomas Engel
Rutherford B. Hayes Presidential Center, Fremont
* Rutherford B. Hayes Papers
Western Reserve Historical Society, Cleveland
Pamphlet Collection

Pennsylvania
Historical Society of Pennsylvania, Philadelphia
Salmon P. Chase Papers
Jay Cooke Papers
* Pennsylvania Historical and Museum Commission, Railroad Museum, Strasburg
Northern Pacific Railroad Photographs

South Carolina
South Carolina Historical Society, Charleston
* Valk Family Genealogy
Clemson University Library
* John C. Calhoun Papers

Tennessee
Andrew Johnson National Historic Site, Greenville
Carleton E. Watkins Photographs

Texas
Southern Methodist University Library
  *De Golyer Collection

Virginia
Swem Library, College of William and Mary, Williamsburg
  *Alumni Files: Archibald Peachy
  *Thomas Roderick Dew Papers
  *Galt Family Papers
  *Nathaniel Beverley Tucker papers
  *William and Mary College Papers
*Carol M. Newman Library, Virginia Tech University, Blacksburg
  Herman Haupt Memoirs
  Northern Pacific Railroad File

Wisconsin
State Historical Society of Wisconsin, Madison
  James Davie Butler Papers
  Joseph Thompson Dodge Papers
  Hercules L. Dousman Papers
  Michael M. Harrington Papers
  *Dwight L. Moody Papers
  John H. Tweedy Papers
  *Daniel L. Wells Papers
*Oshkosh Public Library, Oshkosh
  Allen & Billings File

Foreign Collections

Great Britain
Bodleian Library, Oxford University
  James Bryce Papers
Buckinghamshire County Archives, Aylesbury
  Charles N. Fremantle Papers
Trinity College, Cambridge
  Richard Monckton Milnes Papers
Imperial College, London
  Huxley Papers
*Archives of Northern Ireland, Belfast
  Sir James Emerson Tennent Papers
Scottish Record Office, Edinburgh
  Carron Company Letterbooks
  Edinburgh Chamber of Commerce Minute Books
  Glasgow Chamber of Commerce Minute Books
  Arthur Kinnaird Scrapbooks

Canada
Public Archives of Canada, Ottawa
Sir Wilfrid Laurier Papers
Sir John A. Macdonald Papers

ILLUSTRATION CREDITS

Permission by the following to use items from their collections is gratefully acknowledged.

Billings Mansion Archives, Woodstock, Vermont

Sophia Billings; Oel Billings; Laura Billings Simmons; Watkins 1862 panorama of San Francisco; Britton & Ray lithograph of the Montgomery Block; Henry Wager Halleck and Archibald Cary Peachy; Frederick Billings and Trenor W. Park; Watkins photograph of New Almaden; Watkins photographs of the Yosemite Valley; Wenzler portrait of Julia Parmly; Brady photograph of Frederick Billings; Parmly Billings and Bailey & Billings bank; Haynes photograph of the Grand Canyon of the Yellowstone; Mammoth Hot Springs; George Perkins Marsh; the Marsh estate; Powers Farm; watercolor view of the Billings Mansion; South Park Street, Woodstock; Billings family portrait; Samuel E. Kilner; Julia Parmly Billings and her daughters

Mrs. John H. McDill, Woodstock, Vermont

Bezer Simmons

Wadsworth Atheneum, Hartford, Connecticut
Bequest of Mrs. Clara Hinton Gould

Evening in the Tropics

San Francisco Maritime National Historical Park

San Francisco harbor in 1851 (HDC #A11.7881 and 7881.2)

Southwest Museum, Los Angeles

Jessie Benton Frémont (Photograph #N.30665)

Haynes Foundation Collection of photographs by F. Jay Haynes
Montana Historical Society, Helena, Montana

Dalrymple Farm; Big Cut; Bismarck depot; balcony of the Headquarters Hotel; Gold Creek, Montana

Oregon Historical Society, Portland, Oregon

Henry Villard (Negative #ORHI 55645)

Billings Farm and Museum, Woodstock, Vermont
George Aitken

Woodstock Historical Society, 26 Elm Street, Woodstock Vermont 05091
Woodstock from the slopes of Mount Tom

Index